AF566191

GLUCOSE-6-PHOSPHATE DEHYDROGENASE

Academic Press Rapid Manuscript Reproduction

GLUCOSE-6-PHOSPHATE DEHYDROGENASE

Edited by

Akira Yoshida

Department of Biochemical Genetics
Beckman Research Institute of the City of Hope
Duarte, California

Ernest Beutler

Department of Basic and Clinical Research
Scripps Clinic and Research Foundation
La Jolla, California

1986

ACADEMIC PRESS, INC.
Harcourt Brace Jovanovich, Publishers
Orlando San Diego New York Austin
Boston London Sidney Tokyo Toronto

ACADEMIC PRESS, INC.
Orlando, Florida 32887

United Kingdom Edition published by
ACADEMIC PRESS INC. (LONDON) LTD.
24–28 Oval Road, London NW1 7DX

Library of Congress Cataloging in Publication Data

Glucose-6-phosphate dehydrogenase.

Consists of lectures presented at an international symposium held in Nov. 1985, under the auspices of the National Institutes of Health and the City of Hope National Medical Center.

Includes index.

1. Glucosephosphate dehydrogenase deficiency—Congresses. 2. Glucosephosphate dehydrogenase—Physiological effect—Congresses. I. Yoshida, Akira, Date . II. Beutler, Ernest, Date . III. National Institutes of Health (U.S.) IV. City of Hope National Medical Center (U.S.) [DNLM: 1. Glucosephosphate Dehydrogenase—metabolism—congresses.
2. Glucosephophate Dehydrogenase Deficiency—congresses.
QU 140 G567 1985]
RC632.G55G57 1986 616.3'9 86-45890
ISBN 0–12–772640–3 (alk. paper)

PRINTED IN THE UNITED STATES OF AMERICA

86 87 88 89 9 8 7 6 5 4 3 2 1

CONTENTS

G6PD DEFICIENCY AND HEMOLYTIC ANEMIA

EXPRESSION OF Gd LOCUS

MOLECULAR BIOLOGY OF HUMAN G6PD

PREFACE

Glucose-6-phosphate dehydrogenase (G6PD) was first found in red blood cells and in yeast by Warburg and Christian in 1931. The enzyme is widely distributed, from microorganisms to humans, and its discovery led to the elucidation of the pentose-phosphate pathway by some of the early giants of biochemistry, including Lipmann, Dickens, Horecker, and Racker.

G6PD is a key enzyme in the generation of NADPH and in the production of ribose-5-phosphate, a building block of nucleic acids. About 30 years ago, a deficiency of G6PD activity in red blood cells of persons sensitive to the hemolytic effect of the antimalarial compound primaquine was discovered. G6PD deficiency was soon also found to be one of the causes of infection-induced hemolysis, hereditary nonspherocytic hemolytic anemia, fava bean–induced hemolytic anemia, and, in some populations, severe neonatal jaundice leading to kernicterus.

Many hematologists, geneticists, and biochemists found G6PD to be an invaluable tool to study a variety of fundamental biological problems. These studies have greatly advanced our knowledge of genetics, evolution, molecular structure, enzyme kinetics, and hormonal and nutritional regulation. Recognizing the need for a comprehensive conference on these diverse subjects related to G6PD, an international symposium on G6PD was held in November 1985, under the auspices of the National Institutes of Health (HL–33787) and the City of Hope National Medical Center. The participants, representing a broad range of expertise, reviewed the current scope of our knowledge, critically evaluated previous studies, and discussed future studies of both the basic and practical aspects of the G6PD problem. This volume contains all of the lectures presented by the participants. It is hoped that these proceedings will stimulate future studies and bridge the gap between the disciplines.

We wish to acknowledge the contributions of all participants, and the invaluable planning and editorial assistance of Dian De Sha and Anna Milne.

Akira Yoshida
Ernest Beutler

Paul E. Carson—In Memoriam

Paul E. Carson was born in 1925 and passed away on May 25, 1985, in Chicago. Dr. Carson was Professor and Chairman of the Department of Pharmacology and Senior Attending Physician at Rush–Presbyterian St. Luke's Medical Center in Chicago.

Dr. Carson was a world-renowned scientist, physician, and educator. His pioneering studies on the toxicity of antimalarial drugs led to the establishment of better chemotherapeutic agents and regimens against this scourge, and to the discovery of hereditary glucose-6-phosphate dehydrogenase deficiency. The discovery of this condition, which affects millions of people throughout the world, in turn played a seminal role in the creation of the field of pharmacogenetics.

In his distinguished career, Dr. Carson had served as Research Fellow and Chief, Department of Biochemistry for the Atomic Bomb Casualty Commission in Japan (1950–1952); Physician and Scientist, Department of Biochemistry in Brookhaven National Laboratory (1952–1955); and faculty member at the University of Chicago (1955–1971). The recipient of a Research Career Development Award, Dr. Carson had been honored with the Distinguished Service Award from the Medical Alumni Association of the University of Chicago. He was a member of many societies including the Central Society for Clinical Research, the American Society for Human Genetics, and the American Society of Clinical Pharmacology and Therapeutics. At the time of his death, he was heavily engaged in elucidating the pharmacokinetics of 8-aminoquinolines and other antimalarial compounds.

He is survived by his wife, Mary Carson of Chicago, a son, Jeffrey, and two daughters, Jan and Amy.

G6PD DEFICIENCY AND HEMOLYTIC ANEMIA

DRUG-INDUCED HEMOLYTIC ANEMIA AND NON-SPHEROCYTIC HEMOLYTIC ANEMIA [1,2]

Ernest Beutler

Scripps Clinic & Research Foundation

Department of Basic & Clinical Research

10666 North Torrey Pines Road

La Jolla, California 92037

Glucose-6-phosphate dehydrogenase (G-6-PD) deficiency was discovered as an outgrowth of investigations of the hemolytic effect of the antimalarial compound primaquine. The great interest that the scientific community has manifested in this enzyme deficiency over the past thirty years is due, in large part, to the usefulness of this X-linked enzyme defect in a variety of genetic studies. But a major portion of the interest in G-6-PD deficiency must also be attributed to the fact that this defect also has clinical consequences. These consequences may vary in severity from a trivial shortening of the red cell life span to life threatening hemolytic anemia or kernicterus in

[1]This is publication number 4196 BCR from the Research Institute of Scripps Clinic.
[2]This work was supported in part by Grant #HL 25552 from the National Institutes of Health.

the newborn. It seems appropriate at the beginning of this conference, therefore, to briefly review the clinical manifestations of G-6-PD deficiency and to speculate how they may come about.

I. DRUG INDUCED HEMOLYSIS

In the common, polymorphic forms of G-6-PD deficiency hemolytic anemia occurs only when the erythrocyte is exposed to certain stresses. Most G-6-PD deficient individuals with common forms of G-6-PD deficiency such as G-6-PD A- and G-6-PD Mediterranean go through life without ever being aware that they are enzyme deficient. Their red cell life span is normal, or nearly so, and the hemoglobin level of the blood is maintained at normal levels.

The administration of certain drugs has the capacity to alter this state. The prototype of drugs that precipitate hemolytic episodes in G-6-PD deficiency is primaquine, but there are many others that can also do so. Table 1 presents a list of drugs which have been shown to precipitate hemolytic anemia in G-6-PD deficiency.

The number of drugs that we now acknowledge to have hemolytic potency is much smaller than was once thought (1,2).

Because G-6-PD deficiency was first discovered when the administration of a drug produced hemolytic anemia, there was initially a tendency to believe that hemolysis occurring in G-6-PD deficient individuals was always due to drug ingestion.

TABLE 1. Drugs and chemicals that have clearly been shown to cause clinically significant hemolytic anemia in G-6-PD deficiency.

Acetanilid
Methylene blue
Nalidixic acid (Negram)
Naphthalene
Niridazole (Ambilhar)
Nitrofurantoin (Furadantin)
Phenylhydrazine
Primaquine
Pamaquine
Pentaquine
Phenylazodiaminopyridine (pyridium)(3)
Sulfanilamide
Sulfacetamide
Sulfapyridine
Sulfamethoxazole (Gantanol)
Thiazolesulfone
Toluidine blue
Trinitrotoluene (TNT)

Thus, any drug that had been taken prior to the onset of a hemolytic episode was listed as one which might precipitate hemolysis in G-6-PD deficient persons. It was realized only later that other stresses could precipitate hemolytic anemia in G-6-PD deficient individuals. Prominent among such factors was infection, particularly bacterial pneumonias and typhoid fever (1,4). In many cases it was presumably the stress imposed by infection that had produced the hemolytic anemia and not the drug at all. A list of drugs that were once thought to cause hemolytic anemia in G-6-PD deficiency but that are now known to be innocuous is presented in Table 2.

Acetaminophen (paracetamol, Tylenol, Tralgon, p-hydroxyacetanilide)	Menapthone
Acetophenetidin (phenacetin)	p-Aminobenzoic acid
Acetylsalicylic acid (aspirin)	Phenylbutazone
Aminopyrine (Pyramidon, amidopyrine)	Phenytoin
Antazoline (Antistine)	Probenecid (Benemid)
Antipyrine	Procaine amide hydrochloride(Pronestyl)
Ascorbic acid (vitamin C)	Pyrimethamine (Daraprim)
Benzhexol (Artane)	Quinidine
Chloramphenicol	Quinine
Chlorguanide (Proguanil, Paludrine)	Streptomycin
Chloroquine	Sulfacytine
Colchicine	Sulfadiazine
Diphenylhydramine (Benedryl)	Sulfaquanidine
Isoniazide	Sulfamerazine
L-DOPA	Sulfamethoxypyriazine (Kynex)
Menadione sodium bisulfite (Hykinone)	Sulfisoxazole (Gantrisin)
	Trimethoprim
	Tripelennamine (Pyribenzamine)
	Vitamin K

TABLE 2. Drugs that can probably be given safely in normal therapeutic doses to G-6-PD deficient subjects (without nonspherocytic hemolytic anemia).

Fava beans, too, have the capacity to precipitate hemolysis in some G-6-PD deficient persons. This interesting type of hemolytic anemia, designated as favism, will be discussed later in this symposium.

II. MECHANISM OF HEMOLYSIS

In the many years that have passed since the hemolytic effect of primaquine was first found to be due to G-6-PD deficiency many studies have been carried out to clarify the mechanism by which red cells are destroyed. Even now all the details are not fully understood. The most likely mechanism is that the generation of hydrogen peroxide and free radicals damage hemoglobin and red cell components. Reduced glutathione (GSH) acting through the enzyme glutathione peroxidase serves as a scavenger for low levels of hydrogen peroxide. In the process of removing hydrogen peroxide GSH is oxidized to glutathione disulfide (GSSG). Red cells from G-6-PD deficient individuals lack the capacity to regenerate GSH by reducing GSSG. Thus, they are deprived of the protective effect of GSH and, in addition, the accumulated GSSG may damage red cells. The mechanism by which G-6-PD protects red cells against destruction is schematically illustrated in Figure 1.

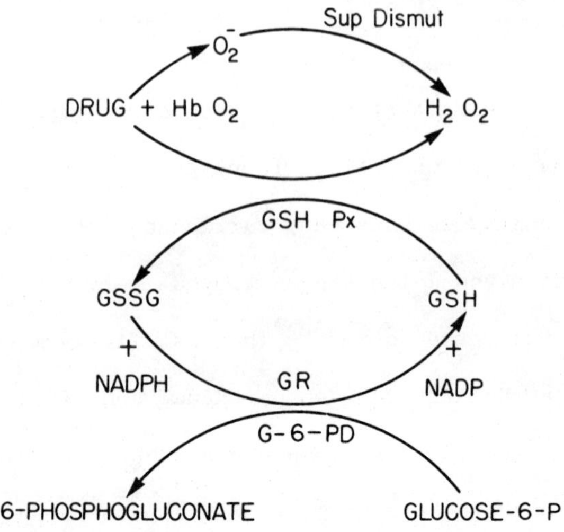

Fig. 1. A schema of the mechanism of protection of red cells against peroxide by G-6-PD.

How the peroxide does its mischief is not entirely clear. One mechanism may be the cross-linking membrane proteins (5). We shall hear more about this later in this symposium. Peroxide and free radicals may also result in denaturation of hemoglobin, and their attachment to the membrane as Heinz bodies impedes the passage of red cells through the narrow slits which separate the splenic pulp from the sinuses (6).

III. HEREDITARY NONSPHEROCYTIC HEMOLYTIC ANEMIA

Drugs, infections, or the ingestion of fava beans are most frequently associated with hemolytic anemia in patients with the common, polymorphic deficient variants of the enzyme. But there is a group of patients that inherit functionally more severe variants, and these individuals have a significantly shortened red cell life span even in the absence of stress. They have a disorder which is designated hereditary non-spherocytic hemolytic anemia (HNSHA). Unlike the polymorphic forms of G-6-PD deficiency, the clinical consequences of these variants is sufficiently severe that the health of the affected individuals is impaired. Thus, selection against these variants has made them sporadic in occurrence and each tends to be different from the other.

When G-6-PD activity is measured in hemolysates from the red cells of such patients with HNSHA, they are sometimes found to have enzyme activity that is only modestly reduced. Not uncommonly, enzyme activities that are considerably higher than those found in polymorphic variants are encountered. Yet, spontaneous hemolysis of these cells occurs. It is clear that from a functional point of view the measurement of enzyme activity is not highly predictive. There may be a number of reasons for this. Yoshida (7) demonstrated that in some variants giving rise to HNSHA the enzyme was extraordinarily sensitive to inhibition by NADPH. Under such circumstances the NADPH/NADP ratio, normally very high, would fall to very low levels at a steady state. But not all variants that give rise to HNSHA have diminished K_i values for NADPH. Another important factor may be the rate at which enzyme activity declines during the erythrocyte life span, and, indeed, frequently the variants giving rise to HNSHA often manifest marked thermal instability. An elevated K_m for G-6-P may also play a role in some instances, but our ability to predict which variants will manifest spontaneous hemolysis and which will result in hemolytic anemia only when a stress is applied is very limited.

There is also a gray area between the polymorphic variants and those in which chronic hemolytic anemia is present that requires further exploration. Cases of chronic hemolytic anemia occurring in association with a G-6-PD variant indistinguishable from G-6-PD Mediterranean have been observed (8,9). In one recently described patient (10) an individual heterozygous for both 6-phosphogluconolactonase deficiency and G-6-PD deficiency was found to have chronic hemolysis. Neither the lactonase deficiency gene nor the G-6-PD deficiency alone appeared to produce hemolytic anemia in other family members. Reinvestigation of an earlier reported case (9) failed to reveal evidence of lactonase deficiency and it is to be presumed that other genetic interactions and possibly environmental factors may be responsible for the occasional occurrence of chronic hemolysis in persons with G-6-PD Mediterranean.

REFERENCES

1. Beutler E: In: Hemolytic Anemia in Disorders of Red Cell Metabolism. Plenum Publishing Company, New York 1978.

2. Beutler E: Chemical toxicity of the erythrocyte. In: Toxicology of the Blood and Bone Marrow. Irons RD (eds.): pp. 39-49. Raven Press, New York 1985.

3. Tishler M: Phenazopyridine-induced hemolytic anemia in a patient with G-6-PD deficiency. Acta Haematol 70:208-209, 1983.

4. Burka ER, Weaver III Z, Marks PA: Clinical spectrum of hemolytic anemia associated with G 6 PD deficiency. Ann Intern Med 64:817-825, 1966.

5. Johnson GJ, Allen DW, Cadman S, Fairbanks VF, White JG, Lampkin BC, Kaplan ME: Red-cell-membrane polypeptide aggregates in glucose-6-phosphate dehydrogenase mutants with chronic hemolytic disease. N Engl J Med 301:522-527, 1979.

6. Rifkind RA: Heinz body anemia: An ultrastructural study. II. Red cell sequestration and destruction. Blood 26:433-448, 1965.

7. Yoshida A: Hemolytic anemia and G-6-PD deficiency. Science 179:532-537, 1973.

8. Ben-Ishay D, Izak G: Chronic hemolysis associated with glucose-6-phosphate deficiency. J Lab Clin Med 63:1002-1009, 1964.

9. Beutler E, Mathai CK, Smith JE: Biochemical variants of glucose-6-phosphate dehydrogenase giving rise to congenital nonspherocytic hemolytic disease. Blood 31:131-150, 1968.

10. Beutler E, Kuhl W, Gelbart T: 6-Phosphogluconolactonase deficiency, a hereditary erythrocyte enzyme deficiency: Possible interaction with glucose-6-phosphate dehydrogenase deficiency. Proc Natl Acad Sci USA 82:3876-3878, 1985.

PHARMACOGENETIC INTERACTION OF GLUCOSE-6-PHOSPHATE DEHYDROGENASE DEFICIENCY WITH ACETYLATION AND HYDROXYLATION

George J. Brewer

Departments of Human Genetics and Internal Medicine,
University of Michigan, Ann Arbor, MI, USA

About 30 years ago, one of our hosts, Ernie Beutler, along with Ray Dern and Alf Alving, conducted a brilliant series of investigations published in seven papers in the Journal of Laboratory and Clinical Medicine (1-7). These studies pinned down the problem of primaquine sensitivity-type hemolysis to a red cell defect involving an abnormality in glutathione metabolism, setting the stage for the late Paul Carson to show that the enzymatic defect was in glucose-6-phosphate dehydrogenase (G-6-PD) of the red cell (8). In the midst of those seven papers, paper V produced a mystery which existed for 25 years (5). About five years ago, we became interested in this mystery, and believe we were fortunate enough to have solved it.

THE MYSTERY

In the process of showing that multiple drugs in addition to primaquine would cause hemolysis of G-6-PD deficient red cells, Dern et al. (5) studied a sulfone called promizole, also called thiazosulfone. In contrast to other drugs, promizole produced a clear bimodal response, either negligible or marked hemolysis. This experiment was carried out by measuring hemolysis of chromium-labeled G-6-PD deficient red cells transfused into normal subjects who then ingested promizole. Promizole produced hemolysis in about half of the cases, the hemolysis depending not on the deficient donor but on the non-deficient recipient who ingested the drug. This suggested to the authors that "it is evident that the observed variability was due to differences in the absorption, excretion, or metabolism of the drug by the individual carriers of the labeled cells. It was observed that the greatest thiazosulfone-induced destruction of primaquine sensitive cells occurred in a normal recipient who developed a marked methemoglobulinemia, abdominal pain, nausea, headache, and a slight anemia during drug administration. This subject, when given only 3.0 Gm. of thiazosulfone daily, developed severe toxic symptoms and hemolyzed transfused primaquine sensitive cells almost as extensively as when 6.0 Gm. were given. On the other hand, sensitive cells from the same donor, when transfused into another recipient, were not hemolyzed by the daily administration of 18.0 Gm. of thiazosulfone. It was noted that the latter recipient did not exhibit any signs of toxicity. It is evident that the primaquine-sensitive cells are highly susceptible to hemolysis by thiazosulfone, but the observed degree of hemolysis in an individual is dependent on extracorpuscular factors characteristic of the particular recipient" (5).

These interpretations, characteristic of the quality of this series of studies, were quite farsighted. As I said earlier, about five years ago we became interested in this mystery and subsequently were able to show that indeed variable metabolism of promizole is the reason for variable hemolysis. Two other pieces of background information are relevant. The first is that promizole has never seen much clinical use, while a structurally related sulfone, dapsone, has seen extensive clinical use, for leprosy, malaria, and dermatitis herpetiformis. Our studies include both drugs, and we have found they behave similarly in our systems. Thus, the hemolytic potential is identical for the two drugs, and what we say about one is applicable to the other. Second, the parent compounds of most of the hemolytic drugs are, in and of themselves, not damaging, but require some type of metabolic transformation to a hemolytic metabolite (9). This is true of primaquine, and is also true of promizole and dapsone.

OUR STUDY OF METABOLIC INTERACTIONS

We were aware that dapsone was a substrate for N-acetylation by the N-acetyl transferase system of the liver. Based on structural similarity (Figure 1) we assumed that promizole would also be a substrate, and in preliminary studies, showed that this was the case. It is well-established, based on extensive studies of isoniazid, that rate of acetylation by N-acetyl transferase is genetically polymorphic due to a single gene, two allele, system (10). Fast acetylation is dominant to slow. About half of both the black and the white populations are homozygous for the slow allele, and are phenotypically slow acetylators. We hypothesized that the reason only half of the recipients hemolyzed G-6-PD deficient red cells was because they were slow acetylators (11).

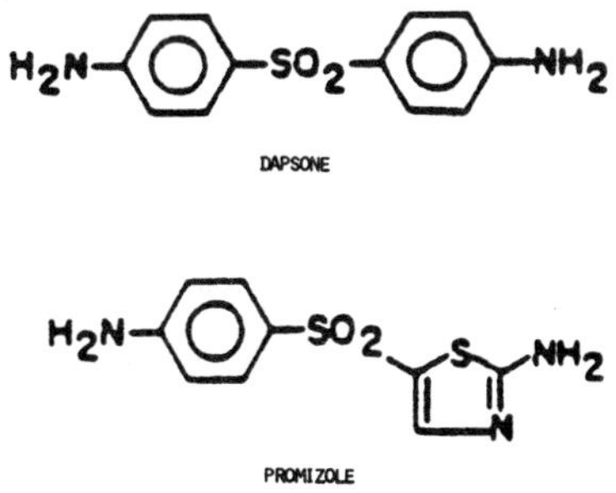

$H_2N-C_6H_4-SO_2-C_6H_4-NH_2$

DAPSONE

$H_2N-C_6H_4-SO_2-$(2-amino-thiazol-5-yl) (NH_2, S, N)

PROMIZOLE

Figure 1. The structures of promizole and dapsone compared.

We decided to carry out an in vitro incubation study, using GSH depletion of G-6-PD deficient cells as an indicator of oxidant damage. To model variation in acetylation, acetylated derivatives of promizole and dapsone were compared to the parent compound. However, we still had the problem that the parent drugs needed some type of metabolic transformation to become active oxidant damaging agents. Based on the literature, we suspected that hydroxylation by the liver's cytochrome p450 system was the metabolic activation required to make promizole and dapsone capable of producing oxidant damage. Israili et al. (12) had shown that the N-hydroxyl derivative of dapsone, but not the parent compound, produced methemoglobin in red cells. Glader and Conrad (13) showed that this same derivative, but not the parent compound, causes glutathione (GSH) depletion in G-6-PD deficient red cells. Scott and Rasbridge (14) demonstrated increased autohemolysis of G-6-PD deficient red cells when they were incubated with hydroxylated dapsone derivative but not with dapsone itself or several other derivatives. To model hydroxylation, mouse liver microsomes, whose microsomal p450 system had been previously induced by phenobarbital, were included in the incubation.

The data for both promizole and dapsone are summarized in Figures 2 and 3. With uninduced microsomes, some nonspecific decrease of GSH occurs. The main points are these: 1) In the complete system, promizole produces a striking loss of GSH (Figure 2). 2) If SKF-525, a specific inhibitor of microsomal p450 hydroxylation is included, almost all of the drug effect is lost. Parallel results are seen with dapsone. 3) If acetylated promizole is used, almost no drug effect is seen (Figure 3). Almost identical results are again seen with dapsone, versus its diacetylated derivative in this system. Normal red cells showed almost no GSH depletion under any of the incubation systems.

These results were all published four years ago in the same journal in which the original primaquine sensitive series, including THE MYSTERY, were published (11). The journal saw fit to republish the original article of Dern et al. (5), along with an editorial on pharmacogenetic interactions by Frischer and the late Paul Carson, discoverer of the enzyme defect (15).

Incubations of Complete System
(Drug, Hydroxylation System,
G6PD Deficient Red Cells),
or Plus SKF-525A,
or with Uninduced
rather than Induced Microsomes

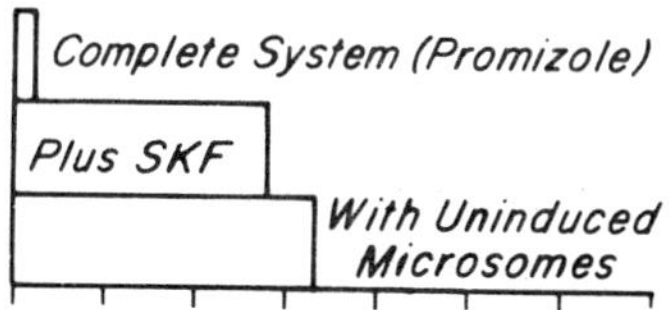

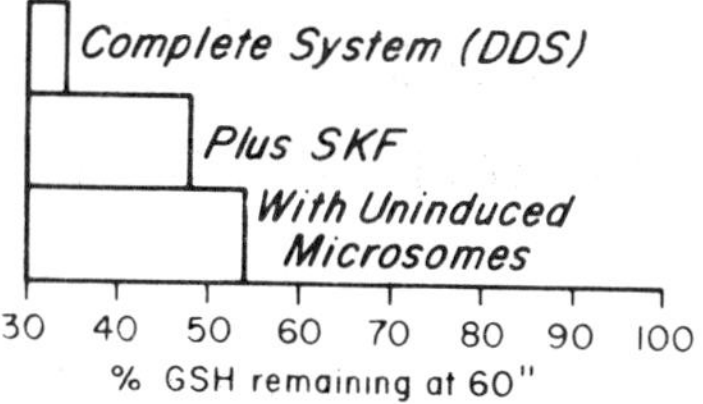

CONCLUSION:

MUCH OF THE GSH LOSS
CAUSED BY A DRUG IS RELATED
TO HYDROXYLATION ACTIVITY

Figure 2. This figure shows the results of incubations with promizole or dapsone on the GSH of G-6-PD deficient red cells (N=3 in each case). Specific loss of GSH is related to hydroxylation. Normal cells showed little effect from any of the incubations. Reprinted with permission from The Red Cell: Fifth Ann Arbor Conference, George J. Brewer, ed.; Alan R. Liss, Inc., 1981.

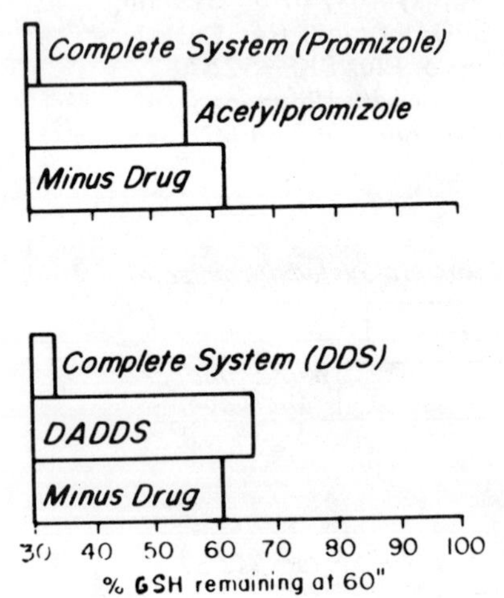

Figure 3. This figure shows that acetylpromizole and diacetylated dapsone cause no specific loss of GSH in the presence of the hydroxylation system, in contrast to the nonacetylated drugs shown in the previous figure. Reprinted with permission from The Red Cell: Fifth Ann Arbor Conference, George J. Brewer, ed.,; Alan R. Liss, Inc., 1981.

We also looked at methemoglobin formation, another measure of oxidant damage, in these same incubates, and saw completely parallel results (Figure 4). Hydroxylation of the parent drug, but not the acetylated derivative, produced methemoglobinemia.

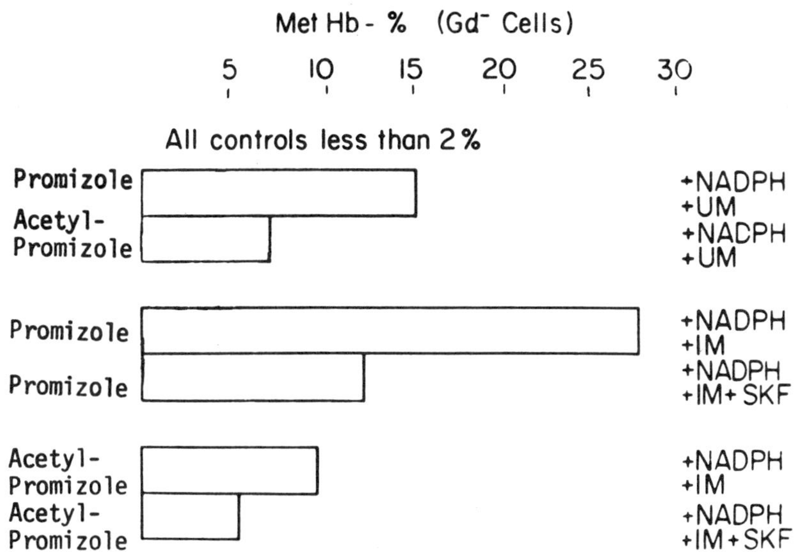

Figure 4. In the upper part of the figure, the nonspecific production of methemoglobin in the presence of the NADPH generating system and uninduced microsomes (UM) is shown. In the middle part of the figure, promizole is shown as producing a considerable amount of methemoglobin in the presence of induced microsomes (IM) above the controls (specific methemoglobin production), and SKF 525, a hydroxylation inhibitor, reduces this number back to the nonspecific range. In the bottom third of the figure it is shown that acetylpromizole does not cause the formation of specific methemoglobin.

Subsequently, we have examined the stability of the damaging agent in these incubates by preincubating induced microsomes with drug, then spinning out the microsomes and incubating the supernatant with deficient cells. We found the supernatant to be fully active in causing GSH depletion, thus leading to the conclusion that the microsomes resulted in a toxic stable intermediate of the drug, no doubt the hydroxylated derivative. I would like to return briefly to the Dern et al. (5) paper to make a clinical guess. I suspect that the individual recipient who had so many symptoms from promizole was not only a slow acetylator, but also had an unusually active hydroxylation system, possibly because of barbiturate ingestion, other environmental causes, or his p450 genotype. In this connection, those non-deficients who suffer side effects from dapsone, such as methemoglobinemia, are probably not only slow acetylators, but have unusually active hydroxylation as well.

A REMAINING MYSTERY

Similar to the situation with the sulfones, only a portion of G-6-PD deficient Mediterraneans are susceptible to favism, the fava bean induced hemolysis. At first blush one might assume that acetylation phenotype could also cause the bimodal susceptibility of favism, but this is unlikely to be the case, because black deficients appear uniformly resistant to favism, yet half of them are slow acetylators. A good bet to explain this mystery is variation in p450 hydroxylation. Four distinct p450 polymorphisms appear to have been identified (Table 1), with probably more to come (16). In each case, a different but partially overlapping subpopulation of people are slow oxidizers. We can postulate that rapid oxidizers of a non-toxic parent compound in the bean, possibly divicine, would be susceptible to favism.

SUMMARY

The mystery of bimodality in hemolysis in G-6-PD deficiency resulting from certain sulfones is explained by acetylation variation. The interaction of three pharmacogenetic systems occurs to determine the hemolytic phenotype (Figure 5). These are the G-6-PD deficiency, acetylation, and p450 hydroxylation. It seems likely that pharmacogenetic and ecogenetic interactions of this type will be seen frequently, because of the redundancy of the detoxifying systems of the body. Bimodality in favism is probably not explained by acetylation; our best guess is that it will be explained by hydroxylation variation.

Table I. Drug Oxidation Polymorphisms

Drug	Frequency of slow oxidizer (%)
1. Debrisoquine (and Sparteine, Phenformin, etc.)	9
2. Tolbutamide	25
3. Mephenytoin (and Nirvanol)	5
4. Carbocysteine (and Penicillamine)	12

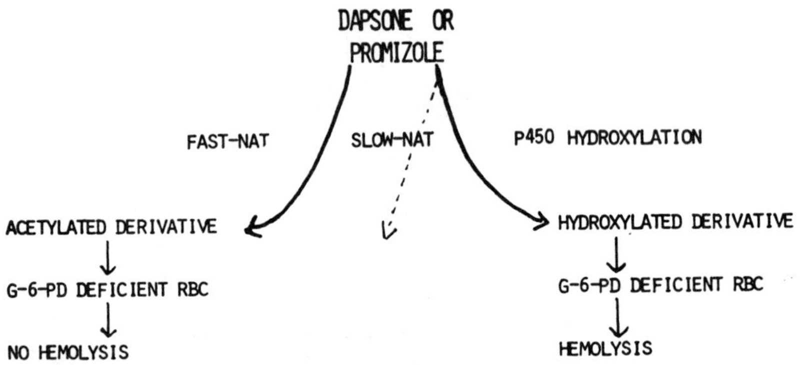

Figure 5. This figure illustrates the pharmacogenetic interactive pathways involved to produce a hemolytic phenotype after ingestion of dapsone or promizole. First, the patient must be G-6-PD deficient. Second, the patient must be a slow acetylator (slow NAT), and last, the drugs have to be hydroxylated to an active intermediate.

REFERENCES

1. Dern, R.J., Weinstein, I.M., LeRoy, G.V., Talmage, D.W., and Alving, A.S. (1954). J. Lab. & Clin. Med. 43 303.
2. Dern, R.J., Beutler, E., and Alving, A.S. (1954). J. Lab. & Clin. Med. 44 171.
3. Beutler E., Dern, R.J., and Alving, A.S. (1954). J. Lab. & Clin. Med. 44 177.
4. Beutler, E., Dern, R.J., and Alving, A.S. (1954). J. Lab. & Clin. Med. 44 439.
5. Dern, R.J., Beutler, E., and Alving, A.S., (1955). J. Lab. & Clin. Med. 45 30.

6. Dern, R.J., Beutler, E., and Alving, A.S., (1955). J. Lab & Clin. Med. 45 40.
7. Beutler, E., Dern, R.J., Flanagan, C.L., and Alving, A.S., (1955). J. Lab & Clin. Med. 45 286.
8. Carson, P.E., Flanagan, C.L., Ickes, C.E., and Alving, A. S. (1956). Science 124 484.
9. Bloom, K.E., Brewer, G.J., Magon, A.M., and Wetterstroem, N. (1983). Clin. Pharmacol. Ther. 33 403.
10. Weber, W.W., and Hein, D.W. (1985). Pharmacological Reviews 37 25.
11. Magon, A.M., Leipzig, R.M., Zannoni, V.G., and Brewer, G.J. (1981). J. Lab. Clin. Med. 97 764.
12. Israili, A.H., Cucinell, S.A., Vaught, J., Davis, E., Lesser, J.M., and Dayton, P.G. (1973). J. Pharmacol. Exp. Ther. 187 138.
13. Glader, B.E., and Conrad, M.E. (1973). J. Lab. Clin. Med. 81 267.
14. Scott, G.L., and Rasbridge, M.R. (1973). Br. J. Haematol. 24 307.
15. Frischer, H., and Carson, P.E. (1981). Interactions in Pharmacogenetics, 97 760.
16. Kupfer, A. (1984). In "Drug Metabolism," Siest, G. (ed), p. 25, Pergamon Press.

FAVISM: EPIDEMIOLOGICAL AND CLINICAL ASPECTS

Christos Kattamis

Thalassemia Unit, 1st Department of Pediatrics, University of Athens, St.Sophie's Children's Hospital, Athens , Greece

1. INTRODUCTION

Favism is an acute, often dramatic hemolytic syndrome, induced by the consumption of fresh or dry beans. The syndrome is common in certain Mediterranean countries, especially Greece, Italy, Israel (among Sephardic Jews) and China; it affects individuals deficient in red cell glucose-6-phosphate dehydrogenase (G-6-PD d.), and mainly those with the Mediterranean variant. (WHO, 1967)

In addition to favism deficient individuals are also predisposed to severe neonatal jaundice, and hemolysis, induced by a series of drugs, by viral or bacterial infections and metabolic disturbances. Of viral infections, infectious hepatitis was shown to induce mild to severe hemolysis in more than 80% of affected deficient individuals. In these patients hepatitis may be also associated with intense and prolonged jaundice. (Kattamis, Tzortzatou, 1970)

Here we shall discuss the clinical, epidemiologic and basic hematologic features of favism. Data concern mainly severely affected patients, with clinical signs of hemolysis, necessitating admission to hospital and treatment.

2. EPIDEMIOLOGICAL DATA

2.1. Morbidity

Favism is known to have existed in Greece since ancient times; Hippocrates described its main clinical features, and introduced the basis for prevention by the aphorism "κυάμων απέχεσθε" which means "avoid fava beans".

Despite this, favism is still common in Greece, representing the most common acute hemolytic anemia in childhood.

Table I. Annual rate of admissions of patients with favism during 1969-1971 in two Children's Hospitals of Athens

Year	Patients admitted		Rate per thousand
	Total	Favism	
1969	16,335	100	6.1
1970	17,695	100	5.6
1971	18,108	77	4.2
1969-71	52,138	277	5.3

The annual morbidity rate is not exactly known; Table I, gives only a rough index, illustrating the rate of admissions of patients with favism in two Children's Hospitals of Athens in three years. From a total of 52, 138 admissions, 277 that is 5.3 per thousand, were admitted for favism.

2.2. Age and Sex Distribution

The age and the sex distribution of 1015 severely affected patients, is illustrated in figure 1.

The disease affects mainly children aged 2-5 years. More than 60% of all patients belonged to this age group. After the age of 6 the incidence fell steadily, the fall being steeper after the age of 10 years. Only 85 (8.4%) were above 10 years while 34 (3.4%) were infants less than twelve months old. Of them 18 were breast fed. Hemolysis in breast fed infants appeared 2 to 6 days after the ingestion of fava beans by the mother who was clinically free of symptoms.

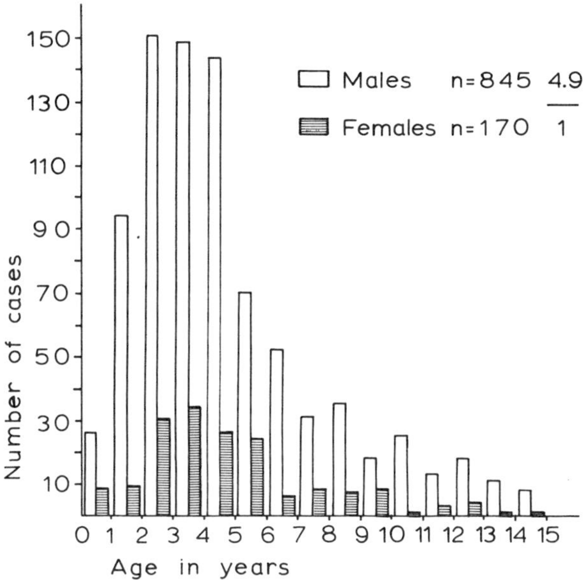

Figure 1. Age and sex distribution of 1015 patients with favism.

The sex ratio in this series was 845 males to 170 females that is 4.9 boys to 1 girl.

The predilection of favism for very young children and infants is difficult to explain. It could be argued that the incidence of favism falls with increasing age, simply because affected children tend to avoid consuming beans after they have had a hemolytic episode. If this was true one would expect all patients to have a hemolytic crisis not only on the first, but also after any subsequent consumption of fava beans. Careful inquiries in a large number of patients disclosed that only about 25% of them were affected following the first consumption of fava beans, and that most of them ingested beans at a later time with no untoward effects. The remaining 75% had their crisis quite unexpectedly although they had previously consumed fava beans several times. On the other hand a second and third episode was noted in a very few cases.

The preponderance of males could be easily explained by the sex linked character of the enzymatic defect which is fully expressed in males. Nevertheless, there are reasons to believe that in heterozygotes hemolysis may be so mild as to escape diagnosis. Therefore the 4.9:1 male female ratio is not the real one but that of selected, severe hospitalized patients.

2.3. Susceptibility to Favism

It should be stressed that consumption of fava beans does not necessarily cause acute hemolysis in all deficient individuals. On the contrary a rather small percentage of deficient individuals is susceptible. This is illustrated by a study on the prevalence of favism among deficient males from different areas of Greece who ingested fava beans more than once.

Table II. Incidence of favism among G-6-PD deficient males in areas of Greece

Areas	No G-6-PD deficient	Cases of favism	Incidence of favism %
Arta	78	5	6.4
Rhodes(North)	43	3	6.9
Rhodes(South)	100	15	15.0
Total	221	23	10.4
SELECTED AREAS OF RHODES			
Without favism	45	0	0
With favism	32	9	28.1

Of a total of 221 male deficient individuals, above 15 years of age only 23 (10.4%) had hemolysis. It is of interest that the incidence of favism among G-6-PD deficient males differs from one area to another and that the higher incidence of 15% was found in southern Rhodes. (Kattamis et al, 1969a)

Difference in susceptibility from one area to another is more evident, in the study of two selected areas of Rhodes, with a high incidence of G-6-PD deficiency, the one was known to have a very low morbidity rate of favism and the other a high one. From a careful questionnaire it was disclosed that none of 45 deficient in the first area had favism in contrast to 9(28%) cases among 32 deficient males in the second area.

In studying further the geographical distribution of G-6-PD deficiency and favism on the island of Rhodes, it was shown that the prevalence of favism was not always proportional to that of G-6-PD deficiency. There were areas with a high incidence of G-6-PD deficiency but very low morbidity for favism and others with rather low incidence of G-6-PD deficiency but with high morbidity rates. (Kattamis et al, 1969a)

These findings suggest that in addition to G-6-PD deficiency and fava beans some other factor or factors enhance the susceptibility of G-6-PD deficient individuals to hemolysis.

These factors are possibly genetic in origin. Stamatoyannopoulos et al, (1966) in a study on the familial predisposition to favism in Greece concluded that "an autosomal gene in heterozygous state enhances the susceptibility to favism of G-6-PD deficient individuals". Also Sartori (1971) accepts the presence of a favic predisposition which behaves as an autosomal recessive trait and seems to be age related. Individuals not deficient in G-6-PD but with a favic predisposition complain of vomiting, headache and fever on eating fava beans, while G-6-PD deficient individuals with favic predisposition are those with hemolytic crisis.

We have no experience with individuals normal in G-6-PD activity who experienced general symptoms without hemolysis, after fava beans consumption.

Though the data collected till now favor the hypothesis that the additional factors which enchance the susceptibility of deficient individuals to favism is genetic in origin environmental factors and acquired conditions could not be easily excluded.

2.4. Seasonal Distribution

Another puzzling point of the epidemiology of favism is its seasonal distribution. Seasonal distribution in the Athens area is shown in figure 2.

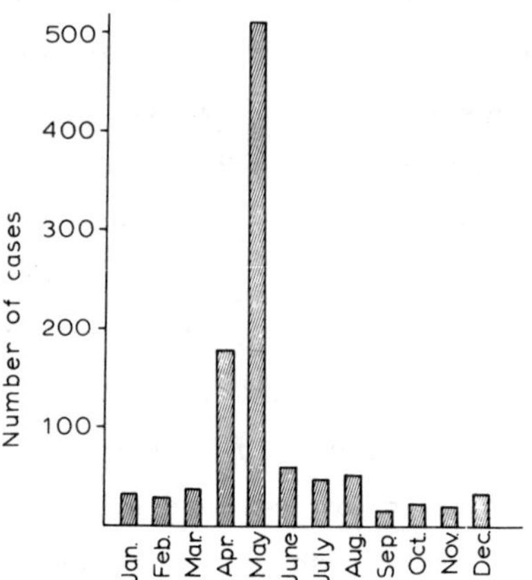

Figure 2. Seasonal distribution of favism.

The disease occurs throughout the year, although its incidence is considerably higher during Spring and especially in May. Fifty per cent of cases were observed during this month by eating exclusively fresh fava beans. During the other months hemolysis was precipitated by dry beans; none of our patients had hemolysis by simple inhalation of pollen from the fava plant, which is reported in the literature as a cause of favism. (Fiorelli et al, 1970) Additional though indirect evidence that inhaled pollen does not precipitate favism at least in Greece, is the very low incidence of favism observed during February and March, when the plants are in blossom. If pollen precipitates hemolysis one would expect even a slight increase in the incidence of favism during these months. In contrast all patients admitted during this period reported consumption of dry beans.

On the other hand the peak in the incidence of favism in May coincides with the ripening of beans.There is a good evidence suggesting that the degree and rate of ripening of beans are of importance for the pathogenesis of favism. Sartori (1971) showed that latitude and altitude influence the seasonal occurrence of favism in proportion to the ripening of beans.

We conducted a similar study in Greece by collecting data for three years from areas with different latitude, the results are shown in figure 3.

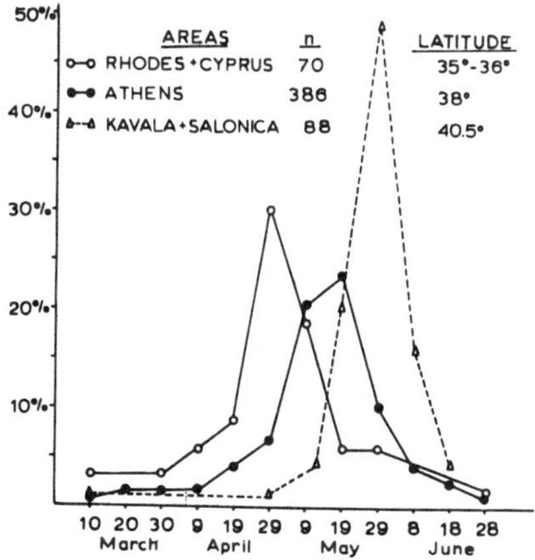

Figure 3. Seasonal distribution of favism in three areas with different latitude.

Based on latitude patients were divided in three groups. The first group includes patients from Cyprus and Rhodes with a latitude of 35-36°; the second group consisted of patients from Athens with a latitude of 38° and the third group of patients from northern Greece (Kavala and Salonica) with a latitude of 40.5^o.

From the percentage seasonal distribution curves is clear that the peak of favism in the first group occurred between 19-29 April, in the second group ten-twenty days later and in the third group even later at the end of May. On the other hand in the same area there were no considerable differences in the seasonal distribution of favism from one year to another.

Differences in the seasonal distribution of favism could be attributed to differences in the nature and the concentration of the noxious agent (whatever it may be) of fresh fava beans which is related to their stage of ripening, which depends on climate and temperature. It may also be argued that consumption is more extensive, during period of ripening.

3. CLINICAL ASPECTS

3.1. Onset of Clinical Symptoms

The time of onset of clinical symptoms after fava beans consumption varied but usually began during the first two days with malaise, abdominal pains, and not infrequently mild fever. The majority of the patients were admitted to hospital when major clinical signs of hemolysis were established. The most common symptom, which alarmed parents and for which they were referred to the hospital was hemoglobinuria. (Kattamis et al 1969)

The time in days which elapsed between the ingestion of fava

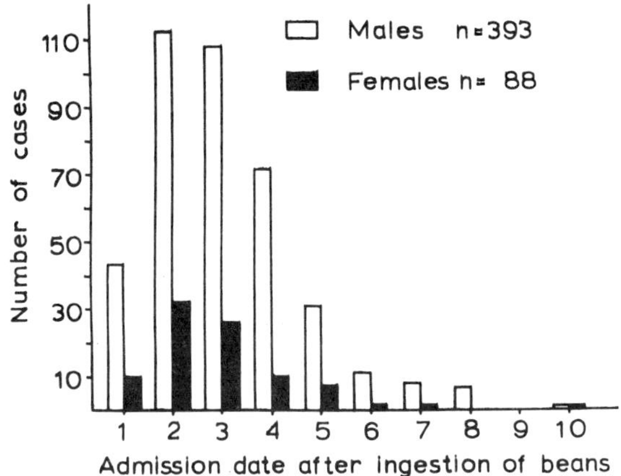

Figure 4. Day of admission after consumption of beans

beans and admission to hospital, is illustrated in figure 4.

The majority of patients entered the hospital as soon as the main clinical symptoms were evident. However some patients were admitted very early with rather high hemoglobin levels, while others after they had become severely anemic. The histogram at admission date is only a rough index of the time of onset and severity of hemolysis were not related to sex or the type of beans (fresh or dry), ingested.

About 75% per cent were admitted between the second and fourth day of consumption and only 6% after the sixth day. The 30 patients with late admission were not severely affected; hemoglobin ranged between 7-8 g/dl, and they managed to compensate hemolysis without transfusion. In contrast all patients admitted before the sixth day were transfused, even those with high initial hemoglobin levels which later dropped to less than 6-7 g/dl.

3.2. Anemia

Anemia varied considerably as illustrated by the histogram of hemoglobin levels on admission (figure 5). More than 80% were admitted with hemoglobin less than 7 g/dl and 36% with less than 5g/dl. In some patients anemia was extremely severe with hemoglobin levels of 1.5-3 g/dl. Nevertheless all patients recovered completely following blood transfusion.

In this series hemoglobin levels were similar in males and females.

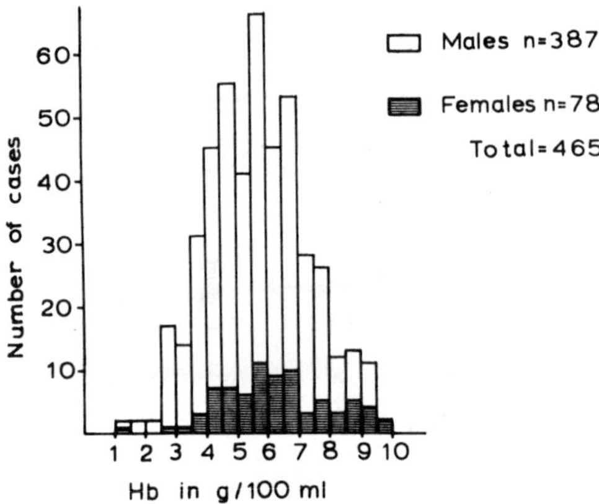

Figure 5. Hemoglobin levels on admission

3.3. Hemoglobinuria

As already mentioned this was the main manifestation initially observed by parents. In more than 80% of patients hemoglobinuria was the prevailing symptom. However in a small proportion and as a rule in patients with late admission, hemoglobinuria was not evident.

Hemoglobinuria, usually lasted for 1-3 days, and frequently was enhanced after transfusion. As a rule hemoglobinuria was a good index for the clinical follow-up of patients. Acute hemolysis seems to be severe and not compensated, as long as hemoglobinuria clears, hemoglobin usually stabilizes and gradually increases.

3.4. Jaundice

Jaundice is referred to as one of the main clinical signs of favism. In examining our patients we were impressed to find that jaundice usually was not prominent and frequently escaped diagnosis.(Kattamis et al, 1976) It was thus considered worth evaluating jaundice clinically and estimating the bilirubin levels of 85 consecutive cases on admission to hospital. The results are shown in figure 6.

In 40% bilirubin was less than 2mg/dl and jaundice was as a rule clinically undetectable; it was mild in 25% (bilirubin 2-4 mg/dl) moderate in 15% (bilirubin 4-6mg/dl) and severe (bilirubin above 6mg/dl) in only 20%.

Bilirubin was mainly of the unconjugated type though in some patients conjugated bilirubin was also found in small proportion.

Bilirubin levels were not related either to anemia and hemoglobin or to reticulocytes levels. (Figure 7)

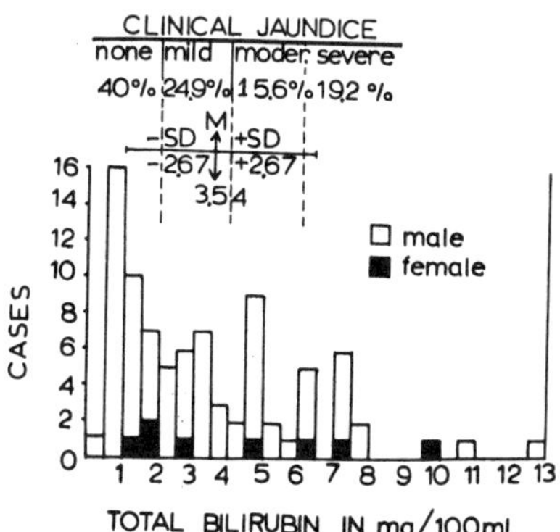

Figure 6. Bilirubin levels and jaundice in 85 patients

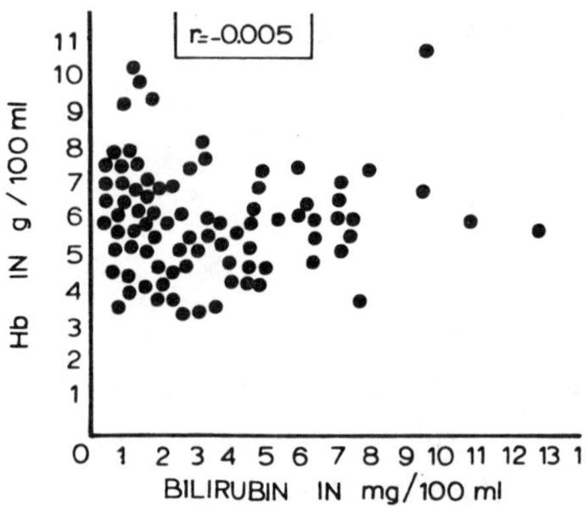

Figure 7. Relation of bilirubin to hemoglobin levels.

Of particular interest were the patients with severe bilirubinemia, for which all other possible causes of bilirubinemia and mainly liver impairment were excluded. The three patients with the higher levels of bilirubin, had also severe neonatal jaundice and were treated with exchange transfusion.

It was assumed that the pathogenesis of bilirubinemia in these severely jaundiced patients could be similar to that of hyper-bilirubinemia of G-6-PD deficient newborns, for which a liver factor, probably genetic in origin, has been incriminated.

3.5. Diagnosis

Diagnosis of favism in its severe form, is not difficult and is based on the history of fava beans consumption, on the clinical signs of acute hemolysis (hemoglobinuria, paleness and jaundice) and also on laboratory findings of hemolysis, mainly hemoglobinuria and extreme anemia, with characteristic red cell abnormalities reticulocytosis and G-6-PD deficiency.

4. LABORATORY DATA

4.1. Red Cell Morphology

During acute hemolysis there are characterestic changes in red cell morphology, pathognomonic for the experienced observer, are the fragmented red cells spherocytes, and erythrocytes with an abnormal distribution of hemoglobin which appeared dense and contracted at one side. (Sansone, Chiappara, 1984).

4.2. Reticulocytes - Serum Ferritin

This is another useful index to establish the diagnosis of hemolytic anemia. The degree of reticulocytosis depends greatly on the time of investigation. The variation of reticulocyte counts on admission is shown in figure 8.

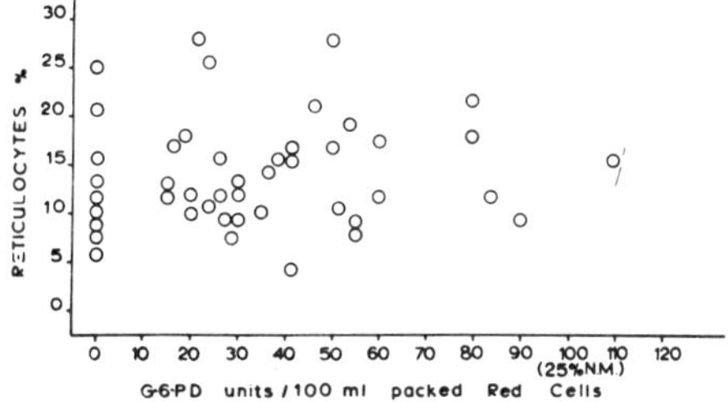

Figure 8. Reticulocytosis and relation to G-6-PD activity in male patients with favism.

Hemolysis is also associated with very high levels of serum ferritin. In a recent study in 30 patients, it was demonstrated that serum ferritin were very high during hemolysis (727±377 ng/ml) as compared to period of rest (28.4±12 ng/ml).

4.3. G-6-PD Activity

Red cell G-6-PD activity could be studied by different methods. During acute hemolysis the most reliable are the quantitative estimation and cyanomethemoglobin elution, technique.

The enzymic activity in 90 male patients during crisis and also during period of rest is illustrated in figure 9.

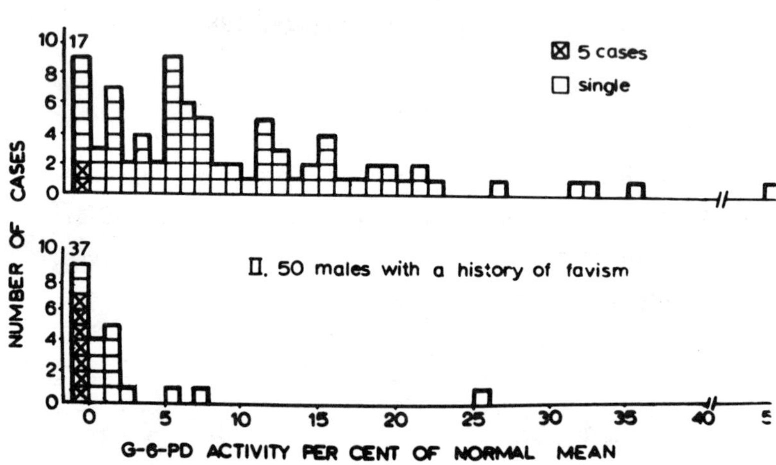

Figure 9. Red cell G-6-PD activity in per cent of normal mean, in 90 male patients one acute crisis and in 50 with a history of favism.

In patients during period of rest the enzymic activity was completely absent in 37 (75%). In the remainder it did not exceed 5-10% of the normal mean; these levels are consistent with those of the common Mediterranean variant, mainly seen in Greece.

Of the male patients studied during crisis only 17 (19%) were totally deficient as compared to 75% at the period of rest; during crisis the majority of patients showed an increase in enzymic activity which never reached normal levels and ranged between 5-35% of the normal mean. The increase of activity during crisis was attributed to the destruction of old cells and overproduction of young erythrocytes. At the peak of the hemolytic process a rather homogeneous red cell population with young erythrocytes is expected to be present. This explains the mild increase of enzymic activity in most of our patients as it is known that the young red cell population of the Mediterranean variant has an activity of about 30% of the normal mean compared to that of 1% in the population of old cells.

Of interest were the findings of the cyanomethemoglobin elution procedure in which deficient erythrocytes appear as ghosts because of the elution of cyanomethemoglobin. In male hemizygotes nearly all red cells are deficient and appear as ghosts. During crisis it was found that a good percentage of red cell appeared to be normal. This means that young red cells which are partially deficient give a falsely normal appearance. The same picture of pseudomosaicism was demonstrated previously in patients with the mild Greek variant. (Papayannopoulou , Stamatoyannopoulos 1964)

G-6-PD activity was also determined during crisis in 22 females with severe hemolysis. Enzymic activity and the presumed genotype of these patients are shown in Figure 10. G-6-PD activity ranged widely from total absence to 95% of the normal mean. This could be explained by the fact that affected females were either homozygotes, as were assumed for six patients or heterozygotes. The presumed genotype in these patients was based on enzymic studies of their parents. Logically any female heterozygote may have hemolysis after eating fava beans; the spectrum of hemolysis is expected to be wide and in relation of the percentage of deficient cells and the degree of enzymic activity in heterozygotes, which during period of rest ranges from zero to normal levels.

Severe hemolysis similar to that of male hemizygotes is expected to occur only in homozygotes or heterozygotes with a severe degree of enzymic deficiency.

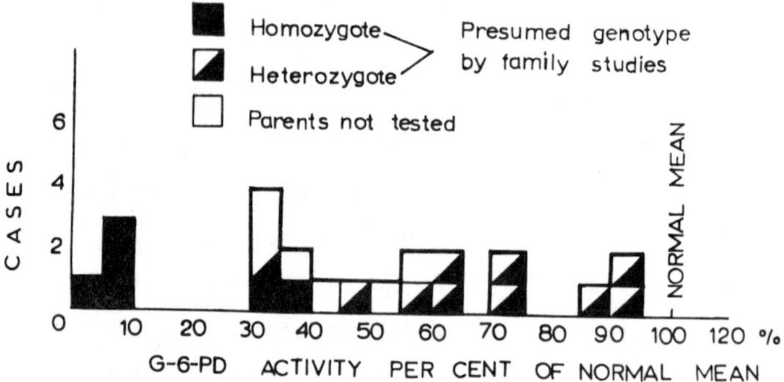

Figure 10. Red cell G-6-PD activity in 22 female patients, during crisis.

Severely deficient heterozygotes may have normal enzymic activity during crisis as was the case with some of our female patients. On crisis, all old deficient cells are destroyed and enzymic activity is the measured on a red cell population consisting of young deficient erythrocytes, and of all normal unhemolysed old and young red cells which also are increased and have a very high activity.

As a matter of fact the severity of hemolysis in heterozygotes is expected to be associated with the degree of enzymic deficiency and the percentage of deficient cells which are prone to hemolysis when affected by the toxic substances of beans. However at present our knowledge on the clinical spectrum of hemolysis in heterozygotes with favism is limited. This assumption was further complicated by the findings of 25 heterozygotes, mothers of male patients who ingested the same type of fava beans which caused severe hemolysis to their sons and were studied simultaneously with their children. None of them had clinical signs of severe hemolysis but hematologic studies disclosed that 5 (20%) of them had mild hemolysis indicated by an increase in reticulocytes (5-7%) and a drop in hemoglobin ranging from 1.8-3.1 g/dl. G-6-PD activity estimated by cyanomethemoglobin elution, quantitative and semi-quantitative determination, in these mothers ranged considerably; anyhow the presence of mild hemolysis was unrelated to the degree of enzyme deficiency. (Kattamis et al, 1972)

It is thus postulated that heterozygotes may experience two types of hemolysis; a mild rather common and compensated one, which is unrelated to the degree of G-6-PD deficiency and a severe one which is proportional to the degree of enzymic deficiency and for which additional predisposing factors are necessary.

A similar compensated mild hemolysis, was disclosed in 8 heterozygous mothers, who consumed fava beans, and their breast fed infant had hemolysis. All had mild increase in reticulocytes (4-6%) and 2 had a mild drop in hemoglobin which restored to normal in 2-3 weeks.

COMMENTS

It is well documented that frequently ingestion of fava beans by G-6-PD deficient individuals is not followed by severe hemolysis with the clinical manifestations already described. Thus the question raised is whether consumption of beans is followed by a mild compensated hemolysis, which may escape diagnosis. To answer this question we examined 10 G-6-PD deficient male siblings of male patients with favism, who consumed fava beans simultaneously with the affected member of the family. None of the siblings had laboratory evidence of hemolysis.

Another point necessitating discussion is the existing evidence for a decline in the prevalence of favism, during the last decade in Greece. In 1983 only 15 patients were treated in our department, with a total of 12.267 admission. The annual rate is only 1.3 per thousand as compared to 5.3 per thousand during the years 1969-1971.

Two main causes are assumed to be responsible for this reduction. The first is related to the considerable changes noted in recent years in the nutritional habits of the urban population of Greece, and the second to intensive health education campaigns.

Prevention of hemolytic anemias and severe neonatal jaundice are included in health education programs while paediatricians play a major role in these programs. To this end G-6-PD screening was added lately as a supplement to the neonatal screening programs of hypothyroidism and hyperphenylalanemia.

The validity of G-6-PD screening is greatly questioned for prevention of severe neonatal jaundice. In contrast, written instructions forwarded to parents of deficient newborns may cause great anxiety in relation to drug administration.

The incidence in a variety of psychological problems among G-6-PD deficient families is expanding. These problems are related to the way of counseling regarding the relation of G-6-PD deficiency to drug induced hemolysis. The list includes drugs which have been incriminated as causing hemolysis in rare cases.

In our experience hemolysis due to drugs commonly used in paediatric practice is not only extremely rare but also mild and well compensated. In a five year period only three patients were treated in our department with drug induced hemolysis. Two had severe infections and received a number of drugs, while the third was a patient with sickle-cell anemia and G-6-PD deficiency, who received analgesics, including novalgin, for thrombotic and probably hemolytic crisis. (unpublished data)

REFERENCES

Fessas, Ph., Doxiadis, SA., and Valaes, I. (1962). Brit. Med. Jour. 2, 1359

Fiorelli, G., Silvetti, M., Binaghi, P., Crisponi, G., and Marras, A. (1970). Abstracts, XIII International Congress of Hematology, Munich, p.180

Kattamis, C.A. (1967). Acta Paed.Scand., Suppl., 172, 303

Kattamis, C.A., Kyriazakou, M., and Haidas, St. (1969). J.Med.Genet., 6, 34

Kattamis, C.A., Haidas, A., and Haidas, St. (1969a). J.Med.Genet., 6, 286

Kattamis, C.A., and Tjortzatou, F. (1970). J.Pediatr., 77, 422

Kattamis, C.A., Haidas, St., Tjortzatou, F., Lagos, P., and Metaxotou-Mavromati, A. (1972). Archieves Greek Society Hematology, i,92, (in Greek)

Meloni, T., Cagnazzo, G., Dore, A., and Cutillo, S. (1973). Journal of Pediatrics, 82, 1048

Papayannopoulou, T., and Stamatoyannopoulos, G. (1964) Lancet, ii, 1215

Sansone, G., and Chiappara, M. (1984). Pathologica, 76, 191

Sartoni, E. (1971). Journal Medical Genetics, 8, 462

Stamatoyannopoulos, G., Fraser, G.R., Motulsky, A.G., Fessas, Ph., Akrivakis, A., and Papayannopoulou, T. (1966). Am.J.Hum. Genet., 18, 253

World Health Organization. Standardization of Precedures for the study of G-6-PD. (1967). No 366 Geneva

ETIOLOGICAL ASPECTS OF FAVISM[1]

P. Arese, L. Mannuzzu, F. Turrini, S. Galiano, and G.F. Gaetani**

Department of Hygiene, University of Torino, Torino, and *Department of Hematology, University of Genova, Genova, Italy

I. THE FAVIC CRISIS

The acute red cell (RC) lysis which occurs in sensitive individuals after ingestion of Vicia faba beans is called favism (1-3). Sensitivity is due to the deficiency of RC glucose-6-phosphate dehydrogenase (G6PD) (4). Only low-activity variants of G6PD confer sensitivity toward fava bean components. Favism resembles drug-induced hemolysis, but differs from it in that only a few G6PD-deficient subjects appear to be sensitive to fava beans (4). It is currently held that favism can be produced in the following ways: 1) by eating fresh raw beans; 2) by eating fresh or frozen cooked beans; 3) through mother's milk in breast-fed children. Favism after eating dry cooked beans is very rare (1). The disease generally starts some hours after ingestion, with chills, weakness and pallor. Hemoglobinuria followed by jaundice appears in some cases as a result of massive hemolysis. Death may occur from acute anemia if blood transfusions are not given in time. Favism is not always a dramatic disease: abortive forms

[1] *Supported by CNR Grants PF 84.00864.51 to P.A. and PF 60/84. 0087.51 to G.F., and by the Thalassemia Research Institute (CNR Cagliari).*

do exist and might be widespread. Females heterozygous for G6PD-deficiency can also experience hemolytic crises of favism, and the severity of hemolysis is directly correlated to the degree of mosaicism (1,3,4).

II. GENETIC ASPECTS

If the G6PD-deficiency is the only factor governing susceptibility to fava beans, all G6PD-deficient subjects should present hemolytic crisis after ingestion of fava beans. As a matter of fact there are numerous deficient subjects who are used to eating fava beans and never experience anemia. For this reason the existence of several additional factors have been considered in the pathogenesis of favism.

Favism is prevalent in the Mediterranean area, in the Middle East and in South-East Asia. Moreover the incidence of favism varies to a considerable extent according to the geographical area, and it does not parallel the rate of prevalence of G6PD-deficiency. Stamatoyannopoulos et al. (5) report that in the Karditsa area in Greece, where G6PD-deficiency is 27 percent, favism is rare; by contrast, in Corfù with a deficiency of G6PD of only 5 percent, favism is more common. Family studies by these authors suggest the possibility of influence of a second autosomal locus in the pathogenesis of favism. Their studies seem to indicate that predisposition is transmissible from one generation to the next and that heterozygosity for the second trait is sufficient to enhance susceptibility. The authors conclude that possession of the postulated trait and of G6PD-deficiency does not necessarily lead to favism after the consumption of fava beans, but susceptibility to acute hemolysis is enhanced. The second postulated genetic trait in the case of favism may involve the red cell itself, increasing its susceptibility to hemolysis, or the absorption, detoxication or excretion of the toxic agents of fava beans.

As far as the RC are concerned, no differences have been detected among the biochemical characteristics of G6PD of subjects suffering or not of favism (6). Association of RC genetic polymorphisms and favism has been investigated by Bottini et al. (7). The authors found that the analysis of the distribution of phenotypes and alleles of RC acid phosphatase are significantly different in G6PD-deficient subjects with favism than in the general population. Another study, related to the absorption, detoxication or excretion of the noxious agents of fava beans has been done by Cassimos et al. (8). The authors demonstrated a clear cut difference in the urine excretion of D-glucaric acid between controls, subjects with favism and subjects with G6PD deficiency only. More than 90 percent of the subjects with a past history of favism had a reduced excretion of D-glucaric acid, where only 33 percent of subjects whose G6PD deficiency was a laboratory finding had the same low values. Only 4 percent of normal children had a reduced excretion of D-glucaric acid. The low urinary excretion of D-glucaric acid could reflect the additional genetic abnormality present in subjects susceptible to fava beans. Unfortunately these authors have not carried out any family study related to the abnormality found.

In a search of other factors that may play a role in the pathogenesis of favism, studies were carried out on beta-glucosidase activity of small intestine biopsies from normal subjects and G6PD-deficient subjects with or without favism (9). Beta-glucosidase might be involved in the absorption and metabolism of toxic agents present in fava beans; a quantitative polymorphism could explain the different susceptibility to fava beans of G6PD-deficient subjects. However no consistent variation of beta-glucosidase activity was evident among the subjects examined (9).

Susceptibility to fava beans is largely confined to the Mediterranean area, although cases have been reported in other countries, and in different G6PD variants. All cases observed had a very low enzyme activity and no cases of hemolytic anemia have been observed in subjects of African origin carrying the G6PD A^- gene (4).

III. HAZARDOUS CONSTITUENTS OF VICIA FABA

Fava beans (*Vicia faba major*, broad bean) represent 13 percent of the world pulse production and is the fourth most important pulse crop in the world (10). It is a staple food in Southern Italy, Sicily, Sardinia, Greece, Egypt, the Middle East and Iran. Fava beans are widely cultivated because of their fast growth in semi-arid climates during the winter-spring time with no need for irrigation. Fava beans are rich in valuable nutrients, such as proteins (22-26 percent, wet weight basis), starch (43-45 percent, wet weight basis), and lipids (2.2-2.7 percent, wet weight basis) (11). Fava beans are also extremely rich in two glycosidic compounds, vicine and convicine (V-C), which generate the redox aglycones divicine (2,6-diamino-4,5-dihydroxypyrimidine) and isouramil (6-amino-2,4,5-trihydroxypyrimidine) upon splitting of the beta-glycosidic bond between glucose and the hydroxyl group at C5. Fava beans also contain high amounts of ascorbate as well as varying concentrations of L-DOPA glycoside (2). Table I illustrates data from the literature on the V-C levels in different cultivars of fava beans. An age-dependent decrease in V-C levels has been constantly observed. Sisini et al. (19,29) have shown that such decay is remarkably constant in all fava bean strains studied sofar. This age-dependent decay is evident either in a variety of strains, or in a single strain studied at different stages of maturation (Fig. 1). Fava bean seeds contain a rather active beta-glucosidase which is able to split the beta-glycosidic

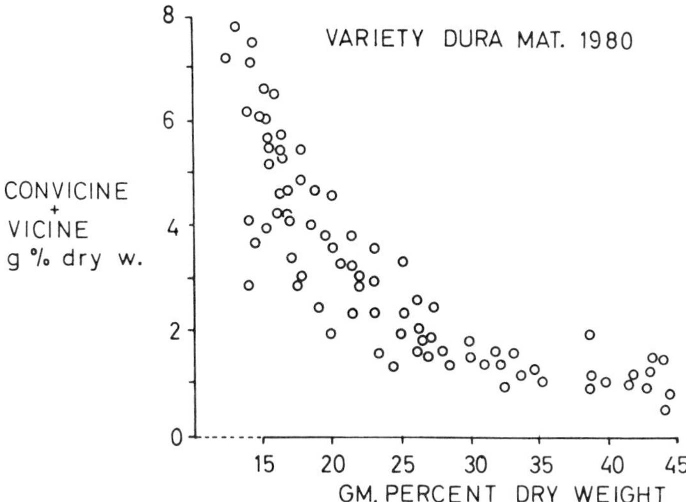

Fig. 1 Age-dependent decay of convicine plus vicine in a pure line of Vicia faba . Data according (19,20).

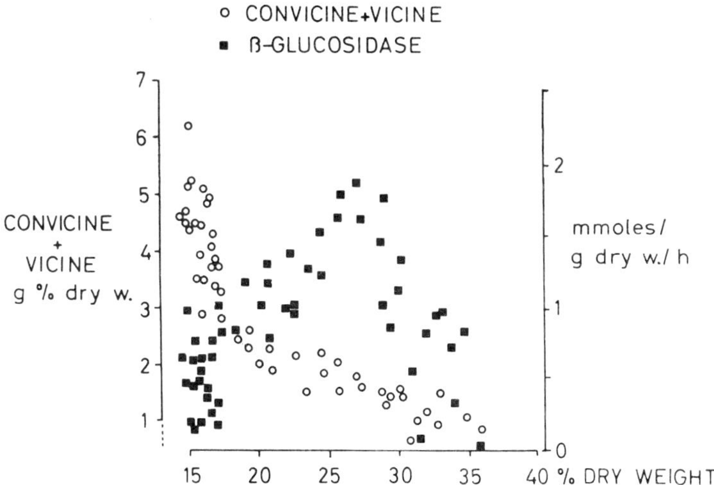

Fig. 2 Age-dependent behavior of convicine plus vicine and beta-glucosidase activity in 17 pure lines of Vicia faba chosen at random. Data according (19,20).

Table I. Levels of vicine and convicine in Vicia faba seeds

	No. of species	Type of sample	Vicine Mean	Vicine Range	Convicine Mean	Convicine Range
Bjerg et al. (12)	170	Dry	0.55	0.19-1.02	0.32	0.11-0.71
Chevion and Navok (13)	1	Fresh	0.48	=	=	
Collier (14)	14	Dry	0.98	0.89-1.07	(vicine+convicine)	
Engel (15)	21	Dry	0.42	0.27-0.68	=	=
Gardiner et al. (16)	78	Dry	0.59	0.45-0.90	0.26	0.15-0.54
Jamalian (17)	58	Dry	0.33	0.23-0.61	=	=
Olsen and Andersen (18)	3	Dry	0.73	0.69-0.77	0.29	0.26-0.32
Sisini (unpublished)	20	Fresh	0.93	0.3-1.45	(vicine+convicine)	

Data are expressed as g percent dry weight

bond of V-C at almost the same rate as the chromogenic substrate p-nitrophenylphosphate. Beta-glucosidase activity is very low in young seeds, increases about six-fold to a maximum in ripe seeds which have 20-27 percent dry weight, and drop again in older seeds (Fig. 2). The strong variability in both V-C levels and beta-glucosidase activity with seed ripening is likely to play a role in triggering the hemolytic crisis. Small children preferentially eat very young beans (dry weight: 15-25 percent), which are tender and sweet. These beans have extremely high levels of V-C, but low beta-glucosidase activity. Riper beans are usually cooked; they have very high beta-glucosidase activity, and still contain between 1 and 2 percent V-C, dry weight basis. It is customary in several Mediterranean countries to dry fava beans and eat them after months of storage. This procedure did not affect the levels of V-C, but reduced to nil ascorbate. Beta-glucosidase too dropped remarkably (19).Boiling fava beans did not lower the levels of V-C. The beta-glycosidic bond was split after few minutes treatment with hot 0.5 N HCl; however, 0.1 N HCl at 37°C was not able to split the beta-glycosidic bond (21). Beta-glucosidase is heat-labile, and is completely inactivated by boiling fava beans for 30 min. Its resistance to an acidic milieu simulating gastric juice was also tested. One hour incubation of homogenized fresh fava beans in 0.1 N HCl, or 0.01 N HCl supplemented with 1 mg/ml pepsin at 37°C led to 80 percent or 30-32 percent inactivation, respectively (A. Sisini, unpublished). Up to now efforts to select fava bean strains devoid of or low in V-C have failed. Hundreds of cultivars grown under very different climatic conditions gave very similar levels of V-C.

IV. RED CELL GLUTATHIONE IN FAVISM

A. Effect of Divicine in a Cell-Free System

The effects and mechanism of action of divicine and isouramil are practically the same. Due to the easier availability of divicine, all the following studies have been performed with this substance. Upon hydrolysis of vicine either by beta-glucosidase or by acid in oxygenated buffered medium, the semiquinoid free radical species of divicine (D-O·) is generated. A possible mechanism is the one-electron reduction of dioxygen to superoxide anion:

$$\text{D-OH} + O_2 \rightarrow O_2^{\bar{\cdot}} + \text{D-O}^{\cdot} + H^+$$

Superoxide anion is spontaneously dismutated to hydrogen peroxide:

$$O_2^{\dot{-}} + O_2^{\dot{-}} + 2\,H^+ \rightarrow H_2O_2 + O_2$$

In this reaction almost equimolecular amounts of superoxide anion are formed and oxygen consumed (22) during vicine hydrolysis at pH 6.0. The free radical species has been detected by ESR spectroscopy (23). The signal was greatly enhanced when the pH value of the incubation mixture was increased from 6 to 9, and almost disappeared when the pH was lowered below 5. The ESR spectrum was relatively stable with time, showing a 20 percent decrease in intensity after 15 min at pH 6.0. No signal was seen if beta-glucosidase was omitted from the incubation mixture or if divicine was maintained in a nitrogen atmosphere, indicating that the release of the aglycone and its reaction with oxygen are essential for radical formation. The addition of GSH completely suppressed the signal while NADPH left it unchanged (23,24). The semiquinoid free radical form of divicine reacts with GSH. A one-electron hydrogen abstraction occurs and oxidized glutathione, GSSG, is generated with the intermediate formation of the tihyl radical $GS^{\bullet}$ (25):

$$GSH + D\text{-}O^{\bullet} \rightarrow GS^{\bullet} + D\text{-}OH$$

$$GS^{\bullet} + GS^{\bullet} \rightarrow GSSG$$

Direct oxidation of GSH by $D\text{-}O^{\bullet}$ is fast and complete within about 1 min. Almost stoichiometric amounts of GSSG are formed. According to Benatti et al. (26) after 60 min incubation in air at 37°C, more than 90 percent of GSH has been converted to GSSG, while about 8 percent was accounted for by two GSH-divicine adducts: one of them ("305 nm peak") was completely reduced by NADPH and glutathione reductase, while the second ("320 nm peak") was a dead-end product. During the fast oxidation a stoichiometry of one mol oxidized glutathione per mol divicine was observed. The fast, stoichiometric oxidation of GSH is followed by a slow oxidation which continues until all GSH has been oxidized (Fig. 3 and 4). The slow and steady oxidation of GSH is due to the redox cycle:

$$a)\ D\text{-}O^{\bullet} + GSH \rightarrow D\text{-}OH + GS^{\bullet}$$

$$b)\ D\text{-}OH + O_2 \rightarrow D\text{-}O^{\bullet} + O_2^{\dot{-}} + H^+$$

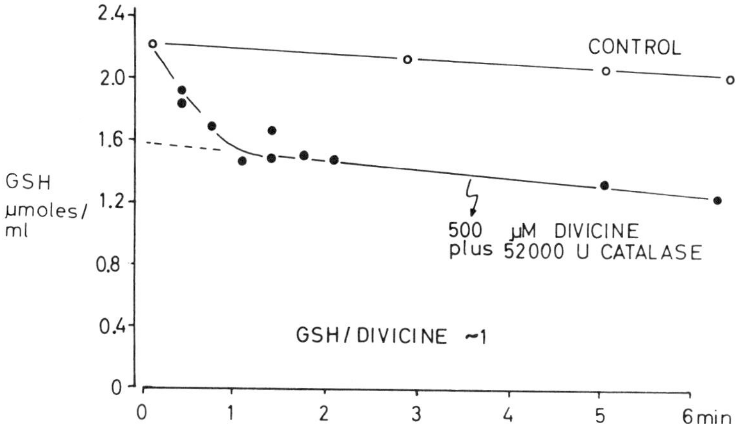

Fig. 3 Short-term kinetics of GSH oxidation by divicine in a cell-free system. The reaction medium was phosphate buffered saline, pH 7.4, 37°C. Data according (28).

The reaction between divicine and GSH is pH-dependent and is accelerated at more alkaline pH-values (Fig. 4). This reflects the higher amount of the free radical species at alkaline pH-values. The formation of the superoxide anion and the cyclic redox

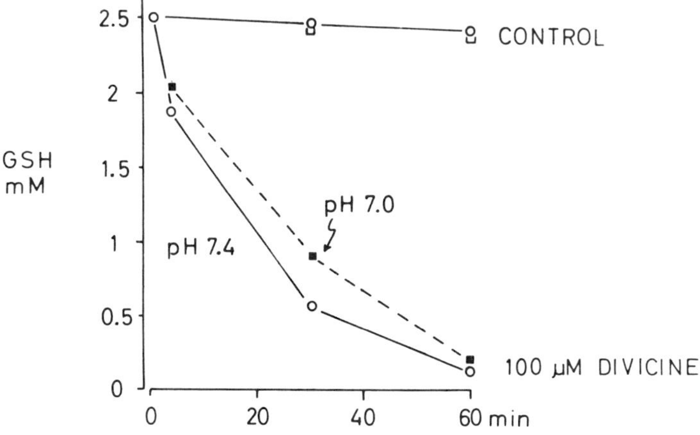

Fig. 4 Influence of pH on GSH oxidation by divicine in a cell-free system. The reaction medium was phosphate buffered saline, pH 7.4, 37°C. Data according (27).

process were demonstrated by the reduction of ferricytochrome c by divicine, and by the regeneration of the reducing divicine species by addition of GSH (27). Spontaneous dismutation of superoxide anion leads to the formation of hydrogen peroxide. Addition of catalase had a minimal inhibitory effect on GSH oxidation by divicine, especially at high divicine concentration. This indicates that although divicine may generate hydrogen peroxide during cyclic oxidation-reduction, the action on GSH is primarily via direct attack. Oxygen radicals other than superoxide anion do not seem to be involved, since scavengers such as mannitol, beta carotene, ethanol, uric acid or alpha-tocopherol had no effect on the reaction kinetics when applied in excess concentrations (27).

B. Effect of Divicine in G6PD-Deficient Red Cells

Divicine added in different final concentrations to suspensions of G6PD-deficient RC always produces end-point kinetics of GSH oxidation. The oxidation comes to a halt in about 20-30 min at 37°C, and only a minimal regeneration of reduced GSH is observed when extending the incubation period up to 24 hours. A stoichiometry of about 1 mol oxidized GSH per mol divicine was constantly observed in the range 50-500 μM divicine. Addition of ascorbate increased the stoichiometry to a maximum value of 2.73 at 100μM divicine, but did not modify the end-point kinetics (Table II). Upon addition of divicine a burst of hydrogen peroxide formation was monitored in the RC suspension. The maximum level attained few nmoles/ml RC suspension in the first 2-3 min, and dropped to almost nil thereafter (28). If divicine was added to de-oxygenated RC suspensions, a more pronounced effect was observed (Fig. 5). A reason for this may be the distinctly higher intracellular pH-value of deoxygenated RC, which favors the formation of the semiquinoid radical form of divicine. Exact stoichiometric amounts of GSSG were formed parallel to GSH oxidation (Fig. 6). Since the

Table II. Stoichiometry between GSH oxidation and divicine in G6PD-deficient red cells. Data according (28).

Divicine µM	Ascorbate µM	GSH oxidized µM	GSH/Divicine [a]
50	=	43	0.86
50	50	90	1.80
50	250	74	1.48
100	=	132	1.32
100	50	215	2.15
100	250	273	2.73
250	=	312	1.24
250	50	372	1.48
250	250	472	1.88
500	=	482	0.96
500	50	502	1.00
500	250	527	1.05

[a] *RC were suspended in isoosmotic, isoionic buffer, pH 7.45, Ht 36-38 percent; 60 min incubation in air atmosphere at 37^oC. Mean of two experiments. GSH assayed in triplicate.*

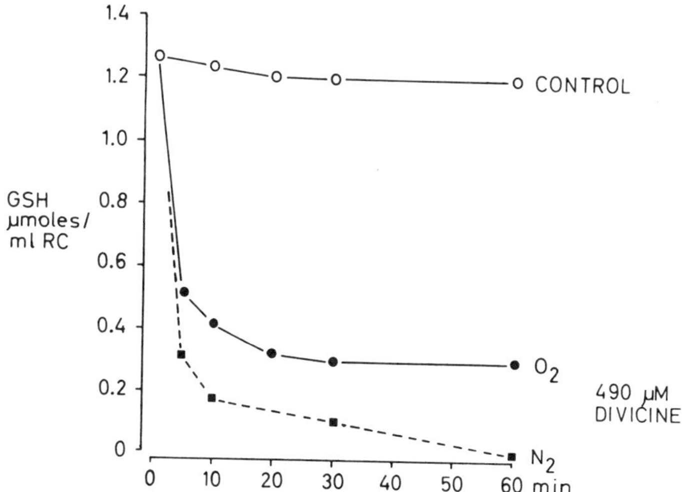

Fig. 5 Influence of oxygenation and deoxygenation on the reaction between GSH and divicine in G6PD-deficient RC. Washed RC were suspended in a isoosmotic, isoionic buffer, pH 7.45, Ht 30 percent, 37^oC and incubated in a IL Tonometer flushed with air (O_2) or nitrogen (N_2). Data according (27,28).

"305 nm peak" complex described by Benatti et al. (26) reacts with NADPH and glutathione reductase, we cannot exclude that a portion of GSSG shown in Fig. 6 would be such a GSSG-divicine adduct.

As shown previously by Srivastava and Beutler (29), GSSG is permeable across the RC membrane. We observed such a leakage in long-term incubations. The rate of GSSG leakage is constant over 6-8 hours and amounts to 0.15-0.17 μmoles /ml RC/h. The effect of divicine on GSH is very similar to that of diamide, although the latter reaction is faster and its stoichiometry is two moles GSH oxidized per mole divicine added (30).

C. Red Cell Glutathione During the Favic Crisis

Lowered GSH and increased GSSG are constant findings during the early stages of favic crisis (31). As a general rule, the level of RC GSH inversely correlates with the precocity and gravity of the crisis, and is lowest in the most dense RC fraction which collects the more damaged cells. As shown in Table IIIA, in three very severe precocious crises characterized by low hematocrit

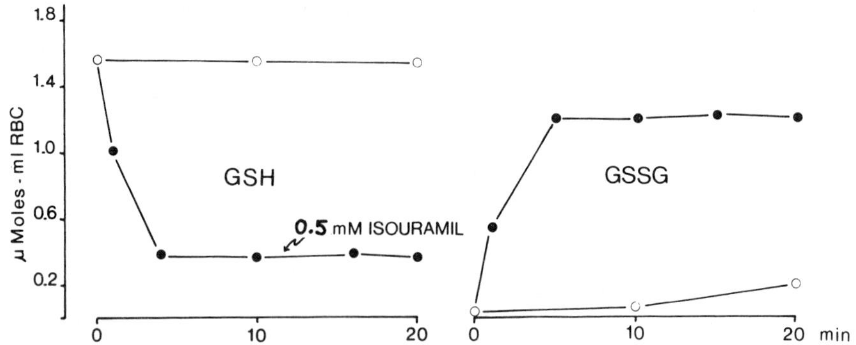

Fig. 6 Stoichiometry between GSH oxidation and GSSG formation in G6PD-deficient RC treated with 0.5 mM isouramil. Washed RC were suspended in isoosmotic, isoionic buffer, pH 7.45, Ht 40 percent, and incubated in air atmosphere at 37°C. Data according (30).

Table III. GSH, cross bonded red cells and Heinz bodies in favic crisis

	Crisis No.	Ht	GSH nmoles/ml RC	Cross bonded RC [a]	Heinz bodies [b]
A) Crisis in progress	6/84	29	113	50	500
	7/84	20	298	80	1280
	19/84	32	342	19	367
	20/84	20	600	63	454
B) Crisis in remission	4/84	27	1592	1	84
	5/84	48	2240	1	=
	9/84	23	1724	1	91
	10/84	35	2102	1	99
	12/84	29	1582	1	125

C) Time-course in a favic crisis

Days after bean consumption	1	2	3	4	5	6
GSH nmoles/ml RC	430	1836	1861	2044	2021	1743
Heinz bodies [b]	362	92	139	77	119	103
Ht	20	19	19	18	26	37
Cross bonded [a] RC	11	1	1	1	1	1

[a] *Cross bonded RC are expressed as percent of total RC*

[b] *Heinz bodies are expressed as turbidity (O.D. x 10^{-3} at 700 nm) of lysed RC before and after centrifugation*

values and high percentages of cross bonded RC (see Section VII,B) extremely low GSH values are measured in the most dense fraction while less severe crises or crises already in the remission phase have distinctly higher GSH levels (Table IIIB). The longitudinal behavior of GSH in a single crisis which was followed from its very beginning is shown in Table IIIC). Restoration of overshooting levels of GSH is very fast.

V. METABOLIC MODIFICATIONS IN FAVISM

In vitro studies based on $^{14}CO_2$ evolution from intact RC incubated in the presence of 1-^{14}C-glucose show that G6PD-deficient RC in unstressed conditions utilize glucose through the pentose phosphate pathway (PPP) at a rate similar to normal RC. On the other hand, in the presence of oxidizing agents, PPP in normal RC is stimulated several fold while in G6PD-deficient RC carrying the Mediterranean variant the stimulation is minimal (32). This behavior may be explained by taking into account the different intracellular metabolic conditions prevailing in normal and in G6PD-deficient RC. In the former almost all $NADP^+$ is present in the reduced form, available $NADP^+$ being less than 1 percent, so that G6PD is operating at a very low rate. By contrast, in G6PD-deficient RC the $NADPH/NADP^+$ ratio is reversed, NADPH being only 25 percent of the total in G6PD-deficient (Mediterranean variant) RC, and PPP is already operating at near maximal rate (32,33). In normal RC under oxidative stimulation, while the flow of glucose through the PPP increases several fold and thus neutralizes the higher levels of hydrogen peroxide generated by the oxidizing agent, the coenzyme ratio remains unchanged and so does the GSH concentration. On the other hand, in G6PD-deficient RC, the glucose flow through the PPP can hardly increase at all, and a fall in NADPH and GSH is observed, with an accompanying increase in hydrogen peroxide.

This has been verified *in vivo* during favic crisis. During this investigation a significantly lower concentration of ATP was also observed (Table IV). The extent of stimulation of the PPP required to cause irreversible damage in deficient RC is not high. By the $^{14}CO_2$ evolution method it was found that when normal RC are incubated in autologous serum, containing the metabolites of ingested primaquine, the stimulation of PPP was only 15-25 percent (34). Yet, under these conditions there would be massive

Table IV. Metabolic differences in G6PD-deficient red cells during favic crisis. Data according (31).

	GSH nmoles/ml RC	$NADPH/NADP^+$	ATP nmoles/ml
Resting conditions	1433	0.39	1566
During the favic crisis	897	0.10	1143

hemolysis <u>in vivo</u> of G6PD-deficient RC. These data are surprising only at first sight if we consider that in G6PD-Mediterranean RC, in spite of maximal intracellular activity of PPP, generation of NADPH is barely sufficient to avoid chronic hemolysis, and that in older RC the $NADPH/NADP^+$ ratio is even more disturbed, so much that even a mild oxidative stress can tip the balance.

The metabolic alterations observed in G6PD-deficient RC during hemolytic crisis after ingestion of fava beans have been confirmed <u>in vitro</u> after incubation in the presence of divicine and isouramil (30,35). Both compounds stimulate the PPP of normal RC even at very low concentration (50 μM), with a considerable potentiation by ascorbic acid. G6PD-deficient RC are functionally unable to increase the rate of PPP, and an irreversible oxidation of GSH and NADPH occurs. It must be pointed out that no difference in the response to divicine or isouramil has been observed between RC of subjects with or without a past history of favism (30).

VI. CALCIUM HOMEOSTASIS IN FAVISM

A. Effects of Divicine in G6PD-Deficient Red Cells

Low passive permeability and high activity of a membrane bound calcium pump keep intraerythrocytic calcium low (about 15 μM/liter cells) (36-38). Increase in intracellular calcium modifies the activity of several enzymes and is considered detrimental to RC function. An influx of calcium has been proposed as a final common pathway for cell death induced by many toxins (39). Increased calcium activates a number of latent enzymes activities (40-43). High intracellular calcium may produce the echinocytic shape change by accumulation of diacylglycerol (43) or by the reversible modification of the spectrin-actin cytoskeleton, which is linked to the membrane _via_ the transmembrane protein glycophorin (44). High calcium, finally, critically affects RC rheology and is constantly associated with decreased deformability (45). Calcium-related cell rigidity appears to be important in the pathogenesis of hemolysis in sickle cell anemia, in uremia and beta thalassemia (46-48).

Quite recently De Flora et al. (49) have studied the effect of divicine plus ascorbate on Ca-ATPase activity in the RC. Three hour incubation of normal and G6PD-deficient RC in buffered saline solution with 2 mM divicine _plus_ 200 μM ascorbate inhibited Ca-ATPase activity by 61-64 percent in both conditions. GSH, which was drastically lowered in the treated G6PD-deficient RC, and ATP, which was unchanged apparently do not influence the pump inactivation.

The effect of divicine on the passive calcium permeability was also studied (Fig. 7). Intact RC were incubated in plasma with 100 μM divicine _plus_ 30 μM ascorbate. This ratio reflects that found in fresh fava beans. After different incubation times (from 30 min to 7 hours) RC were isolated, ATP-depleted by addition of iodoacetate and incubated further with ^{45}calcium for 15-240 min.

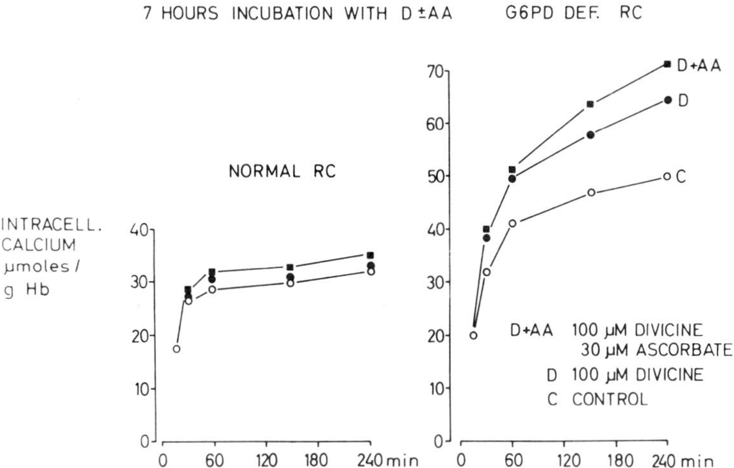

Fig. 7 Effect of divicine and ascorbate on the passive calcium influx into ATP-depleted normal and G6PD-deficient RC. Washed RC were suspended in own plasma (Ht 45 percent) and incubated for 7 h at 37°C. Divicine plus/minus ascorbate was added at time 0. RC were then washed three times with a isoosmotic, isoionic buffer, pH 7.4, and suspended in the same buffer containing 0.5 mM iodoacetate. After 90 min incubation and three washes the RC were suspended in the same buffer without iodoacetate, and supplemented with 1.5 mM calcium plus ^{45}calcium. At selected times aliquots were freed of extracellular radioactivity by passage through Ficoll, and cell-associated radioactivity counted (F. Turrini, L. Mannuzzu, A. Naitana, unpublished results).

Extracellular ^{45}calcium was removed by centrifuging the cells through a Ficoll cushion and the cell-associated ^{45}calcium counted. The rapid exponential increase of cell-associated radioactivity reflects isotopic equilibration with an easily accessible (membrane)compartment , while the asymptotic part of the curve corresponds to net influx. This is extremely low in normal, treated or untreated RC, and is markedly increased in divicine treated G6PD-deficient RC. Increase in passive calcium permeability was also observed after 30 min incubation of deficient RC with 500 µM divicine (not shown). In this case, however, a distinctly higher rate of isotope penetration was observed.

B. Calcium Homeostasis During the Crisis

Two independent reports (49,50) have recently shown remarkable increase in RC calcium level during the favic crisis. The values reportad by De Flora et al. (49) in children (four cases) range from 195 to 555 μmoles per liter cells, while Turrini et al. (50) in four cases (mostly adults) found lower values, ranging from 143 to 244 μmoles per liter cells in the lowest fraction of density separated RC. Except for their age, the probands were comparable in that all of them came from North Sardinia and were examined in the same period. Both groups found lowered Ca-ATPase activity. The activity was lowered by 56 percent (mean of 7 crises) according to De Flora et al. (49), and 57 percent (mean of 4 crises) according to Turrini et al. (50). This last result is referred to the most dense fraction, while the top fraction had constantly increased Ca-ATPase activities.

The calcium extruding capacity of human RC is very high, so that decrease in maximal capacity of the enzyme cannot explain the drastical increase in intracellular calcium levels observed in the favic crisis. By contrast, combination of increased passive permeability and decreased outpumping might offer a plausible explanation. The mechanism by which divicine opens the calcium channels is unknown. A working hypothesis could be the following: diamide and divicine are sulfhydryl reagents that stimulate phagocytosis of G6PD-deficient RC at comparably low concentrations. Provisional data (H.U. Lutz, F. Bussolino, R. Flepp, P. Arese, unpublished results) show that diamide enhances the binding of naturally circulating antibodies to the RC. If these results are confirmed, it is conceivable that opening of calcium channels occurs via binding of specific immunoglobulins in analogy to recent evidence (51,52).

Probably connected with increased calcium levels are modifications in the SDS-PAGE electrophoretic pattern of membrane proteins (30,50). In the bottom fractions of crisis RC, the following findings were constant: (1) the appearance of high molecular weight aggregates. In some cases the aggregates could be reduced by dithioerythritol (DTE). In other cases only a partial reduction was possible; (2) the decrease of band 3 protein, which does not reappear after reduction with DTE; (3) the presence of multiple new bands scattered over the whole range. Similar membrane protein modifications are reported in beta thalassemia (53) and G6PD-deficiency due to variants associated with chronic hemolytic anemia (54). The same changes were observed after incubation of G6PD-deficient RC with 500 μM divicine or diamide (not shown).

VII. RHEOLOGY AND MEMBRANE CROSS BONDING IN FAVISM

A. Effect of Divicine on Rheology and Membrane Cross Bonding in G6PD-Deficient Red Cells

Two rheological parameters have been measured in divicine-treated normal and G6PD-deficient RC: filterability across Nuclepore membranes and the elongation index by the rheoscope technique of Schmid-Schönbein (55). After four hour treatment with 250 μM divicine the filtration pressure almost doubled relative to untreated controls (30). Elongation of RC at different shear rates was measured in the rheoscope. Elongation, which is a way to assess RC deformability, was already diminished 4 hours after addition of 500 μM divicine, and remained almost unchanged after 8 and 24 hours (56).

Membrane cross bonding is a tight bond which connects opposing cytoplasmic faces of RC membrane (57). This feature is produced by incubating G6PD-deficient RC for 6 to 9 hours in autologous plasma in presence of 0.5-1 mM divicine and shrinking the cells

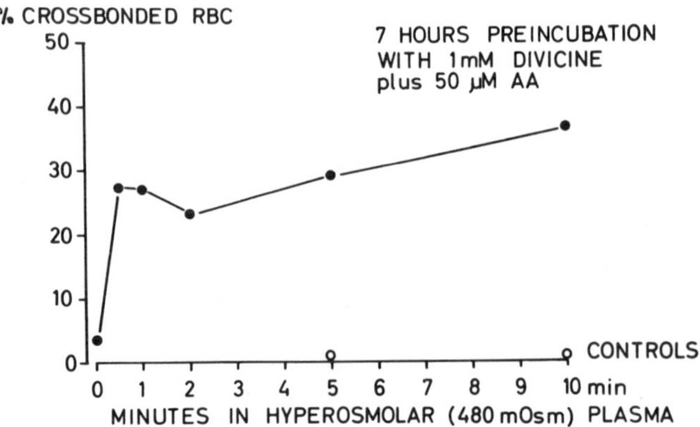

Fig. 8 Formation of cross bonding in divicine-treated G6PD-deficient red cells. Data according (57).

in hypertonic plasma of 400-700 mOsm. Shrinkage is necessary because cytoplasmic faces of the membrane must come into contact for cross bonding to occur. Cross bonding in hypertonic plasma is fast: appreciable percentages of cross bonded RC are seen after 30 sec, and the process is complete after 3 min (Fig. 8). Plasma is necessary for cross bonding. When hypertonic saline was used, the speed of the process was one order of magnitude lower. The nature of the plasmatic factor is at present unknown. Disulfide bonds are not involved, since dithioerythritol, a penetrating reductant, did not reverse cross bonding. Non-covalent interactions of cytoskeletal proteins of the two touching membrane areas seem therefore responsible for cross bonding. Heinz body formation was very scanty in these experiments. This is in contrast with the very high number of Heinz bodies observed during the crisis. The reason for this difference is unclear. Nevertheless, at least in vitro, Heinz bodies are not necessary for membrane cross bonding to occur (57).

B. Red Cell Rheology and Membrane Cross Bonding During the Crisis

Red cells obtained during the crisis were sheared in the rheoscope and studied morphologically (55,57). In all crises still in progress, a varying number of completely indeformable, heavily altered cross bonded RC was constantly found (Table III). The percentage of cross bonded RC varied from 80 to 19 percent. Cross bonded RC were cleared from the circulation in the early phases of the crisis and could not be observed any longer after 24-48 hours from its onset (Table IIIB,C). The shape of cross bonded RC is variable, but they all have two common features: (1) hemoglobin is confined to one side of the cell, leaving the other part as transparent as a ghost (2) in the transparent part of the cell, the opposing parts of the membrane are touching. Swelling of the cells in buffered saline of 200 mOsm makes the hemoglobin containing part of the cell more rounded or even spherical, but does not separate the cross bonded areas. Although the stress at the boundary between the swollen and the crosslinked areas must be very high, cross bonding is strong enough to resist such stress for there was no evidence of peeling apart.

For membrane cross bonding to occur, oxidant membrane lesion in sensitive RC and short periods of hypertonicity are necessary. In peritubular capillaries of the kidneys hypertonicity as high as 1000 mOsm is attained. These hypertonic regions receive about 0.25 percent of the cardiac output, and RC spend there about 30 sec (57). Combining these figures with the data obtained *in vitro*, a cross bonding rate between 0.16 and 2 percent of the total RC per hour can be calculated. This is in keeping with observed rates of hemolysis in the favic crisis. The different morphology of cross bonded RC *in vitro* and *in vivo* is due to the centric contact occurring in the dimple *in vitro*, while membrane contact during the crisis can be due either to osmotic shrinking in the kidney or squeezing of the cells in the spleen or other narrow environments of the microcirculation.

The same type of RC deformation was described in a number of oxidant-elicited hemolytic conditions occurring mostly in G6PD-deficient subjects after ingestion of sulphpyridine, naphthalene, lysol, phenacetin, sulfanilamide, antimalarials, phenylhydrazine, and in the course of favism and fulminant hepatitis (58,59). These cells, which are identical with cross bonded RC, were called "double colored RC" (60), "eccentrocytes" (58), or "hemighosts" (59). They were studied in some detail in acetylphenylhydrazine-treated dogs and found to be completely rigid, highly viscous and containing higher amounts of hemoglobin (58). In the dog, after three days of treatment, top levels of "eccentrocytes" were found (25-30 percent of total RC population), and RC disappearance paralleled the decline of "eccentrocytes". Loss of RC stopped when all "eccentrocytes" were removed. A second challenge with acetylphenylhydrazine produced Heinz bodies but no "eccentrocytes" and no further RC loss occurred (58). These experiments clearly demonstrate that eccentrocytes are removed while the presence of Heinz bodies does not _per se_ cause RC removal.

VIII. ERYTHROPHAGOCYTOSIS IN FAVISM

A. Effect of Divicine on Phagocytosis of G6PD-Deficient Red Cells

Extremely low amounts of divicine stimulate phagocytosis of G6PD-deficient RC by human adherent monocytes. As low as 25 μM divicine added to G6PD-deficient RC elicited a sharp increase in phagocytosis. As shown in Table VA, stimulation by 25 μM divicine is near to maximum, since deficient RC treated with 500 μM divicine or 1 mM diamide or opsonized by anti-D antibodies (not shown) were phagocytized at comparable rates. Phagocytosis was

Table V. Erythrophagocytosis by human monocytes in favism [a]

A) G6PD-deficient red cells treated *in vitro* with divicine or diamide

Subject	Control	Divicine 25 μM	Divicine 500 μM	Diamide 20 μM
M.A.	0.3	3.1	2.9	2.4
C.G.	0.3	3.0	3.8	2.3

B) Favic crisis in progress

Days after bean consumption	1-3	120
Crisis No.		
6/84	3.5	0.3
7/84	2.0	0.4
19/84	4.0	0.5
20/84	2.4	0.4

C) Favic crisis in remission

Days after bean consumption	5-6	120
Crisis No.		
2/84	0.39	0.33
14/84	0.32	0.35

[a] *Phagocytosis is expressed as red cells per phagocyte. RC were radioactively labelled and phagocytized for 60 min at pH 7.4 and 37°C by adherent human monocytes (F. Turrini, F. Bussolino, P. Arese, unpublished results)*

complement-dependent, and completely abolished by heat-inactivated or DFP-inactivated serum, or when RC were pretreated with a monoclonal antibody specific for the C3b receptor (not shown).

Phagocytosis stimulated by micromolar concentrations of divicine is remarkably similar to that elicited by comparable amounts of diamide. Stimulation of phagocytosis of normal and G6PD-deficient RC by micromolar diamide is also complement-dependent and is the result of a complex interplay between the minimally altered RC membrane, complement factors and naturally occurring antibodies (H.U. Lutz, F. Bussolino, R. Flepp, P. Arese, unpublished results). The membrane lesion leading to recognition by macrophages is still unknown. None of a number of biochemical, rheological or morphological parameters was measurably modified by such low concentrations of diamide or divicine (not shown).

B. Erythrophagocytosis During the Crisis - Extravascular vs. Intravascular Hemolysis

Erythrophagocytosis by adherent human monocytes was studied in a number of favic crises. As shown in Table VB, a constant and remarkable increase in phagocytosis particularly evident in the most dense RC fraction (not shown) was observed in all crises. Phagocytosis was almost absent during remission (Table VC).

Signs of intense phagocytosis have been noted in the past by Sansone (1). In a systematic study of sternal marrow smears in 22 favic crises he constantly remarked on the presence of macrophages "literally engorged with red cells and any type of cellular debris". The same author also reported the presence of ingested red cells and cell debris in circulating monocytes (1).

The nature and site of favic hemolysis are basically unknown. The common observation of intense hemoglobinuria and hemoglobinemia has led to the opinion that hemolysis is vastly intravascular. We have performed a balance study between RC disappearance and hemoglobin elimination in urine in a small series of crises which were delivered to the hospital shortly after outbreak. A typical example is presented in Table VI. In this case, where only RC disappearance and urinary hemoglobin excretion are balanced, total heme eliminated in the urine accounted for only about 5 percent of the RC which disappeared during the crisis. In other cases not presented here (T.H. Fischer, A. Naitana, F. Turrini, P. Are-

Table VI. Balance between hemoglobin disappearing from blood and hemoglobin eliminated in the urine during a favic crisis

Days from admission	1	2	3	4	5	6	7	8	120
Hb g/dl	13.5	13.5	9.8	7.9	8.1	10.4	11.6	11.9	13.5
Ht	45	43	30	28	25	32	33	33	44
Total heme in urine g/day	(7)[a]	5.74	6.19	1.13	2.45	0.85	(0.7)[a]	0.52	=

TOTAL HEMOGLOBIN LOSS : 433 g

TOTAL HEMOGLOBIN IN URINE : 24.6 g

PERCENT OF TOTAL HEMOGLOBIN LOSS IN URINE : 5.6 Percent

Patient F.L., male, aged 77, weight 64 Kg. Fava bean consumption 17 hours before admission. Transfusion of packed RC equivalent to 126 g hemoglobin on admission. (F. Turrini, L. Mannuzzu, A. Naitana, P. Arese, unpublished results)

[a] *estimated*

se, unpublished results), in which contributions by haptoglobin, hemopexin, plasmatic hemoglobin and hemoglobin catabolites were also considered, figures not exceeding 10 percent intravascular hemolysis were obtained. This is in keeping with the massive accumulation of radioactivity in spleen and liver of chromium-labelled, G6PD-deficient RC transfused into normal subjects who received primaquine medication (61). The time course of radioactivity accumulation is also very similar to the kinetics of RC destruction in the favic crisis. The crisis is never fulminant but rather slow, the lapse between fava bean consumption and appearance of dark urine being never shorter than 12-24 hours.

Furthermore we never observed in vitro hemolysis of G6PD-deficient RC treated with up to 1 mM divicine or isouramil even after prolonged incubation at constant pH-value and physiological glucose concentrations.

The dimension of reticulo-endothelial system in man (about 2 x 10^{11} phagocytes (62)) may account for the disappearing RC in the favic crisis. Disappearance of about 6.75 x 10^{12} RC, corresponding to the loss of about 1.5 liters of blood, which is a common observation during favic crisis, would mean that each macrophage in the body should ingest about 35 red cells. Each adherent human monocyte can ingest up to 4 RC within 40-60 min. The realistic, yet unproven assumption has to be made that macrophages can repeat the ingestion/digestion cycle several times in 24-48 hours. The concept of predominantly extravascular hemolysis also goes along with the extremely low divicine concentrations which stimulate phagocytosis by human monocytes, and the analogy with the diamide-stimulated, complement and IgG-dependent erythrophagocytosis.

No data are available on the divicine-isouramil levels in blood after fava bean consumption. Assuming the ingestion of 100 g fresh beans (20 percent dry weight), containing about 6 percent divicine plus isouramil, and assuming 10 percent absorption of the active principles, about 400-500 μM concentration can be reached in blood. This is enough to elicit maximal erythrophagocytosis. Finally, an old observation by Sansone (1) should also be kept in mind. In the systematic study of sternal marrow smears of favic patients he constantly observed evidence of "extraordinarily intense" phagocytic activity.

IX. CONCLUDING REMARKS

Several lines of evidence point to a causal role of divicine and isouramil in the pathogenesis of favism. A very peculiar and early effect of both substances is the irreversible oxidation of GSH and other protein-bound -SH groups. The consequences are manyfold and mainly encompass membrane alterations, such as passive permeability to calcium, inhibition of Ca-ATPase, formation of large polypeptide aggregates and diminution of band 3 protein. Intracellular calcium levels are increased while RC deformability and filterability are decreased. A peculiar alteration consisting in the stable linkage of the opposing cytoplasmic faces of the membrane (membrane cross bonding), elicited by the combination of still undefined membrane modifications and osmotic RC shrinkage, has been described. Heinz bodies are formed and the level of methemoglobin is increased. All these effects were found in RC isolated at early stages of the favic crisis, and were elicited <u>in vitro</u> by treating G6PD-deficient RC with 100-500 μM divicine or isouramil. However, the only parameter that was enhanced by very low (25 μM) concentrations of divicine was the specific phagocytosis of G6PD-deficient RC by human adherent macrophages. The mechanism for this is still unknown. Provisional experiments show that the favic crisis is predominantly extravascular. If this observation is extended and confirmed, a challenging task for the near future will be to understand what the minimal alterations in the membrane are that induce phagocytosis. A further unsettled point which awaits clarification are the determinants of blood levels of divicine and isouramil which are high enough to elicit the observed modifications. Since the human intestine is unable to split beta-glycosides in the normal absorption process (G. Semenza, personal communication), we put forward the working hypothesis that the inactive glycosides are split into the active aglycones by the beta-glucosidase of the fava bean itself. In this respect it is interesting to note that crises are observed mostly in small children, where the gastric transit is rapid, and in old people, where achilia is frequent. Furthermore, the beta-glucosidase is very heat-sensitive and does not resist boiling which is a customary procedure in order to make large beans, particularly dry ones palatable. Due to the low thermal conductivity of the beans, short frying does not inactivate the enzyme. On the other hand, beta-glucosidase is moderately acid-sensitive and resists prolonged treatment with 0.01 N HCl plus pepsin. Finally, an old yet very comprehensive study by Marcolongo (63), performed on 592 cases of documented favism showed that 80 percent of all cases was caused by ingestion of raw beans, 19 percent by cooked beans (cooking specifications were not given) and only 1 percent by consumption of dry beans.

ACKNOWLEDGEMENTS

P.A. would like to thank Anna Naitana, Sassari, for expert technical assistance, the patients and physicians (Dr. G. Palomba, Oristano; Dr. Trincas, Cabras; Dr. Paolini, Sassari) for their cooperation, and Prof. A. Sisini, Sassari for the communication of unpublished results

REFERENCES

1. Sansone, G., Piga, A.M., and Segni, G. (1958). "Il favismo". Minerva Medica Editrice, Torino.
2. Mager, J., Chevion, M., and Glaser, G. (1980). In "Toxic Constituents of Plant Foodstuffs, Ed. 2" (I.E. Liener, ed), p. 265. Academic Press, New York.
3. Arese, P. (1982). Rev. Pure Appl. Pharmacol. Sci. 3, 123.
4. Beutler, E. (1983). In "The Metabolic Basis of Inherited Disease, Ed. 5" (J.B. Stanbury, J.B. Wyngaarden, D.S. Fredrickson, J.L. Goldstein, and M.S. Brown, eds.), p. 1629. McGraw-Hill, New York.
5. Stamatoyannopoulos, G., Fraser, G.R., Motulsky, A.G., Fessas, P., Akrivakis, A., and Papayannopoulou, T. (1966). Am. J. Hum. Genet. 18, 253.
6. Gaetani, G.F., unpublished results.
7. Bottini, E., Lucarelli, P., Agostino, R., Palmarino, R., Businco, L., and Antognoni, G. (1971). Science 171, 409.
8. Cassimos, C., Malaka Zafirius, K., and Tsiures, J. (1974). J. Pediatr. 84, 871.
9. Mareni, C., Repetto, L., Forteleoni, G., and Gaetani, G.F. (1984). J. Med. Genet. 21, 278.
10. F.A.O. (1976). Production Yearbook. Rome.
11. Bressani, R. (1973). "Nutritional Improvement of Food Legumes by Breeding", p. 15. Protein Advisory Group, United Nations, New York.
12. Bjerg, B., Poulsen, M.H., and Sorensen, H. (1980). Fabis 2, 51.
13. Chevion, M., and Navok, T. (1983). Anal. Biochem. 128, 152.
14. Collier, H.B. (1976). Can. Inst. Food. Sci. Technol. J. 9, 155.
15. Engel, A.B. (1970). In "Rapport No. R3195". Central Institute for Nutrition and Food Research. Zeist. The Netherlands.
16. Gardiner, E.E., Marquardt, RR, and Kemp, G. (1982). Can. J. Plant Sci. 62, 589.
17. Jamalian, J. (1978). J. Sci. Food Agric. 29, 136.
18. Olsen, H.S., and Andersen, J.H. (1978). J. Sci. Food Agric. 29, 323.
19. Sisini, A., Spanu, A., and Arese, P. (1981). Boll. Soc. Ital. Biol. Sper. 57, 1496.
20. Sisini, A., Virdis-Usai, R., and Arese, P. (1981). Boll. Soc. Ital. Biol. Sper. 57, 15o3.
21. Bendich, A., and Clements, G.C. (1953). Biochim. Biophys. Acta 12, 462.
22. Chevion, M., Navok, T., Glaser, G., and Mager, J. (1982). Eur. J. Biochem. 127, 405.
23. Albano, E., Tomasi, A., Mannuzzu, L., and Arese, P. (1984). Biochem. Pharmacol. 33, 1701.

24. Albano, E., personal communication.
25. Kosower, N.S., and Kosower, E.M. (1976). In "Free Radicals in Biology, Vol. 2" (W.A. Pryor, ed.), p. 55. Academic Press, New York.
26. Benatti, U., Guida, L., and De Flora, A. (1984). Biochem. Biophys. Res. Comm. 120, 747.
27. Baker, M.A. (1983). "Favism: Studies on the Biochemical Mechanisms by Which the Fava Bean Toxin, Divicine, Causes Oxidative Damage in the Glucose-6-Phosphate Dehydrogenase Deficient Erythrocyte", M.S. Thesis. Columbia University, New York.
28. Baker, M.A., Bosia, A., Pescarmona, G.P., Turrini, F., and Arese, P. (1984). Toxicol. Pathol. 12, 331.
29. Srivastava, S.K., and Beutler, E. (1969). J. Biol. Chem. 244, 9.
30. Arese, P., Bosia, A., Naitana, A., Gaetani, S., D'Aquino, M., and Gaetani, G.F. (1981). In "The Red Cell. Fifth Ann Arbor Conference" (G. Brewer, ed.), p. 725. Alan R. Liss, Inc., New York.
31. Gaetani, G.F., Mareni, C., Salvidio, E., Galiano, S., Meloni, T., and Arese, P. (1979). Br. J. Haematol. 43, 39.
32. Gaetani, G.F., Parker, J.C., and Kirkman, H.N. (1974). Proc. Natl. Acad. Sci. USA 71, 3584.
33. Kirkman, H.N., Gaetani, G.F., Clemons, E.H., and Mareni, C. (1975). J. Clin. Invest. 55, 875.
34. Welt, S.I., Jackson, E.H., Kirkman, H.N., and Parker, J.C. (1971). Ann. N.Y.Acad. Sci. 179, 625.
35. Mager, J., Glaser, G., Razin, A., Izak, G., Bien, S., and Noan, M. (1965). Biochem. Biophys. Res. Comm. 20, 235.
36. Sarkadi, B. (1980). Biochim. Biophys. Acta 604, 159.
37. Larsen, F., Katz, S., Roufogalis, B.D., and Brooks, D.E. (1981). Nature 294, 667.
38. O'Rear, E.A., Udden, M.M., McIntire, L.V., and Lynch, E.C. (1981). Am. J. Hematol. 11, 283.
39. Schanne, F.A.X., Kane, A.B., Young, E.E., and Farber, J.L. (1979). Science 206, 700.
40. Melloni, E., Sparatore, B., Salamino, F., Michetti, M., and Pontremoli, S. (1982). Biochem. Biophys. Res. Comm. 106, 731.
41. Lorand, L., Weissman, L.B., Epel, D.L., and Bruner-Lorand, J. (1976). Proc. Natl. Acad. Sci. USA 73, 4479.
42. Ferrell, J.E., Jr., and Huestris, W.H. (1984). J. Cell Biol. 98, 1992.
43. Allan, D., and Michell, R.H. (1977). Biochem. J. 166, 495.
44. Anderson, R.A., and Lovrien, R.E. (1981). Nature 292, 158.
45. Rogausch, H. (1978). Pflügers Arch. 373, 43.
46. Eaton, J.W., Skelton, T.D., Swofford, H.S., Kolpin, C.E., and Jacob, H.S. (1973). Nature 246, 105.
47. Forman, S., Bischel, M., and Hochstein, P. (1973). Ann. Int. Med. 79, 841.
48. Rachmilewitz, E.A., Shinar, E., Shalev, O., Milner, Y.,

Erusalimsky, J., and Schrier, S.L. (1983). Biomed. Biochim. Acta 42, 27.
49. De Flora, A., Benatti, U., Guida, L., Forteleoni, G., and Meloni, T. (1985).Blood 66, 294.
50. Turrini, F., Naitana, A., Mannuzzu, L., Pescarmona, G.P., and Arese, P. (1985). Blood 66, 302.
51. Young, J.D.-E., Ko, S.S., and Cohn, Z.A. (1984). Proc. Natl. Acad. Sci. USA 81, 7258.
52. Kanner, B.I., and Metzger, H. (1984). J. Biol. Chem. 259, 10188.
53. Kahane, I., Shifter, A., and Rachmilewitz, E.A. (1978). FEBS Lett. 85, 267.
54. Johnson, G.J., Allen, D.W., Cadman, S., Fairbanks, V.F., White, J.G., Lampkin, B.C., and Kaplan, M.E. (1979). N. Engl. J. Med. 301, 522.
55. Schmid-Schönbein, H., von Gosen, J., Heinrich, L., Klose, H.J., and Volger, E. (1973). Microvsc. Res. 6, 366.
56. Arese, P., Naitana, A., Mannuzzu, L., Turrini, F., Haest, C.W.M., Fischer, T.M., and Deuticke, B. (1983). In "Advances in Red Cell Biology" (D.J. Weatherall, S. Gorini, and G. Fiorelli, eds.), p.375. Raven Press, New York.
57. Fischer, T.M., Meloni, T., Pescarmona, G.P., and Arese, P. (1985). Br. J. Haematol. 59, 159.
58. Ham, T.H., Grauel, J.A., Dunn, R.F., Murphy, J.R., White, J.G., and Kellermeyer, R.W. (1973). J. Lab. Clin. Med. 82, 898.
59. Chan, T.K., Chan, W.C., and Weed, R.I. (1982). Br. J. Haematol. 50, 575.
60. Sansone, G. (1957). Minerva Medica 48, 3317.
61. Salvidio, E., Pannacciulli, I., Tizianello, A., and Ajmar, F. (1967). N. Engl. J. Med. 276, 1339.
62. Wintrobe, M.M. (1975). "Clinical Hematology, Ed. 7", p.256. Lea & Febiger, Philadelphia.
63. Marcolongo, F. (1953). Minerva Medica 44, 1963.

DIVICINE AND G6PD-DEFICIENT ERYTHROCYTES: AN INTEGRATED MODEL OF CYTOTOXICITY IN FAVISM

A. De Flora

U. Benatti

L. Guida

E. Zocchi

Institute of Biochemistry, University of Genoa, Genoa, Italy

I. INTRODUCTION

The biochemical picture of G6PD-deficient erythrocytes in favism and in drug-induced hemolytic anemia is characterized by severe oxidant damage, as shown by hematological, morphological and metabolic evidences (1-4). However, a number of facts tend to complicate a correct interpretation of experimental findings. Among these are the attendant reticulocytosis (masking the

[1]Supported in part by grants from the Special Project "Ingegneria Genetica e Basi Molecolari delle Malattie Ereditarie", C.N.R., Rome, and from the Ministry of Education, Rome.

genuine picture of damaged erythrocytes whose clearance from circulation may be picked up variably by the observer), the lack of adequate controls (because of obvious ethical reasons) and the occurrence of still unidentified pathogenetic factors interacting with severe deficiency of G6PD activity, with ingestion of fava beans, or with administration of potentially hemolyzing drugs.

For the above reasons, discrimination between causes and effects in the wide array of biochemical and cellular abnormalities of the affected erythrocytes during acute hemolysis may prove to be exceedingly difficult. In spite of this, it seems correct, from a standpoint of methodology, to focus on oxidant challenge exerted by toxic components of fava beans (or by oxidizing drugs or by metabolites thereof) as the triggering event for a phenomenologically known yet mechanistically undefined chain of discrete steps that result in eventual hemolysis. This approach is supported by the well established fact that a still uncontradicted requirement for the erythrocyte to hemolyze *in vitro* following ingestion of fava beans is severe deficiency of G6PD activity, *i.e.*, of the enzyme activity that starts and regulates the integrated chain of redox reactions resulting in reduction of glutathione.

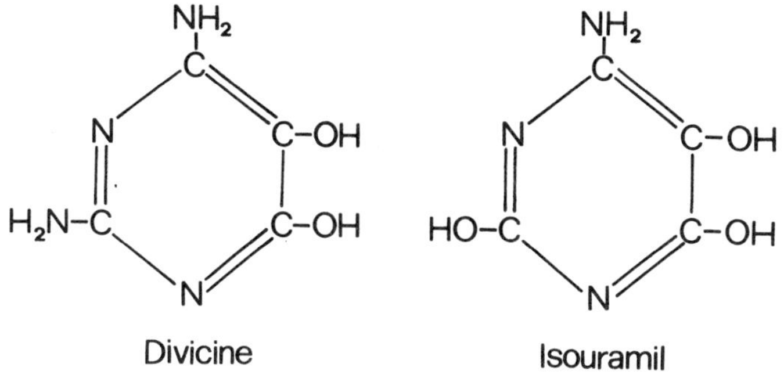

Divicine

Isouramil

Figure 1. Structures of Divicine and Isouramil.

Search for toxic components of fava beans endowed with oxidizing properties was highlighted twenty years ago by the observation that the two pyrimidine aglycones shown in Figure 1, *i.e.* divicine and isouramil (resulting from hydrolysis of the two corresponding β-glucosides vicine and convicine, respectively, that are stored as inactive components in broad beans), display an unusually high reactivity in re-oxidizing GSH (5,6). Since then, a wealth of experimental observations have added support to the contention that these *o*-hydroquinone species are important, if not the exclusive, aethiopathogenetic factors in favism (7-15). Such a conclusion is mostly based upon the remarkable similarities between specific damaging effects induced by either aglycone on G6PD-deficient erythrocytes *in vitro* and the biochemical and morphological features of erythrocytes drawn during hemolytic crisis in favism. As already stressed, proper allowance should be made for the variegate and superimposed events that characterize the *in vivo* situation, thereby complicating the sharp and typically oxidative abnormalities that are found in the erythrocytes challenged with divicine or isouramil in reconstructed systems.

II. AUTOOXIDATION OF DIVICINE

Although most data observed in our laboratory concern divicine (obtained either by acidic or β-glucosidase-catalyzed hydrolysis of commercial vicine), closely comparable results were generally reported using isouramil, confirming that a common pattern of autooxidation of either pyrimidine aglycone (DH_2 or IH_2) leads to formation of the two corresponding quinone forms (D_{ox} and I_{ox}, respectively). In addition, most damaging effects were obtained in the presence of both DH_2 and ascorbate, the latter compound acting, however, to magnify these intracellular events, probably by means of an accelerated redox cycling between DH_2 and D_{ox}.

There is in fact little doubt that most cytotoxic effects that are induced *in vitro* or *in vivo* by divicine and isouramil are strictly inherent to the autooxidation mechanism undergone by either native aglycone: indeed, removal of oxygen results in complete abrogation of these effects.

The patterns of divicine autooxidation have recently received considerable attention. Chevion *et al.* (9) reported the chemical properties, as well as the interconversion, of the two major forms (*i.e.*, hydroquinone and quinone, respectively) and described the rapid decomposition of the latter species in the presence of oxygen. Albano *et al.* (16) obtained ESR evidence for a semiquinone free radical intermediate the formation of which is strictly dependent on availability of oxygen.

Kinetic analysis of divicine autooxidation, based on disappearance of the 280 nm peak (hydroquinone) and on appearance of the 240 nm-absorbing quinonic species (9), showed a variety of autooxidation pathways (17). The *in vitro* predominant mechanism is O_2-dependent, yet it seems to be of little physiological relevance since it should be effectively inhibited by superoxide dismutase intracellularly. A second mechanism of divicine autooxidation depends on H_2O_2 and hemoglobin as a result of formation of ferrylhemoglobin, the latter species being endowed with very strong oxidizing properties (18). Finally, if both the above mechanisms are properly prevented (*e.g.*, by superoxide dismutase and catalase), a third autooxidation pathway, requiring build-up of an autocatalytic intermediate, becomes effective (17).

Therefore, it may be significant, from a pathophysiological standpoint, that GSH counteracts operation of the third autooxidation pathway of divicine, this representing an additional defense system of the normal erythrocytes whereby divicine oxidation and its deleterious consequences can be efficiently prevented. Accordingly, the impaired metabolic competence of G6PD-deficient erythrocytes to regenerate their GSH would be expected to enhance their susceptibility to oxidant damage resulting from uncontrolled occurrence of divicine autooxidation through the third pathway.

III. INTERACTIONS BETWEEN DIVICINE AND THE HEMOGLOBIN-METHEMOGLOBIN SYSTEM

Divicine and its by-products can interfere with the hemoglobin (Hb)-methemoglobin (MetHb) system in a variety of ways (17). In the presence of oxygen and at low DH_2/Hb ratios, H_2O_2 will produce MetHb and ferrylhemoglobin. Ferrylhemoglobin, or MetHb-peroxide complex, is known to be one of the most reactive oxidizing species in the erythrocyte (18) and to catalyze several peroxidation reactions. Among other effects, ferrylhemoglobin is responsible for a feedback mechanism whereby autooxidation of divicine is significantly accelerated (and

MetHb formation consomitantly increased), thus enhancing the detrimental consequences it entails. An extrapolation to the intracellular conditions that characterize the hemolytic crisis of favic patients is clearly still impossible; yet, it is interesting to remember that remarkably increased MetHb concentrations are invariably observed both upon *in vitro* incubation of G6PD-deficient erythrocytes with divicine (11) and in favism as well (T. Meloni, G. Forteleoni, U. Benatti, and A. De Flora, unpublished data). The consistency of the latter observation makes assay of MetHb levels the most convenient parameter for adequately monitoring progress of the hemolytic crisis in favic patients and for deciding about transfusion therapy accordingly.

On the other hand, we have recently reported that, under anaerobic conditions especially, native divicine (*i.e.*, its hydroquinone species) reduces MetHb efficiently (19). This represents an alternative pathway to the autooxidation process, whose extent of intracellular occurrence is completely unknown. Quinoid divicine (D_{ox}) is still formed, this triggering a redox cycle that will drain both NADPH and GSH (19). G6PD-deficient erythrocytes fail to reconvert the quinone to the hydroquinone form and seem therefore to be more susceptible than the normal cells to accumulating MetHb and, by consequence, ferrylhemoglobin.

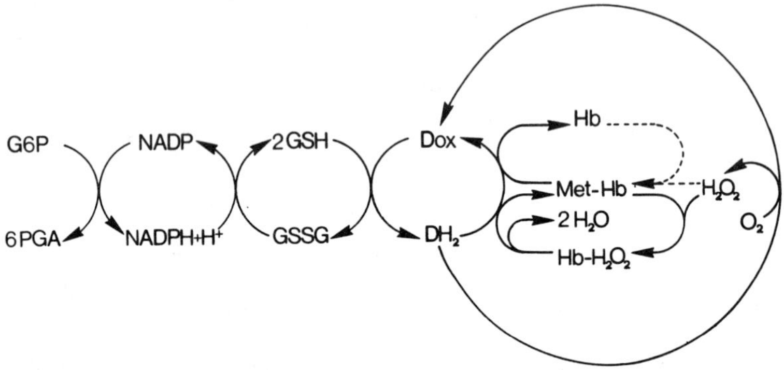

Figure 2. Metabolic correlations between divicine oxidation and hemoglobin forms (from ref. 19, with permission).

Figure 2 provides a tentative picture of the multiple interactions between the DH_2/D_{ox} system and the hemoglobin forms, as well as of the metabolic driving force (*i.e.*, the hexose monophosphate shunt and glutathione reductase activities) that seems to perpetuate redox cycling of divicine itself in the normal erythrocytes.

IV. INTRACELLULAR DAMAGING EFFECTS OF DIVICINE ON G6PD-DEFICIENT ERYTHROCYTES

As already emphasized, all cytotoxic effects induced by divicine are due to its autooxidation. A variety of noxious consequences follow the interaction of divicine with normal and even more with G6PD-deficient erythrocytes (for a review see Ref. 20). These effects include draining of cellular reducing equivalents like GSH and NADPH, inactivation of important enzyme proteins (both cytosolic and membrane-bound), electrolyte imbalances, alterations in membrane cytoskeleton, membrane cross bonding (21) and enhanced erythrophagocytosis (22).

It has been pointed out that it is still impossible to trace a detailed time sequence of the above intracellular events following divicine autooxidation (20). This is the main reason that delays an adequate investigation of those plasma factors that modulate eventual hemolysis of divicine-damaged G6PD-deficient erythrocytes (20) in the test tube. Specifically, these include: a) low M_r, thermostable molecules that increase hemolysis of divicine-pretreated erythrocytes considerably and that are found in all individuals; b) protein molecules, with apparently variable expression in G6PD-deficient subjects, that seem to counteract the hemolytic activity of the low M_r plasma factors.

The well known erratic feature of hemolytic episodes in G6PD-deficient subjects might bear close relevance to the above network of functionally opposite plasma factors, thus stimulating further attempts at identifying them as an important area of investigation in G6PD deficiency. However, until a better knowledge of the integrated events that follow divicine autooxidation in the affected cells is available, any advances in the chemical characterization of plasma factors seems to be remote.

Among the various abnormalities divicine elicits in G6PD-deficient erythrocytes, impairment of Ca^{2+}-ATPase and of Na^{+}/K^{+}-ATPase activities (14,15) and markedly disordered calcium and potassium homeostasis seem to be important for the functional disturbances they entail. Like for other *in vitro* effects of divicine, comparable patterns have been recently observed in favism (15,23). Specifically, the alteration of calcium homeostasis either as induced in reconstructed systems or as experimentally observed in erythrocytes from favic patients, is expected to result in significant stimulation of usually latent enzyme activities (15,23). Recent results (24) support this expectation as far as activation of a cytosolic Ca^{2+}-activated neutral proteinase (CANP) is concerned in irreversibly damaged erythrocytes from favic patients. Since intracellular stimulation of CANP is not devoid of consequences on the integrity of specific membrane proteins (24-26), a vicious cycle might take place, thereby impairing electrolyte homeostasis further and making it irreversible (*e.g.*, through uncontrolled opening of the calcium channels or further inactivation of Ca^{2+}-ATPase). Accordingly, abnormally elevated intraerythrocytic Ca^{2+} levels could represent both a cause and an effect of irreversible cytotoxicity: it may not be fortuitous, from this standpoint, and closely related to functional impairment of the Ca^{2+} pump, that erythrocytes during favic crisis have normal ATP concentrations in spite of the unusually high Ca^{2+} they should actively extrude under these circumstances.

In conclusion, the network of detrimental effects triggered by divicine within G6PD-deficient erythrocytes is exceedingly complex and still uninterpretable in its details. Fig. 3 attempts to summarize the basic alterations that have been observed and their possible correlation with the process of hemolysis. Although emphasizing the role of plasma in oxidative erythrocyte destruction, this picture cannot with certainty exclude the view of extravascular mechanisms of hemolysis being involved (21,22). The unequivocal membrane damage occurring in these conditions has been shown on the contrary to be an adequate signal for erythrophagocytosis to be stimulated (22).

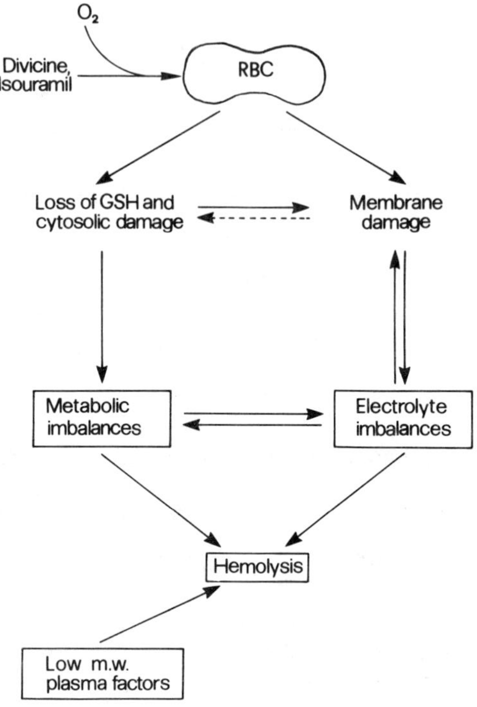

Fig. 3. Tentative mechanisms of hemolysis in divicine-or isouramil-damaged erythrocytes.

Therefore available evidence seems to favor the idea that the individual abnormalities elicited by divicine in the susceptible erythrocytes are compatible with both intravascular and extravascular mechanisms of hemolysis.

V. FUTURE PERSPECTIVES

Research on divicine, mostly concerned with its high chemical reactivity and with the plethora of intracellular effects it induces, should be directed toward two main objectives:

1) Elucidating the still undefined reasons of variability of hemolytic processes (not only inter-individual, but also during the life of G6PD-deficient subjects). As anticipated, adequate consideration should be given to the nature and to the mechanisms of plasma factors yet, plain interaction of divicine with erythrocytes is *per se* open to several sources of such variability. The variety of pathways and of factors involved in the process of divicine autooxidation is just an example of the pleomorphic patterns of the divicine-erythrocyte relationship. The obvious relevance of this heterogeneity to the still unpredictable nature of hemolytic crises in susceptible individuals justifies this area of research critically, in terms of prevention.

2) Extending present and future knowledge on divicine reactivity and toxicity to other cell systems. This might represent the basis for selective studies aimed at designing new cytotoxic molecules potentially useful in cancer chemotherapy. Divicine may induce cytotoxicity for several reasons, to be

properly investigated and elucidated to this purpose: a) formation of bursts of H_2O_2, b) formation of potentially cell-damaging semiquinone species, c) possible antimetabolite activity in nucleated cell types because of its pyrimidine structure. An integration of these partially verified and partially predicted mechanisms in specific cellular targets may be very fruitful in the future.

ACKNOWLEDGMENTS

We are indebted to Dr. C. Winterbourn for helpful discussions and for allowing us to quote results of collaborative experiments before publication.

REFERENCES

1. Sansone, G., Piga, A.M., and Ségni, G., (1958). Il Favismo, Minerva Medica, Torino

2. Beutler, E., (1983). In "The Metabolic Basis of Inherited Disease" (J.B. Stanbury, J.B. Wyngaarden, D.S. Fredrickson, J.L. Goldstein, M.S. Brown, eds.), 5th edition, p. 1629, McGraw-Hill, New York

3. Kattamis, C.A., Kyriazakou, M., and Chaidas, S. (1969). J. Med. Genet. 6, 34

4. Gaetani, G.F., Mareni, C., Salvidio, E., Galiano, S., Meloni, T., and Arese, P. (1979). Br. J. Haematol. 43, 39

5. Lin, J.Y. (1963). J. Form. Med. Assn 62, 777

6. Mager, J., Glaser, G., Razin, A., Izak, G., Bien, S., and Noam, M. (1965). Biochem. Biophys. Res. Commun. 20, 235

7. Mager, J., Chevion, M., and Glaser, G. (1980). In "Toxic Constituents of Plant Foodstuffs (L.I. Liener, ed.), 2nd edition, p. 265, Academic Press, New York

8. Arese, P., Bosia, A., Naitana, A., Gaetani, S., D'Aquino, M., and Gaetani, G.F. (1981). In "Red Cell: Fifth Ann Arbor Conference", p. 725, Alan R. Liss, New York

9. Chevion, M., Navok, T., Glaser, G., and Mager, J. (1982). Eur. J. Biochem. 127, 405

10. Arese, P. (1982). Rev. Pure Appl. Pharmacol. Sci. 3, 123

11. De Flora, A., Benatti, U., Morelli, A., and Guida, L. (1983). Biochem. Int. 7, 281

12. Benatti, U., Guida, L., and De Flora, A. (1984). Biochem. Biophys. Res. Commun. 120, 747

13. Mavelli, I., Ciriolo, M.R., Rossi, L., Meloni, T., Forteleoni, G., De Flora, A., Benatti, U., Morelli, A., and Rotilio, G. (1984). Eur. J. Biochem. 139, 13

14. Benatti, U., Guida, L., Forteleoni, G., Meloni, T., and De Flora, A. (1985). Arch. Biochem. Biophys. 239, 334

15. De Flora, A., Benatti, U., Guida, L., Forteleoni, G., and Meloni, T. (1985). Blood 66, 294

16. Albano, E., Tomasi, A., Mannuzzu, L., and Arese, P. (1984). Biochem. Pharmacol. 33, 1701

17. Winterbourn, C.C., Benatti, U., and De Flora, A. (1986). Biochem. Pharmacol., in press

18. Winterbourn, C.C. (1985). Environmental Health Perspectives, in press

19. Benatti, U., Guida, L., Grasso, M., Tonetti, M., De Flora, A., and Winterbourn, C.C. (1985). Arch. Biochem. Biophys., in press

20. De Flora, A., Benatti, U., and Guida, L. (1985). Free Radical Res. Commun., in press

21. Fischer, T.M., Meloni, T., Pescarmona, G.P., and Arese, P. (1985). Br. J. Haematol. 59, 159

22. Baker, M., Bosia, A., Pescarmona, G.P., Turrini, F., and Arese, P. (1984). Toxicol. Pathol. 12, 331

23. Turrini, F., Naitana, A., Fischer, T.M., Pescarmona, G.P., and Arese, P. (1985). Blood 66, 302

24. De Flora, A., Morelli, A., and Grasso, M. (1986). This volume

25. Pontremoli, S., Melloni, E., Sparatore, B., Michetti, M., and Horecker, B.L. (1984). Proc. Natl. Acad. Sci. 81, 6714

26. Pontremoli, S., Melloni, E., Sparatore, B., Salamino, F., Michetti, M., Sacco, O., and Horecker, B.L. (1985). Biochem. Biophys. Res. Commun. 128, 331

G6PD-RELATED NEONATAL JAUNDICE

Sergio Piomelli, M.D.

Division of Pediatric Hematology-Oncology,
Columbia University College of Physicians and Surgeons,
New York, NY, USA

The bilirubin concentration in the newborn is regulated by a delicate equilibrium between the rate of hemolysis and the ability of the liver to convert the by-products of heme into water-soluble conjugated bilirubin to be excreted in the bile. This balance is so fragile that significant accumulation of bilirubin occurs in every newborn, leading to the so-called "physiological jaundice" that usually peaks at 3-4 days of age. Any factor that either increases hemolysis or decreases bilirubin conjugation may easily shift this precarious equilibrium and result in additional hyperbilirubinemia.

The red cells of the newborn are particularly defective in the enzymatic machinery needed to maintain an intracellular reduced state: NADH-dependent diaphorase [41] and glutathione

The work of the author referred to in this article was supported by NIH Grant #AM26793-06.

peroxidase are less active than in adult red cells [13]. In the newborn liver, uridine-diphosphate-glucuronyl transferase is defective, and bilirubin-carrying proteins (protein Y and Z, ligandin) are also low [26]. All neonatologists are fully aware of the implications of this state of affairs and are watchful for the rapid development of massive hyperbilirubinemia, a very common event in the nursery. Hyperbilirubinemia (when extreme) leads to kernicterus; this (when not fatal) results in severe neurological sequelae. A variety of factors, such as prematurity, infections, acidosis, drugs that compete with bilirubin for binding protein sites and drugs that increase red cell destruction may contribute to the hyperbilirubinemia, on one hand, and to the development of kernicterus, on the other.

Severe jaundice secondary to G6PD deficiency in the newborn may occur through several mechanisms:

1) Neonatal Jaundice in the Common Variants of G6PD Deficiency.
2) Neonatal Jaundice in Congenital Non-Spherocytic Hemolytic Anemias Secondary to G6PD Deficiency.
3) Acute Hemolysis in G6PD Deficiency During the Perinatal Period.

Underlying all of these syndromes is the intrinsic abnormality of the G6PD-deficient red cells: the common variants lead to a moderate degree of chronic hemolysis in the steady state and to acute extreme hemolysis under the appropriate trigger; the congenital non-spherocytic hemolytic variants lead to severe chronic hemolysis. What makes the perinatal period one of great risk for jaundice is the inability of the liver to handle bilirubin efficiently, probably aggravated by the G6PD deficiency itself. Therefore the G6PD-deficient newborn is particularly prone to develop increased jaundice of varying degree and at varying frequency, depending on both the genetic make-up and the environmental circumstances.

1) Neonatal Jaundice in the Common Variants of G6PD Deficiency.

In 1959, Segni reported two cases of kernicterus in G6PD-deficient newborns [45]. On May 21, 1960, Lancet published a letter by G.D. Smith and F. Vella from Singapore on kernicterus and G6PD deficiency [46]. On the same day, S.A. Doxiadis, P. Fessas and T. Valaez read a paper at the Athens Medical Society on the high incidence of G6PD deficiency in severe unexplained neonatal jaundice [8]. In the same year, Panizon reported similar observations in Sardinia [34,35]. It is interesting that, from the very beginning, the severity and the worldwide distribution of the problem were obvious and a debate on the cause of this severe jaundice was started that still remains open. The group in Singapore felt that some extrinsic factor was responsible for precipitating hemolysis; the group in Athens felt that the enzyme deficiency by itself was enough to explain the jaundice, in absence of an extrinsic precipitating cause. After these first reports, observations of increased incidence of jaundice in G6PD-deficient newborns have followed from all over the world: in Europe, from Greece [7,23,50,51,54] and Sardinia [43]; in the Middle East, from Turkey [44], and Israel [1,11,28]; in the Far East, from China [16,20,21], India [2,22], Singapore [4,47], Malaysia [18] and Thailand [10,36]; in Africa, from Ghana [29], Nigeria [3,6,31] and Dakar [48]. In North America, neonatal jaundice was noticed among Afro-American babies [5,9,12,30,33,52], particularly among premature infants.

From all these studies, a picture emerges of a marked increase in frequency of neonatal jaundice among G6PD-deficient male newborns; however, it is also obvious that there is a tremendous variability in this frequency, even among populations that are ethnically very close. For instance, among G6PD-deficient hemizygous males in Cagliari, in Southern Sardinia, the incidence of jaundice (bilirubin >16 mg/dl at 96 hours) was reported as 10.2% [43], while in Sassari, in northern Sardinia, the incidence of jaundice (bilirubin >18mg/dl) was found to be 37% [25,26]. The cause for this discrepancy is not known; it is worth noticing, however, that, in Sassari, besides the common

Gd^{Mediterranean} defect, there is also a similar, but distinct defect: Gd^{Sassari} [49]. Thus it is conceivable that the different incidence of hyperbilirubinemia in Sardinia may reflect a true genetic difference.

In another study, in Greece, there was an increased frequency of jaundice among G6PD-deficient male newborns in three locations studied, in comparison to normal male newborns. However, there was a large difference among the three different locations in the frequency of severe jaundice (bilirubin >16mg/dl) observed even among normal newborns. (Severe jaundice occurred in Alexandria in 1% of normal and 4% of G6PD-deficient newborns; in Rhodes, in 2% of normal and 11% of G6PD-deficient newborns; in Lesbos, in 11% of normal and 43% of G6PD-deficient newborns.) The very high frequency of severe jaundice in Lesbos even in normal newborns suggests that some local "icterogenic factor" may be operative there [51].

Several studies have been directed at analyzing such factors, both genetic and extrinsic, with varying degrees of success. A reduced red cell acid phosphomono-esterase was observed in G6PD-deficient Caucasian newborns [33], but not in Chinese newborns [20]. Vitamin K was excluded as a cause of jaundice in one study [6] and mentholated powders incriminated in another [31]; prematurity, acidosis and hypoxia were found to be icterogenic in another [19].

It appears clear, at this point, that G6PD deficiency strongly predisposes to neonatal jaundice. This enzymatic defect results in a significant reduction in red cell life span [38] and is associated with a significant degree of compensated chronic hemolysis in the adult [37,39]. In the newborn, the same degree of reduction in red cell life span certainly increases the burden of bilirubin on the poorly functioning neonatal liver. On the other hand, it is not inconceivable that the G6PD deficiency in the liver cells may itself additionally interfere with the bilirubin conjugation process [24]. It has been suggested that this mechanism may be mediated by the glutathione-S-transferase activity of ligandin and to the need for GSH of the ligandin-mediated uptake of bilirubin by the liver [24,32]. This mechanism is entirely speculative; however it has been shown that in newborns with severe jaundice (both of unknown etiology and associated with G6PD deficiency) salicylamide glucuronide formation was impaired [23].

A substantial reduction in the frequency of exchange transfusion has been reported in Sassari, after the use of preventive phenobarbital, that induces the UDP-glucuronyl transferase [25]. This observation, together with the demonstration that in cases of severe jaundice there is no marked drop in hemoglobin, has been taken as evidence that the jaundice in G6PD deficiency is nearly exclusively of hepatic, rather than hemolytic nature. Yet, a reduction in newborn red cell life span from 80 to 60 days may result in severe hyperbilirubinemia, but cause only a negligible decrease in hemoglobin by day 4. Any anemia may be also rapidly compensated by a modest increase in reticulocytes. In any event, raising the UDP-glucuronyl transferase activity toward its normal adult level would decrease the hyperbilirubinemia, even in face of increased hemolysis. In adults, in fact, even in cases of extreme chronic hemolysis, the bilirubin level never reaches the values observed in the newborn [39].

Most of the studies reported above have focused on hemizygous males, in whom the defect is most severe and more easily diagnosed [39]. In heterozygous females the establishment of the diagnosis is much more difficult, in part because of the randomness of the "lyonization" and even more so in the newborn heterozygote where the young red cell population results in a shift toward normal in the frequency of G6PD-positive cells [43]. This is mostly an effect of the marked age dependency of the mutant G6PD defective enzymes [37] and only in part of the selective removal from the circulation of G6PD-deficient cells [43]. The extreme difficulty of classification of heterozygous females, even with the most sensitive slide elution tests has been emphasized [24,43]. The frequency of severe jaundice in heterozygous females is much lower than in hemizygous males and, as expected, it is higher among those females who have a more marked enzyme deficiency [24,43].

2) Neonatal Jaundice in Congenital Non-Spherocytic Hemolytic Anemias Secondary to G6PD Deficiency.

Besides the very common variants of G6PD deficiency associated with drug induced hemolysis, several extremely rare mutants with abnormal kinetics result in a much more pronounced chronic severe hemolysis. These syndromes are associated with intense reticulocytosis and result in extreme hyperbilirubinemia at birth, requiring often multiple exchange transfusions. These syndromes may be discriminated at birth from acute episodes of hemolysis only on the basis of family history (when present). Otherwise the true chronic hemolytic nature can be clarified only later, when hemolysis persists. At time of birth, the diagnosis is made even more difficult by the fact that the mothers of these males usually have normal G6PD activity as their short-lived defective red cells are selectively and rapidly removed from the circulation [40].

3) Acute Hemolysis in G6PD Deficiency During the Perinatal Period.

The G6PD-deficient individual throughout life is at risk of acute hemolysis; this, when acute and severe enough, can be life threatening and even fatal. In the perinatal period, the inability of the newborn's liver to handle bilirubin adequately turns even minor episodes of acute hemolysis into catastrophic hyperbilirubinemia, that often leads to kernicterus.

Acute hemolysis may be triggered in utero, by the administration of hemolysis-inducing drugs to the heterozygous mother, during the last part of pregnancy. While the mother herself is usually unaffected, transplacental passage of these agents may, however, induce acute neonatal hemolysis (and jaundice) if the male offspring is enzyme deficient [14]. In one case, fetal hydrops fetalis occurred in a Chinese infant with G6PD deficiency whose mother ingested fava beans and ascorbic acid [27].

After birth, upon returning home, the G6PD-deficient infant can be suddenly exposed to hemolysis inducing agents, often under totally unsuspected circumstances [4]. The most common and, in a sense, the saddest event takes place when the newly born is affectionately wrapped into a blanket that had been carefully stored in naphthalene, to preserve it for the anticipated arrival home of the new heir. Unfortunately, this warm parental gesture results in disastrous consequences. The naphthalene fumes induce in the sensitive G6PD-deficient infant an episode of acute hemolysis; the still immature liver is unable to conjugate bilirubin when exposed to this sudden bolus; the indirect bilirubin reaches rapidly values well above 30 mg/dl (often in the 50s) and kernicterus, in this situation, is practically unavoidable, despite exchange transfusion. This author has seen this sad scenario develop under nearly identical circumstances four times: 2 times in a Chinese and a Mediterranean male newborn, once in a Chinese heterozygous female and once in an Afro-American male child. In all of these cases, of course, these tragedies could have been avoided had the parents been warned of the risk of naphthalene exposure to a G6PD-deficient infant. This would have happened if a regular program of newborn screening were in effect in New York and if pediatricians were more familiar with the possibility of these occurrences. At least in New York City, the incidence of severe perinatal jaundice secondary to G6PD deficiency is too low to stimulate the need for an extensive screening program. In other areas of the world, where G6PD deficiency is present at greater incidence and higher degree, screening programs have been both feasible and successful [25].

4) Social Relevance of G6PD-Related Neonatal Jaundice.

It is obvious that such a severe clinical syndrome, associated with severe kernicterus, represents a major public health problem, particularly for those areas of the world where G6PD deficiency is quite common. Recently, [50] Valaez et al. have critically reviewed the experience with severe jaundice in

Athens, during the 11 year period 1962 to 1973, examining the records of 2976 affected infants. From this extensive review they have concluded:

a) G6PD deficiency is an independent cause of neonatal jaundice;
b) among Greek male newborns, G6PD deficiency is the cause of 30% of all cases of kernicterus;
c) the gene loss due to kernicterus among G6PD-deficient males is 10 times greater than among normal male newborns;
d) among G6PD-deficient males, the greatest gene loss is due to kernicterus;
e) despite the introduction of exchange transfusion and improved management, *late kernicterus* persists among G6PD-deficient males, at a rate 44 times what it is in the general population. G6PD deficiency was the most prevalent cause among infants admitted to the hospital with established kernicterus.

These studies indicate clearly the public health relevance of neonatal jaundice secondary to G6PD deficiency and underscore the need for screening of newborns, at least in those areas of the world where the defect is most common and severe.

5) Prevention and Management of G6PD Related Neonatal Jaundice.

Screening of newborns has been recommended for those areas of the world where the defect is most frequent [24,25]. While screening is relatively simple for hemizygous males, it is fraught with difficulties and uncertainty for heterozygous females [39]. On the other hand, severe jaundice occurs at greater frequency only among those females who have the highest percentage of deficient red cells and are therefore most likely to be correctly identified even by the least sensitive screening tests [24,43].

Once G6PD-deficient newborns have been identified, prophylactic treatment with phenobarbital has been utilized with success in Sardinia and has resulted in a substantial reduction in the need for exchange transfusion [24,25] More recently, phototherapy has been utilized with success in several institutions, including ours. Despite sporadic reports of possible failures at phototherapy [15], this procedure is now successfully utilized around the world, including Sassari (Meloni, personal communication).

None of these procedures has completely removed the need for exchange transfusion. In fact, in G6PD deficiency (as in other hemolytic diseases of the newborn), exchange transfusion removes not only toxic bilirubin, but also, more importantly, the red cells whose hemolysis is the source of bilirubin. Thus, although it may be the last resort, exchange transfusion provides a definitive solution to the problem.

Once a jaundiced newborn has been identified as G6PD-deficient, the best practical advice is in the statement by Bienzle et al [3]:

> *"It is important to emphasize that the finding of G6PD deficiency in a jaundiced newborn does not exonerate the pediatrician from the task of identifying possible precipitating causes of jaundice, such as infections, which require treatment in their own right."*

ACKNOWLEDGMENTS

The author would like to thank Ms. C. Seaman for critical review of this manuscript, Ms. D. Bouras for assistance in careful bibliographical search, and Ms. L. Apelis for patient editing of this manuscript.

REFERENCES

1. Ashkenazi F, Mimouni F, Merlob P, Reisner SH: Neonatal bilirubin levels and glucose-6-phosphate dehydrogenase deficiency in preterm and low-birth-weight infants in Israel. Israel J Med Sciences 19: 1056, 1983.
2. Baxi, AJ, Undevia JV, Bhatia HM: Role of glucose-6-phosphate dehydrogenase deficiency in acute hemolytic crisis and neonatal jaundice. Indian J Med Sci 18: 574, 1964.
3. Bienzle U, Effiong C, Luzzatto L: Erythrocyte glucose 6-phosphate dehydrogenase deficiency (G6PD type A$^-$) and neonatal jaundice. Acta Paediatr Scand 65: 701, 1976.
4. Brown WR, Wong HB: Hyperbilirubinemia and kernicterus in glucose-6-phosphate dehydrogenase-deficient infants in Singapore. Pediatrics 41: 1055, 1968.
5. Calvert AF, Trimble GE: Glucose-6-phosphate dehydrogenase in an Afro-American population. Hum Hered 30: 271, 1980.
6. Capps FPA, Gilles HM, Jolly H, Worlledge SM: Glucose-6-phosphate dehydrogenase deficiency and neonatal jaundice in Nigeria - their relation to the use of prophylactic vitamin K. Lancet II: 379, 1963.
7. Doxiadis SA, Fessas PH, Valaes T, Mastrokalos N: Glucose-6-phosphate dehydrogenase deficiency - a new aetiological factor of severe neonatal jaundice. Lancet I: 297, 1961.
8. Doxiadis SA, Fessas PH, Falaes T: Erythrocyte enzyme deficiency in unexplained kernicterus. Lancet II: 44, 1960.
9. Eshaghpour E, Oski FA et al: The relationship of erythrocyte glucose-6-phosphate dehydrogenase deficiency to hyperbilirubinemia in Negro premature infants. J Pediatr 70: 595, 1957.
10. Flatz G, Springam S, Komkris V: Neonatal jaundice in glucose-6-phosphate dehydrogenase deficiency. Lancet I: 1382, 1963.
11. Freier S, Mayer K, Levene C, Abrahamov A: Neonatal jaundice associated with familiar G-6-PD deficiency in Israel. Arch Dis Child 40: 280, 1965.

12. Gibbs WN, Gray R, Lowry M: Glucose-6-phosphate dehydrogenase deficiency and neonatal jaundice in Jamaica. Brit J Haemat 43: 265, 1979.
13. Gross RT, Bracci R, et al: Hydrogen peroxide toxicity and detoxification on erythrocytes of newborn infants. Blood 29: 481, 1967.
14. Ifekwunigwe AE, Luzzatto L: Kernicterus in G-6-P-D deficiency. Lancet I: 667, 1966
15. Kopelman AE, Ey JL, Le H: Phototherapy in newborn infants with glucose-6-phosphate dehydrogenase deficiency. J Pediatr 93: 497, 1978.
16. Lee T-C, Shih L-Y, Huang P-C, Lin C-C, Blackwell B-N, Blackwell RQ, Hsia DY-Y: Glucose-6-phosphate dehydrogenase deficiency in Taiwan. Amer J Hum Genet 15: 126, 1963.
17. Levin SE, Charlton RW, Freiman I: Glucose-6-phosphate dehydrogenase deficiency and neonatal jaundice in South African Bantu infants. J Pediat 65: 757, 1964.
18. Lie-Injo LE, Virik HK, Lim PW, Lie AK, Ganesan J: Red cell metabolism and severe neonatal jaundice in West Malaysia. Acta Haemat 58: 152, 1977.
19. Lopez R, Cooperman JM: Glucose-6-phosphate dehydrogenase deficiency and hyperbilirubinemia in the newborn. Amer J Dis Child 122 :66, 1971.
20. Lu T-C, Wei H, et al: Erythrocyte acid phosphomonoesterase activity in newly born Chinese deficient in glucose-6-phosphate dehydrogenase. Nature 213: 707, 1967.
21. Lu T-C, Wei H, Quentin Blackwell R: Increased incidence of severe hyperbilirubinemia among newborn Chinese infants with G-6-P D deficiency. Pediatr 37:994, 1968.
22. Majid JD, Hasan MI: Neonatal jaundice due to glucose 6-phosphate dehydrogenase deficiency. J Indian Med Assoc 44: 143, 1965.
23. Malaka-Zafiriu K, Tsiures I, Danielides B, Cassimos C: Salicylamide glucuronide formation in newborns with severe jaundice of unknown etiology and due to glucose-6-phosphate dehydrogenase deficiency in Greece. Helv Paediat Acta 28: 323, 1973.

24. Meloni T, Forteleoni G, Dore A, Cutillo S: Neonatal hyperbilirubinaemia in heterozygous glucose-6-phosphate dehydrogenase deficient females. Brit J Haemat 53: 241, 1983.
25. Meloni T, Costa S, Cutillo S: Three years experience in preventing severe hyperbilirubinemia in newborn infants with erythrocyte G-6-PD deficiency. Bio Neonate 28: 370, 1976.
26. Meloni T, Giorgio C, Dore A, Cutillo S: Phenobarbital for prevention of hyperbilirubinemia in glucose-6-phosphate dehydrogenase-deficient newborn infants. J Pediatr 82: 1048, 1973.
27. Mentzer WC Jr, Collier E: Hydrops fetalis associated with erythrocyte G-6-P-D deficiency and maternal ingestion of fava beans and ascorbic acid. J Pediatr 86: 565, 1975.
28. Milbauer B, Peled N, Svirsky S: Neonatal hyperbilirubinemia and glucose-6-phosphate dehydrogenase deficiency. Israel J Med Sci 9: 11, 1973.
29. Nkrumah FK: A severe neonatal jaundice - analysis of possible associated factors in infants from Accra. Ghana Med J 1: 160, 1973.
30. O'Flynn MED, Hsia DY-Y: Serum bilirubin levels and glucose-6-phosphate dehydrogenase deficiency in newborn American Negroes. J Pediat 63: 160, 1963.
31. Olower SA, Ransome-Kuti O: The risk of jaundice in glucose-6-phosphate dehydrogenase deficient babies exposed to menthol. Acta Paediatr Scand 68: 1, 1979.
32. Oluboyede OA, Esan JF, Francis TI, Luzzatto L: Genetically determined deficiency of glucose-6-phosphate dehydrogenase (type A$^-$) is expressed in the liver. J Lab Clin Med 93: 783, 1979.
33. Oski FA, Shahidi NT, et al: Erythrocyte acid phosphomonoesterase and glucose-6-phosphate dehydrogenase deficiency in Caucasians. Science 139: 409, 1963.
34. Panizon F: L'ictere grave du nouveau-ne associe a une deficience en glucose-6-phosphate deshydrogenase. Bio Neonat 2: 167, 1960.

35. Panizon F: Erythrocyte enzyme deficiency in unexplained kernicterus. Lancet II: 1093, 1960.

36 Phornphutkul C, Whitaker JA, Worathumrong N: Severe hyperbilirubinemia in Thai newborns in association with erythrocyte G6PD deficiency. Clin Pediatr 8: 275,1969.

37. Piomelli S, Corash LM, Davenport DD, Miraglia J, Amorosi EL: In vivo lability of G6PD in Gd^{A-} and $Gd^{Mediterranean}$ deficiency. J Clin Invest 47: 940, 1968.

38. Piomelli S, Spada U, Bernini L, Siniscalco M, Adinolfi M, Mollison PL, Latte B: Survival of ^{51}Cr-labelled red cells in subjects with thalassemia-trait or G6PD deficiency or both abnormalities. Brit J Hematol 10: 171, 1964.

39. Piomelli S, Vora S: G6PD deficiency and related disorders of the pentose pathway. Chapter in: Hematology of Infancy and Childhood. (eds) Nathan D, Oski F. WB Saunders & Co, Philadelphia, PA. p 566, 1981.

40. Rattazzi MC, Corash LM, van Zanen GE, Jaffe ER, Piomelli S: G6PD deficiency and chronic hemolysis: Four new mutants - relationships between clinical syndrome and enzyme kinetics. Blood 38: 205, 1971.

41. Ross JP: Deficient activity of DPNH-dependent methemoglobin diaphorase in G6PD-deficient blood. Blood 21: 51, 1963.

43. Sanna G, Frau F, De Virgiliis S, Piu P, Bertolino F, Cao A: Glucose-6-phosphate dehydrogenase red blood cell phenotype in Gd Mediterranean heterozygous females and hemizygous males at birth. Pediatr Res 15: 1443, 1981.

44. Say B, Ozand P, et al: Erythrocyte glucose-6-phosphate dehydrogenase deficiency in Turkey. Acta Paediatr Scan 54: 319, 1965.

45. Segni G: Su due casi di ittero nucleare in neonati con difetto enzimatico eritrocitario. Minerva Pediat 11: 1420, 1959.

46. Smith GD, Vella F: Erythrocyte enzyme deficiency in unexplained kernicterus. Lancet I: 1133, 1960.

47. Tan KL: Glucose-6-phosphate dehydrogenase status and neonatal jaundice. Arch Dis Childhood 56: 874, 1981.
48. Tchernia PG, Zucker J-M, Oudart J-L, Boal MR, Kuakuvi N: Frequence et incidence du deficit en glucose-6-phosphate deshydrogenase erythrocytaire chez le nouveau-ne africain a Dakar. Nouv Rev Franc Hemat 11: 145, 1971.
49. Testa U, Meloni T, Lania A, Battistuzzi G, Cutillo S, Luzzatto L: Genetic heterogeneity of glucose 6-phosphate dehydrogenase deficiency in Sardinia. Hum Genet 56: 99, 1980.
50. Valaes T, Karaklis A, Doxiadis S: Gene loss due to neonatal jaundice in glucose-6-phosphate dehydrogenase deficiency. in: Physiologic Foundations of Perinatal Care. L Stern, M Xanthou, B Friis-Hansen (eds). Praeger Publishers, N.Y. p 89, 1985.
51. Valaes T, Karaklis A, Stravrakakis D, Bavela-Stravrakakis B, Perakis A, Doxiadis SA: Incidence and mechanism of neonatal jaundice related to glucose-6-phosphate dehydrogenase deficiency. Pediat Res 3: 448, 1969.
52. Wolff JA, Bertram HG, Paya K: Neonatal serum bilirubin and glucose-6-phosphate dehydrogenase - relationship of various perinatal factors to hyperbilirubinemia. Amer J Dis Child 113: 251, 1967.
53. Yue PCK, Strickland M: Glucose-6-phosphate-dehydrogenase deficiency and neonatal jaundice in Chinese male infants in Hong Kong. Lancet I: 350, 1965.
54. Zannos-Mariolea L, Kattamis C: Glucose-6-phosphate dehydrogenase deficiency in Greece. Blood 18: 34, 1961.

REGULATION OF GLUCOSE-6-PHOSPHATE DEHYDROGENASE IN NORMAL AND VARIANT RED BLOOD CELLS[1]

*Henry N. Kirkman and Gian F. Gaetani**

Department of Pediatrics, University of North Carolina, Chapel Hill, North Carolina and *Department of Hematology, University of Genoa, Genoa, Italy

The two forms of nicotinamide adenine dinucleotide (NAD and NADP) have nearly opposite roles in metabolism. NAD occurs in cells largely in the oxidized form (NAD^+) and is utilized in catabolic steps, especially ones making the energy of foods available to cells. NAD^+ and NADH do not seem to have a prominent role in the generation or disposal of oxygen radicals. By contrast, NADP occurs largely in the reduced form (NADPH) and is utilized for anabolism through a variety of steps in reductive biosynthesis. NADPH, moreover, is now recognized as the essential component of a pathway that destroys peroxides in mammals (Fig. 1).

[1] *This work was supported by National Institutes of Health Grants AM-29864 and HD-03110 and by Consiglio Nazionale delle Ricerche Grants PF 83.01005.51 and 82.02374.51*

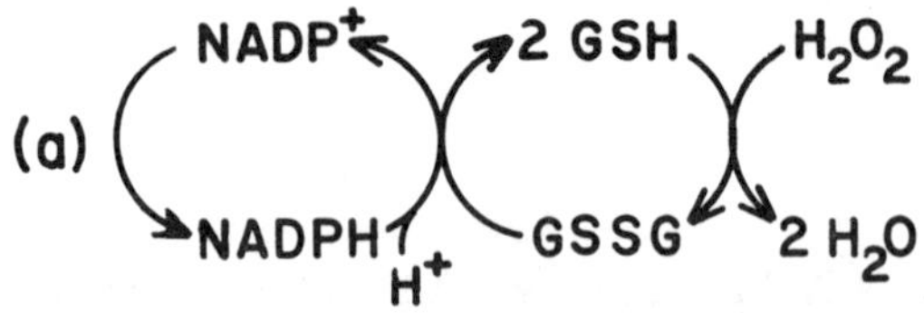

(a) $NADP^+ \rightleftarrows NADPH + H^+$; $2\ GSH \rightleftarrows GSSG$; $H_2O_2 \rightarrow 2\ H_2O$

Fig. 1. Sequence leading to the destruction of hydrogen peroxide. (a) Glucose-6-phosphate dehydrogenase and 6-phosphogluconate dehydrogenase.

Paradoxically, NADPH is also essential for the generation of oxygen radicals in phagocytizing white cells. As exemplified by the lethality of chronic granulomatous disease, this function is essential for the killing of bacteria and for the survival of human beings.

I. THE HEXOSE MONOPHOSPHATE SHUNT AND GLUCOSE-6-PHOSPHATE DEHYDROGENASE.

The principal source of NADPH in many species and cells is the hexose monophosphate shunt, also called the pentose phosphate pathway (Fig. 2). Through the action of two dehydrogenases in tandem, or in series, this pathway generates two molecules of NADPH for each molecule of glucose (specifically, glucose-6-phosphate) entering the pathway. Moreover, the products of the pathway are able to enter the Embden-Meyerhof pathway, where they can be carried up to glucose-6-phosphate (G6P) and enter the shunt again. In the absence of peroxidative/oxidative stress, the shunt accounts for only about 5% of the glucose metabolized in the human red cell (1), yet the passing of glucose through this pathway is required for the survival of the red cell and certain other cells.

One of the two NADPH-generating enzymes of the shunt turns out also to be the initial enzyme catalyzing the entrance of glucose into the pathway: glucose-6-phosphate dehydrogenase (G6PD). Evidence now exists that the step catalyzed by this enzyme is also the committed and rate-limiting step of the pathway. The immediate products of G6PD are NADPH and D-glucono-δ-lactone 6-phosphate. At a physiological pH, the lactone undergoes irreversible hydrolysis to 6-phosphogluconate both spontaneously and through the action of a lactonase. The two actions cause the hydrolysis to be rapid (2). Moreover, assays on lysates of certain cells indicate that the concentrations of 6-phosphogluconate are much less than those of G6P. Since subsequent steps in the pathway are reversible, the assumption must be made that G6PD is the rate-limiting enzyme of the shunt.

The shunt also generates ribulose-5-phosphate and ribose-5-phosphate. The latter becomes phosphoribosyl pyrophosphate, which is utilized for the synthesis of nucleotides and consequently RNA and DNA. Together with the role of the shunt in producing NADPH for reductive biosynthesis, the shunt could be regarded as the "growth" pathway. But care must be taken not to regard G6PD necessarily as catalyzing the rate-limiting step for synthesis of ribose-5-phosphate. As may be seen in Fig. 2, ribose-5-phosphate can be generated from intermediates of the Embden-Meyerhof pathway through reversal of the distal steps in the shunt. The evolving of independent regulation of synthesis of NADPH and ribose-5-phosphate would seem to be necessary. In mammary and fat cells, where lipogenesis occurs, the need for NADPH is relatively greater than the need for ribose-5-phosphate. In contrast, the need for NADPH is relatively low in muscle cells, where little if any fat is synthesized. G6PD therefore should be regarded as the regulated step for synthesis of NADPH but probably not the regulated step for synthesis of ribose-5-phosphate.

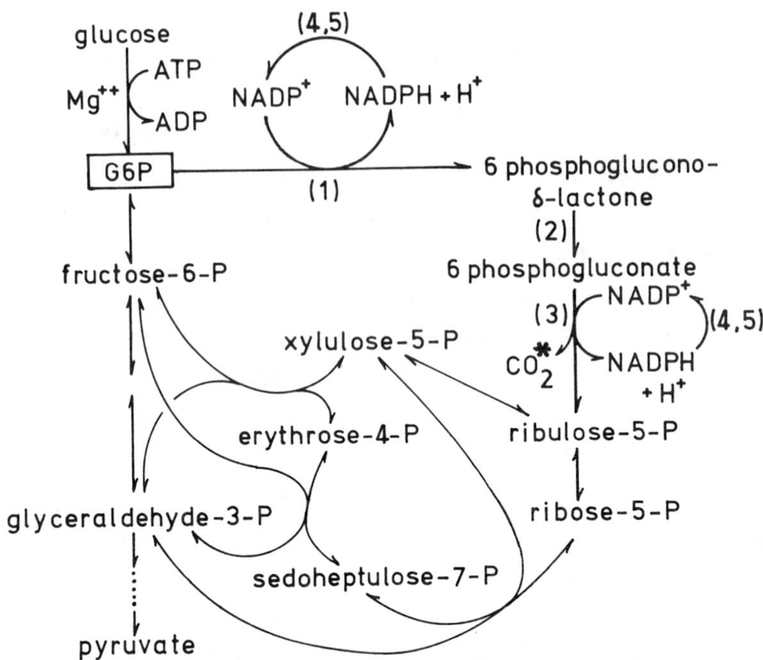

*Fig. 2. The hexose monophosphate shunt (pentose phosphate pathway). 1) G6PD. 2) 6-phosphogluconolactonase and spontaneous hydrolysis. 3) 6-phosphogluconate dehydrogenase. 4) Oxidation of NADPH by methylene blue, coupled to oxygen. 5) Use of NADPH for reduction of GSSG or for reductive biosynthesis. * From the first carbon of G6P. This figure is from Kirkman and Wilson (3).*

Knowledge of the physical and kinetic properties of normal human G6PD is necessary, but not sufficient, for an understanding of the intracellular regulation of the enzyme. At physiological concentrations, temperatures, pH, and ionic strength, human G6PD exists as a dimer (4), possibly with some of the enzyme being in the tetrameric form (5). Both forms of the enzyme are active. When stripped of $NADP^+$, diluted, and allowed to stand at 0–2 ^{0}C, however, the enzyme reversibly becomes an inactive monomer (4). Reactivation occurs when the enzyme is warmed in the presence of $NADP^+$. The subunit has been estimated to have a molecular weight of 52,000 - 59,000 (4,5).

In the course of studies of normal and mutant G6PD's in many parts of the world, the kinetics of the human enzyme have been evaluated probably more often than have the kinetics of any other enzyme in the field of biochemistry. Several different buffers have been employed, but Tris (chloride), pH 8.0, is commonly used. G6PD has two substrates and two products. The vast majority of comparative kinetic studies of normal and variant enzymes have been observations of the initial rate when G6P was held at a constant concentration (well above the Km for G6P) while the concentration of $NADP^+$ was varied, and vice versa. In the hands of nearly all investigators, classical Michaelis-Menten relationships are observed: i.e., the double reciprocal plots form a straight line. When NADPH is added to the reaction, it acts as a competitive inhibitor relative to $NADP^+$.

Although the results of such assays are valuable in phenotyping variants of G6PD, an understanding of how the enzyme might behave at physiological concentrations of substrates requires much more complex analysis of the kinetics of purified G6PD. Soldin and Balinsky studied the behavior of the enzyme under conditions that included simultaneously low concentrations of both $NADP^+$ and G6P (6). They concluded that the mechanism was of the ordered, Bi Bi type although the results could not rule out a random order mechanism with the formation of a dead end enzyme-G6P-NADPH complex. Evidence for the latter was obtained by Kirkman, Wilson and Clemons, who purified the enzyme to homogeneity, as determined by electrophoresis in SDS acrylamide gel. They then determined the initial rate of G6PD in 110 reactions in which G6P, $NADP^+$, and NADPH were present at various concentrations, including concentrations above and below the anticipated Km's for the two substrate (3). The buffer was a Krebs Ringer Tes buffer, pH 7.4, matching serum in ionic strength and concentration of divalent ions. Statistical methods of Wilkinson and Cleland provided the best-fitting estimates of the kinetic constants, as well as the standard errors of the estimates.

$$\frac{1}{V} = \frac{1}{Vmax}\left[1 + \frac{Ka}{A}\left(1 + \frac{Q}{Kii}\right) + \frac{Kb}{B} + \frac{Kia\,Kb}{AB}\left(1 + \frac{Q}{Ki}\right)\right]$$

A : $NADP^+$ Q : NADPH
B : G6P E : Enzyme (G6PD)

Ka : [EB] [A] / [EAB] = 6.51 ± 0.61 µM
Kb : [EA] [B] / [EAB] = 38.3 ± 3.6 µM
Kia : [E] [A] / [EA] = 7.91 ± 1.45 µM
Ki : [E] [Q] / [EQ] = 7.11 ± 1.29 µM
Kii : [EB] [Q] / [EBQ] = 16.98 ± 2.63 µM

Fig. 3. Kinetic equation and constants (± standard error of the estimate) for G6PD assayed in Krebs Ringer/ Tes buffer, pH 7.4 (3).

The equation and constants for the mechanism are given in Fig. 3. The equation is that of a random Bi Bi mechanism in which NADPH functions as a dead end competitive inhibitor (or leads to the formation of an abortive ternary complex).

Kosow earlier obtained the same equation with studies in 0.1 M triethanol amine buffer, pH 8.0, of G6PD from human platelets (7). With the use of alternative substrates, however, Kosow concluded that the mechanism is an ordered Bi Bi mechanism in which $NADP^+$ combines with the free enzyme, and NADPH is the last product to be released. At millimolar concentrations, ADP and ATP functioned as inhibitors (7)

II. A DISCREPANCY BETWEEN OBSERVED AND EXPECTED ACTIVITIES.

For several decades following the recognition that NADPH was an inhibitor (8), the post-translational regulation of G6PD was thought to be largely that of depletion of dinucleotide substrate ($NADP^+$) with simultaneous inhibition from the dinucleotide product (NADPH). The conversion of $NADP^+$ to NADPH causes the function of the enzyme to be doubly restricted, since NADPH functions as a competitive inhibitor relative to $NADP^+$. Conversely, the need of the cell for NADPH is signaled by low concentrations of NADPH and correspondingly higher concentrations of $NADP^+$. Over a 14-year period beginning in 1967, however, evidence began to mount for a more complex system of post-translational regulation of the enzyme. Kosower, Vanderhoff, and London (9) and Rieber and Jaffe (10) found that intact, G6PD-deficient red cells could not regenerate GSH effectively after exposure to azoester, but the corresponding hemolysates could do so. Sapag-Hagar, Lagunas, and Sols noticed that the concentrations of 6-phosphogluconate in extracts of rat liver were considerably less than the concentrations of G6P (11), yet the activities and Km's of the two dehydrogenase were such that 6-phosphogluconate should accumulate. They interpreted their results to mean either that 6-phosphogluconate dehydrogenase is more active within the cell than expected or that some unknown, additional mechanism must exist for disposing of 6-phosphogluconate. An alternative explanation, however is that G6PD functions less well in the cell than expected. Studies of *Escherichia coli* by Orthner and Pizer suggested that G6PD is under some type of intracellular inhibition also in that organism (12).

Further evidence for some unknown restraint or control of G6PD came from measurements of activity of the shunt and of concentrations of NADP and NADPH in normal and G6PD-deficient human red cells. Before those findings are reviewed, some description of the methods should be given.

The historically important studies of Brin and Yonemoto led to useful methods and to an understanding of the shunt in human red cells (13). They found that $^{14}CO_2$ from carbons 1 and 2 of glucose accounted for 95% of the oxidation of glucose and that 85% of the oxygen consumed by the human red cell could be accounted for by $^{14}CO_2$ from ^{14}C-labeled glucose. In the absence of methylene blue most of the $^{14}CO_2$ came from carbon 1 of glucose. The addition of methylene blue, however, increased the rate of oxidation of glucose and caused the second carbon of glucose to become a contributor of $^{14}CO_2$. These results can be interpreted as indicating that the oxidation of glucose in human red cells occurs largely through the initial portion of the shunt, that methylene blue stimulates the oxidation by hastening the conversion of NADPH to $NADP^+$, and that such stimulation causes the products of the shunt to re-enter the Embden-Meyerhof, where they can be converted back to G6P and allowed to enter the shunt again. As a consequence of these features, the intracellular rate of G6PD can be calculated from the rate of evolution of $^{14}CO_2$ from [1-^{14}C]glucose and [2-^{14}C]glucose.

The development of techniques for measuring NADP in red cells is more recent. NADPH is destroyed when heated in acidic solutions whereas $NADP^+$ is destroyed when heated in alkaline solutions. The Lowry and Passonneau method (14) for enzymic determination of NADP, NADPH, and $NADP^+$ has two stages: 1) preliminary dilution of the sample and differential destruction of NADPH or $NADP^+$ and 2) cycling (enzymic) assay of the remaining NADP. Burch, Bradley and Lowry demonstrated that NADPH, in the presence of hemoglobin, has a tendency to undergo oxidation to $NADP^+$ at acid pH or when heated at alkaline pH (15). They found that this post-sampling oxidation could be prevented by: the addition of ascorbic acid to the acidic solutions; the presence of cysteine in the alkaline solutions; and the extensive dilution of the sample in these solutions before they were heated. In the adaptation of these methods to studies of human red cells (3,16), the sample is diluted at least 1,050 fold (2,100 fold relative to packed red cells) in very freshly prepared, cold (0 ^{o}C) 0.04 M

NaOH/0.5 mM cysteine before being heated for destruction of $NADP^+$. Dilution of the sample in acidic solutions containing ascorbic acid must be so extensive that direct determination of [$NADP^+$] (as opposed to taking the difference between [NADP] and [NADPH]) is inaccurate in normal human red cells without oxidative stress (16). Nevertheless, estimations of red cell [NADP] and [NADPH] are routinely possible and estimations of [$NADP^+$] are possible under certain conditions (3). Internal standards of $NADP^+$ and NADPH are used except under experimental conditions in which earlier assays have shown consistent agreement between the results with internal and external standards. The degree of dilution can be less when the hemoglobin/NADP ratio of the sample is less than that of human red cells (14).

The sensitivity of the cycling step makes possible the relatively extensive, preliminary dilution of the sample. In the cycling reaction, the NADPH becomes $NADP^+$ in the presence of glutamate dehydrogenase, α-ketoglutarate, ammonium chloride, and ADP. The $NADP^+$, in turn, becomes NADPH in the presence of G6P and G6PD. Each turn of the cycle results in one molecule of 6-phosphogluconate. The cycling rate is 16,000/hr with the reagents used in our laboratories (3,16). That is, each molecule of NADP results in 16,000 molecules of 6-phosphogluconate during an incubation of one hour. The concentration of NADP in the cycling mixture is so far below the Km of each dehydrogenase that the rate of cycling is essentially proportional to the concentration of NADP (14). The reaction is stopped by heat, and 6-phosphogluconate is determined fluorometrically (14) with reagent 6-phosphogluconate dehydrogenase and $NADP^+$. Various precautions are listed in the book by Drs. Lowry and Passonneau (14). The use of disposable plasticware, glassware, and pipette tips is preferable to extensive acid-washing and alkali-washing of glassware. Contamination is otherwise often a problem in a laboratory where NADP is heavily used and where the cycling assay is capable of detecting $1X10^{-12}$ to $1X10^{-11}$ moles of NADP.

The sensitivity of the cycling assay matches or exceeds that of radioisotopic methods. With the preliminary differential destruction of $NADP^+$ or NADPH, moreover, the cycling assay tells the investigator how much of the NADP has become $NADP^+$ or NADPH during an experiment.

Welt, *et al.* found an average of only about 10% stimulation of the shunt in normal red cells drawn approximately 12 hours after the ingestion of the second of two daily doses of 30 mg of primaquine base (as primaquine phosphate) (17). This increase corresponds to only about 1/1000 of the maximal activity of the G6PD of G6PD A^- red cells. In the same laboratory, red cells of G6PD B, G6PD A^- and G6PD Mediterranean cells were incubated for 6 hours at 37 ^{0}C with α-naphthol at concentrations ranging from a level (20 μM) that caused only a slight drop in GSH to a level (80 μM) that caused a decided drop in GSH in the G6PD-deficient cells (18). Results at 40 μM are shown in Table 1. Each molecule of G6P passing through the shunt results in two molecules of NADPH, which are sufficient to generate four molecules of GSH from oxidized glutathione.

TABLE I. Response of normal and G6PD-deficient red cells to incubation with α-naphthol

Phenotype	Increase in rate of HMS[a] μmol $L^{-1} h^{-1}$	Rate of decrease in GSH[b] μmol $L^{-1} h^{-1}$
Normal	72	6
A^-	16	29
Mediterranean	6	38

[a] Hexose monophosphate shunt. The α-naphthol concentration was 40 μM. The HMS rate for normal red cells without α-naphthol was 90-100 μmol $L^{-1} h^{-1}$.

[b] Divided by 4. These results are calculated from the findings of Gaetani *et al.* (18).

So that the rate of decline in GSH could be compared to the increase in activity of the shunt, the μmol $L^{-1}h^{-1}$ decline in GSH is divided by four. The various increases in shunt activity and rates of decline in GSH are less than the resting rate of the shunt in normal red cells and less than 1% of the maximal rate of G6PD from G6PD A^- red cells.

If such minimal additional requirements for NADPH cannot be met by the G6PD-deficient (A^-) red cell, then one might expect that the G6PD A^- red cell has difficulty keeping NADP in the reduced form even when the cell is not under oxidative or peroxidative stress. Exactly that situation was found when fresh, G6PD-deficient red cells were assayed for NADPH and NADP (Table II) (16). At assumed (unbound) concentrations of G6P and NADP of 30-60 μM, the G6PD in the unstressed G6PD A^- red cell is working at only about 1-2 % of the rate at which it is capable, given the known Km's and Vmax of the enzyme.

A similar discrepancy was found with normal red cells that were placed under oxidative stress by incubation with methylene blue at concentrations of the dye ranging from 0 to 100 μM (3). The observed rate of the intracellular enzyme was determined from the rate of evolution of $^{14}CO_2$ from [1-^{14}C]glucose and [2-^{14}C]glucose.

TABLE II. NADP content of normal and G6PD-deficient red cells[a]

Phenotype	Number of subjects	NADP μM	% of NADP that is NADPH
Normal	8	30.7 ± 2.8	98.3 ± 4.1
A^-	6	50.3 ± 5.8	39.2 ± 6.7
Mediterranean	6	59.1 ± 2.6	24.5 ± 5.0

[a] From Kirkman, *et al.* (16).

The expected rate of the intracellular enzyme was calculated from the observed concentrations of G6P, $NADP^+$ and NADPH, the observed Vmax of the enzyme (in hemolysate), and the known kinetic constants of normal G6PD. Comparison of the observed rate with the expected rate indicated that the intracellular enzyme was under unexplained restraint ranging from 200-fold, when most of the NADP was reduced, to 5-fold, when most of the NADP was oxidized (3). Moreover, a plot of intracellular rate against the observed $NADP^+$/NADP ratio revealed a sigmoid curve (Fig. 4). In contrast, both the expected and observed

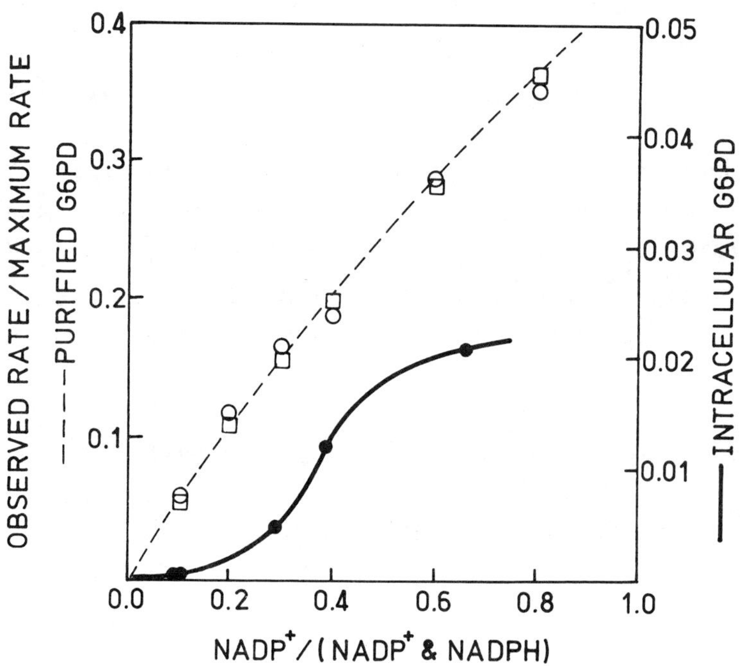

Fig. 4. Comparison of rate of intracellular and purified G6PD. Solid line: Rate of G6PD in normal red cells incubated with methylene blue at concentrations of 0, 0.5, 5 and 100 μM. Observed (○, □) and predicted (broken line) rates of purified G6PD were at concentrations of 40 μM for both G6P and NADP. The predicted rates were from the equation of Fig. 3. From Kirkman, Wilson, and Clemons (3).

activity of the purified enzyme gave a diagonal line at various $NADP^+$/NADP ratios. (G6P and NADP concentrations were constant and comparable to those observed in the cell.)

The initial observations were with normal red cells incubated in Krebs Ringer / Tes buffer, but similar results were obtained with incubation of human red cells in Krebs Ringer / bicarbonate buffer (19). With the latter incubation, moreover, the concentration of 6-phosphogluconate was considered in the calculation. Since $^{14}CO_2$ is the product of 6-phosphogluconate dehydrogenase, rather than of G6PD, the true intracellular rate of G6PD is given by the amount of evolved $^{14}CO_2$ plus the rise, if any, in concentration of 6-phosphogluconate. No sigmoidicity was observed when the rate of $^{14}CO_2$ evolution was plotted against concentrations of methylene blue, either in these experiments or in repetitions (at both of our laboratories) of the experiments in phosphate buffer by Bonsignore, *et al* (19). Nor could sigmoidicity be observed (in either laboratory) when kinetics of the purified, normal enzyme where studied in borate buffer according to the protocol of Afolayan and Luzzatto (19). The intracellular inhibition seemed unlikely to be the consequence of inhibition by intracellular ATP, since the technique of lysis and resealing of red cells allowed relatively large amounts of ATP to be trapped in resealed red cell ghosts without demonstrable intracellular inhibition of G6PD by the ATP (20).

As mentioned, numerous workers have been unable to demonstrate sigmoid kinetics of the purified human enzyme, and the purified enzyme undergoes reversible dissociation into inactive monomers under conditions (4) that seem too extreme to exist within the cell. Intracellular transition of human G6PD between a dimer and monomers was refuted by Kahler and Kirkman, who prepared resealed red cell ghosts from a heterozygous woman who had two electrophoretically distinguishable forms of the enzyme: G6PD A and G6PD B (21). The reasoning behind the experiment is similar to that in using these variants to see if cultured and fused fibroblasts have two active X chromosomes. By the Lyon mechanism, each red cell of the

G6PD A/B woman has only G6PD A or G6PD B. With lysis and resealing of the red cells, however, both variants are present in each cell. If the enzyme is undergoing dissociation into monomers and reassociation into a dimer, a heterodimer (AB) should form. As with somatic cell hybrids containing two active X chromosomes, the heterodimer should be apparent as a prominent, third band that is intermediate in location to the positions of the AA and BB bands on electrophoresis. That is, there should be three bands corresponding to AA, AB, and BB in the proportion 1:2:1. After incubation of the resealed red cells in the presence of glucose, which keeps much of the NADP in the reduced form, only two electrophoretic bands were seen, G6PD A and G6PD B (21).

III. EXTENSIVE BINDING OF INTRACELLULAR NADP.

A major explanation for the intracellular restraint and sigmoid kinetics of G6PD in human red cells was found when efforts were made to identify an inhibitor of the enzyme among the metabolites of red cells. That effort required separating the proteins of the red cell from the components of lower molecular weight. Three methods of separation were used: molecular exclusion column chromatography, dialysis, and ultrafiltration-washing (22). The last method consisted of ultrafiltering hemolysates by centrifuging them in Amicon CF-25 ultrafiltration cones, adding buffer to the protein concentrate, and centrifuging again. Although molecular exclusion chromatography and dialysis required 12-72 hours, the ultrafiltration-washing could be accomplished within one hour. All three methods, however, revealed an unexpected phenomenon: most of the NADP of the human red cell is retained with the protein fraction (22).

Work on the binding of NADP by proteins of hemolysates required preparation of the hemolysate in a way that reduced or eliminated enzymic destruction of NADP during the experiment. White cells and platelets were removed from the heparinized blood by filtration through cellulose particles by the method of Beutler, West and Blume (23). Stroma (membranes) of the cells, which contain proteinases and NADPase, were removed by centrifugation of the 1:10 hemolysate of washed, packed red cells at 16,000 X g for 20 - 30 min. With these precautions, very little destruction of NADP was noticed in experiments requiring studies of the hemolysate at 0 °C - 4 °C for 12 to 24 hours. The ratio of NADPH to NADP, however, underwent gradual change. After removal of G6P, which interacts with G6PD to keep most of the NADP reduced, about half of the NADPH in hemolysates became $NADP^+$ over a period of 12 - 24 hours.

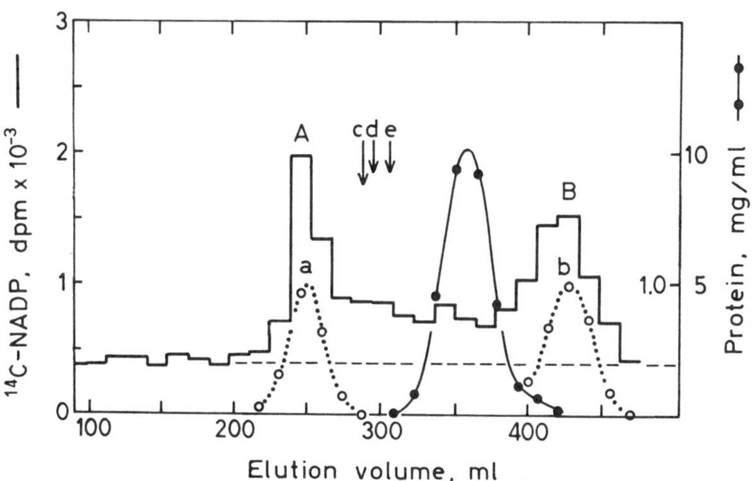

Fig 5. Distribution of [^{14}C]NADP after molecular exclusion chromatography of a mixture of [^{14}C]NADP and stroma-free hemolysate on a 2.6 X 96 cm column of Sephadex G-200. Dotted lines represent relative activities of catalase (a) and NADPH diaphorase (b). Arrows indicate the centers of activities of G6PD (c), glutathione reductase (d), and 6-phosphogluconate dehydrogenase (e). Cycling analysis revealed that Peak A consisted of NADPH whereas Peak B consisted largely of $NADP^+$. From Kirkman and Gaetani (24).

The tendency of proteins of human red cells to bind NADP could be demonstrated either by enzymic (cycling) assays of the protein fraction (22) or by the binding of ^{14}C-labeled NADPH and $NADP^+$ (24). An example of the latter can be seen in Fig. 5. When the proteins emerging from a Sephadex G-200 column or G-150 column were equilibrated with $NADP^+$ (5µM) or with an equimolar mixture of $NADP^+$ (2.5 µM) and NADPH (2.5 µM), the principal region of NADPH-binding was found to be coincident with the catalase peak, whereas the principal region of $NADP^+$ binding was coincident with the NADP diaphorase peak (22,24). Highly purified human catalase was found to contain NADPH and to bind four molecules of NADPH per (tetrameric) molecule of catalase (24). Crystalline catalase from bovine liver was found not only to bind NADPH but also to contain four molecules of tightly bound NADPH per molecule of catalase as the catalase is removed "from the bottle" from biochemical supply companies. The NADPH-binding peak was missing in the hemolysate of a patient with the Swiss type of acatalasemia (24). As was noted by earlier workers, $NADP^+$ binding was found with highly purified NADPH diaphorase. It seems almost more than coincidental that the two principle NADP-binding proteins of the human red cells are enzymes for which the biological function is unclear. But binding of NADP also occurs among proteins with molecular weights between those of catalase (240,000) and NADPH diaphorase (20,000), and binding by these proteins is particularly noticeable when the proteins are equilibrated with $NADP^+$ or NADPH at concentrations higher than 2.5 µM (22). That is, the affinity of these proteins for NADP is less than that of catalase or NADPH diaphorase.

Equilibration of hemolysate proteins with $NADP^+$ or NADPH by ultrafiltration-washing revealed that 1.5 - 2.0 µM concentrations of either dinucleotide were sufficient to cause the binding of 33 - 38 nmol of NADP per 340 mg of hemoglobin (22). In packed cells, where the concentration of hemoglobin is about 340 g/L, this relationship would result in a total red cell NADP concentration of 35.5 - 40 µM, with only 1.5 - 2.0 µM of NADP representing free or unbound NADP.

Equilibration of the hemolysate proteins against NADP at a constant concentration, but at various $NADP^+$/NADP ratios, moreover revealed sigmoid binding. That is, a plot of the fraction, f, of unbound NADP that was oxidized (the $NADP^+$/NADP ratio) against the corresponding fraction, f_b, for bound NADP, resulted in a sigmoid curve (22).

As revealed by computer analysis with dissociation equations, such sigmoidicity is the expected result when two or more proteins are binding most of the NADP, one protein having a 10-fold or greater affinity for NADPH than for $NADP^+$ and another protein having the opposite affinities. Fig. 6 allows an intuitive appreciation for why the sigmoidicity occurs. The left-hand panel has an f_b vs. f plot for each of two hypothetical proteins. The convex curve is that of a protein having a dissociation constant for the protein-$NADP^+$ complex that is 0.1 times the dissociation constant for the protein-NADPH complex. As a consequence of a greater affinity for $NADP^+$ than for NADPH, the fraction of bound NADP that is $NADP^+$ (f_b) tends to lead the corresponding fraction (f) in the solution to which the protein is exposed. Conversely, the concave curve is obtained with a protein having a dissociation constant for the protein-$NADP^+$ complex that is 10 times the dissociation constant for the protein-NADPH complex. The middle panel illustrates the graphic average of the two curves,

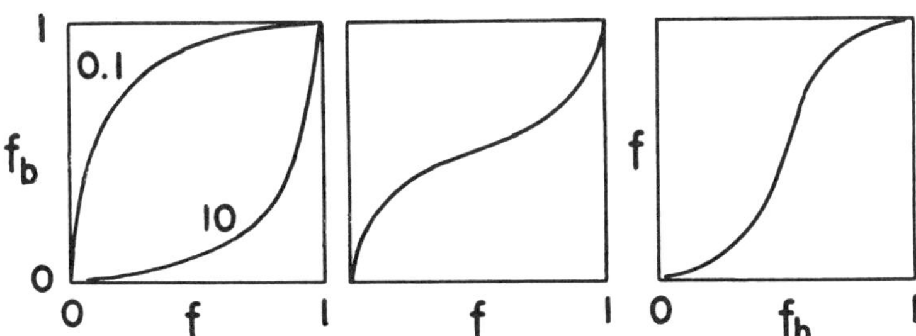

Fig. 6. The $NADP^+$/NADP ratio (f_b) of NADP bound to two hypothetical proteins (text) exposed to NADP at various $NADP^+$/NADP ratios (f) for unbound NADP.

such as would be obtained with an equimolar mixture of the two proteins. Inversion of the ordinate and abscissa (right-hand panel) reveals a sigmoid curve comparable to that seen in plots of intracellular G6PD rates against the $NADP^+$/NADP ratio of the whole cell (Fig. 4). When over 90% of the NADP of the cell is bound, as is the case with human erythrocytes, f_b is similar to the $NADP^+$/NADP ratio of the whole cell.

Thus the sigmoid kinetics of intracellular G6PD are a reflection of the sigmoid nature of the binding and release of $NADP^+$ and NADPH in the red cell. In terms of the $NADP^+$/NADP ratio of the whole cell, the G6PD is able to respond as if the enzyme had sigmoid kinetics. More than would be the case with classical kinetics, the G6PD decreases in rate when most of the (whole cell) NADP is NADPH and increases in rate when very little of the NADP is NADPH. As may be seen in the right-hand panel of Fig. 6, the $NADP^+$/NADP ratio of unbound NADP tends to be an exaggeration (about f=0.5) of the ratio for bound NADP: when the ratio for bound NADP is high, the ratio for unbound NADP is even higher; when the ratio for bound NADP is low, the ratio for unbound NADP is even lower. One implication of this relationship is that caution must be exercised in assuming that inhibition by NADPH can increase the severity of the impairment in G6PD variants that have an unusually low Ki for NADPH. The ratio of NADPH to $NADP^+$ is low in the G6PD-deficient red cell and still lower in that fraction of NADP that is unbound.

The binding of over 90% of the NADP by proteins within the human red cell, coupled with the sigmoid nature of the binding, can account for most of the 5-fold to 200-fold discrepancy between expected and observed activity of intracellular G6PD in normal red cells (22). The discrepancy occurs largely because the concentrations of $NADP^+$and NADPH in the whole cell are very different from the corresponding concentrations of the unbound dinucleotides. More recently, the use of these techniques has suggested that much of the NADP in human leukocytes and rat liver is also bound. Rat liver was homogenized in five volumes of Krebs Ringer / Tes buffer, then the homogenate was centrifuged at 16,000 g for 30 minutes. As determined by either the cycling assay for NADP or the binding of added ^{14}C-labeled NADP, over 90% of the NADP of the supernatant fluid remained with the protein fraction during ultrafiltration. These findings suggest that extensive binding of NADP by intracellular proteins may not be just a peculiarity of human red cells.

IV. UNANSWERED QUESTIONS

At least two features of G6PD and the shunt in human red cells remain puzzles.

A. G6PD Mediterranean.

In theory, characteristics of G6PD variants that should predispose to severe hemolytic problems (congenital or chronic non-spherocytic disease) are: (i) very low Vmax for the enzyme, (ii) high Km's for the substrates (25), $NADP^+$ and G6P, and (iii) severe intracellular lability of the enzyme. The importance of the last item is easily overlooked. Intracellular decay in activity of the enzyme may cause the G6PD in one variant to reach a critically low level in 60 days whereas the lability with another variant might cause it to reach the same critically low level in 12 days. The mean activity of the red cells is the area under the activity-age curve divided by the age at which the red cells are destroyed. As a consequence, activities of G6PD in the red cell samples from both variants might be identical.

A noteworthy exception to these three expectations is G6PD Mediterranean. To be sure, the Mediterranean variant has a *lower* than normal Km for its substrates, and this is important in view of the finding that the concentration of unbound NADP is considerably lower than that of total (bound and unbound) NADP. But the Mediterranean variant has a Vmax that is essentially zero in cells beyond the reticulocyte stage. This point was made by Piomelli *et al*, who removed white cells and platelets with the use of a discontinuous density gradient and found much less than the 3% normal activity for G6PD that had been reported by investigators using cruder methods for removing white cells and platelets (26). Workers using filtration through cellulose particles for removal of white cells now report that the activity of G6PD in Mediterranean G6PD-deficient red cells is essentially non-detectable, even with the use of ^{14}C-labeled glucose or G6P (27).

B. The 6-phosphogluconate paradox.

The concentrations of substrate routinely used in assays for G6PD and 6-phosphogluconate dehydrogenase are many times the Km's for these substrates. As a consequence, measurements of the activity of the two dehydrogenases in lysates of cells provide an estimate of the Vmax of each enzyme. In the following paragraphs, the use of the term "Vmax" will denote the result of such an assay.

Sepag-Hagar, Lagunas, and Sols called attention to low concentrations of 6-phosphogluconate in rat liver relative to the concentrations of G6P. The Vmax and Km values of the two dehydrogenases seemed to indicate that 6-phosphogluconate dehydrogenase would have difficulty utilizing 6-phosphogluconate as rapidly as it might be generated by G6PD.

As in rat liver, the concentration of 6-phosphogluconate in the human red cell is less than the concentration of G6P (Table III)). Yet the Vmax of G6PD is about 1.5 times the Vmax of 6-phosphogluconate dehydrogenase in human red cells. Using 6-phosphogluconate dehydrogenase that had been carried through many purification steps, Pearse and Rosemeyer found the red cell enzyme to have a Km for 6-phosphogluconate of 20 μM (28). This is 1/2 to 1/3 of the Km of G6PD for G6P. Their estimate of the Ki for NADPH was 30 μM, which is 2.3 times the corresponding Ki for G6PD. In the sense of the Km for

TABLE III. Concentrations of glucose-6-phosphate (G6P) and 6-phosphogluconate (6PG) in rat liver and human red cells .

Cells	G6P	6PG
Rat liver[a]	187 nmol/g	44 nmol/g
Human red cells[b]		
no methylene blue	70 nmol/ml	4 nmol/ml
methylene blue (100 μM)	32 nmol/ml	19 nmol/ml

[a] Sapag-Hagar, M., Lagunas, R. & Sols, A. (1973).
[b] Kirkman, H. N. & Gaetani, G. F. (1986).

6-phosphogluconate (versus G6P) and the Ki for NADPH, therefore, the 6-phosphogluconate dehydrogenase of human red cells would seem to have a 2-fold or 3-fold advantage over G6PD.

Pearse and Rosemeyer reported (28), however, that the the Km for $NADP^+$ of the 6-phosphogluconate dehydrogenase was 30 μM, a value four to five times the corresponding Km of G6PD in Krebs Ringer/Tes buffer, pH 7.4 (Fig. 3) and nearly 10 times the Km for $NADP^+$ of G6PD in Tris (chloride) buffer, pH 8.0. Thus the relative values for Vmax and Km (NADP) of human G6PD and 6-phosphogluconate dehydrogenase indicate that the intracellular concentration of 6-phosphogluconate should exceed that of G6P, yet the opposite occurs (Table III). Although binding and sigmoid release of $NADP^+$ explains much of the intracellular inhibition of G6PD, the binding only intensifies the paradox regarding 6-phosphogluconate.

As is pointed out by Professor J. H. Williamson elsewhere in this symposium, mutations that result in accumulations of 6-phosphogluconate in *Drosophila melanogaster* are lethal. The accumulation of large amounts of this acidic substance seem to be harmful to the cell. It is likely therefore that strong selective pressures serve to prevent the development of relative values for Vmax and Km that would allow high concentrations of 6-phosphogluconate to occur within the cell. Eggleston and Krebs reported the presence of a factor in rat liver that decreased the extent to which G6PD and 6-phosphogluconate dehydrogenase were inhibited by NADPH (29). Rodriquez-Segade, Carrion and Freire reported the presence of a similar factor, with molecular weight of 15,000, in the hepatopancreas of a mussel (30). Levy and Christoff, however, recently offered evidence that the original observations of Eggleston and Krebs were the result of an artifact (31).

The paradox would seem to justify a renewed search for a de-inhibiting factor or for a second pathway by which 6-phosphogluconate could be utilized. Before a search is undertaken, however, perhaps the determinations of Vmax, Km's and Ki's of the two dehydrogenase of rat liver and human cells should be repeated. Yoshida and Davé reported a Km for $NADP^+$ (3.3 µM) of 6-phosphogluconate dehydrogenase that was only 1/9 of the estimate of Pearse and Rosemeyer (32). One of us (H.N.K.) found the Km for $NADP^+$ to be only 2.4 ± 0.5 µM for assays with fresh stroma-free hemolysates in the same Tris (chloride) / KCl, pH 8.0, buffer that was used by Pearse and Rosemeyer. The result was 3.9 ± 0.4 µM in the Krebs Ringer / Tes buffer, pH 7.4, used in the studies (3) of Fig. 3. Determinations of Km's and Ki's on cell lysates are generally inadvisable. But stroma-free hemolysates have very little activity of enzymes that consume or generate substrates or (NADPH) product of 6-phosphogluconate dehydrogenase during assays for that enzyme.

Three possible explanations can be formulated for the finding of a lower Km for $NADP^+$ with hemolysates than was obtained by earlier workers with extensively purified 6-phosphogluconate dehydrogenase: 1) Extensive purification of the enzyme results in damage to the dehydrogenase. 2) A de-inhibiting factor is removed during purification. 3) 6-phosphogluconate is contaminated, causing different results with different batches of the substrate. The third possibility is suggested by information from at least one company supplying 6-phosphogluconate. That information implies that results of assays for 6-phosphogluconate dehydrogenase vary from one batch of the substrate to the next.

REFERENCES

1. Rose, I. A., and O'Connell, E. L. (1964). J. Biol. Chem. 239, 12-17.
2. Schofield, P. J., and Sols, A. (1976). Biochem. Biophys. Res. Comm. 71, 1313-1318.
3. Kirkman, H. N., Wilson, W. G., and Clemons, E. H. (1980). J. Lab. Clin. Med. 95, 877-887.
4. Kirkman, H. N., and Hendrickson, E. M. (1962). J. Biol. Chem. 237, 2371-2376.
5. Cohen, P., and Rosemeyer, M. A. (1969). European J. Biochem. 8, 8-15.
6. Soldin, S. J., and Balinsky, D. (1968). Biochem. 7, 1077-1082.
7. Kosow, D. P. (1974). Archiv. Biochem. Biophys. 162, 186-193.
8. Negelein, E., and Haas, E. (1935). Biochem. Z. 282, 206-220.
9. Kosower, N. S., Vanderhoff, G. A., and London, I. M. (1967). Blood 29, 313-319.
10. Rieber, E. E. and Jaffe, E. R. (1970). Blood 35, 166-172.
11. Sapag-Hagar, M., Lagunas, R., and Sols, A. (1973). Biochem. Biophys. Res. Comm. 50, 179-185.
12. Orthner, C. L., and Pizer, L. I. (1974). J. Biol. Chem. 249, 3750-3755.
13. Brin, M., and Yonemoto, R. H. (1958). J. Biol. Chem. 230, 307-317.
14. Lowry, O. H., and Passonneau, J. V. (1972). "A Flexible Sytem of Enzymatic Analysis". Academic Press, New York.
15. Burch, H. B., Bradley, M. E., and Lowry, O. H. (1967). J. Biol. Chem. 242, 4546-4554.
16. Kirkman, H. N., Gaetani, G. D., Clemons, E. H., and Mareni, C. (1975). J. Clin. Invest. 55, 875-878.
17. Welt, S. I., Jackson, E. H., Kirkman, H. N., and Parker, J. C. (1971). Ann. N. Y. Acad. Sci. 179, 625-635.

18. Gaetani, G. F., Parker, J. C., and Kirkman, H. N. (1974). Proc. Natl. Acad. Sci. USA 71, 3584-3587.
19. Kirkman, H. N., and Gaetani, G. F. (1986). J. Biol. Chem. in press.
20. Wilson, W. G., Kirkman, H. N., and Clemons, E. H. (1980). J. Lab. Clin. Med. 95, 888-896.
21. Kahler, S. G., and Kirkman, H. N. (1983). J. Biol. Chem. 258, 717-718.
22. Kirkman, H. N., Gaetani, G. F., and Clemons, E. H. (1986). J. Biol. Chem. in press.
23. Beutler, E., West, C., and Blume, K. G. (1976). J. Lab. Clin. Med. 88, 328-333.
24. Kirkman, H. N., and Gaetani, G. F. (1984). Proc. Natl. Acad. Sci. USA 81, 4343-4347.
25. Kirkman, H. N., Schettini, F., and Pickard, B. M. (1964). J. Lab. Clin. Med. 63, 726-735.
26. Piomelli, S., Corash, L. M., Davenport, D. D., Miraglia, J., and Amorosi, E. L. (1968). J. Clin. Invest. 47, 940-948.
27. Morelli, A., Benatti, U. Lenzerini, L., Sparatore, B., Salamino, F., Melloni, E., Michetti, M., Pontremoli, S., and DeFlora, A. (1981). Blood 58, 642-644.
28. Pearse, B. M. F., and Rosemeyer, M. A. (1974). Eur. J. Biocem. 42, 213-223.
29. Eggleston, L. V., and Krebs, H. A. (1974). Biochem. J. 138, 425-435.
30. Rodriquez-Segade, S., Carrion, A., and Freire, M. (1979). Biochem. Biophys. Res. Comm. 89, 148-154.
31. Levy, H. R., and Christoff, M. (1983). Biochem. J. 214, 959-965.
32. Yoshida, A., and Davé, V. (1975). Arch. Biochem. Biophys. 169, 298-303.

INTRAERYTHROCYTIC STABILITY OF NORMAL AND MUTANT G6PD[1]

De Flora, A., Morelli, A., and Grasso, M.

Institute of Biochemistry, University of Genoa, Genoa, Italy

I. INTRODUCTION

"In vivo" operation of enzyme proteins rests on two basic properties, both being the result of a complex regulation by a number of either intrinsic or environmental factors: catalytic competence (including

---

[1]Supported in part by grants from the Special Project "Ingegneria Genetica e Basi Molecolari delle Malattie Ereditarie", C.N.R., Rome, and from the Ministry of Education, Rome.

affinities for substrates, activators and inhibitors) and protein turnover. G6PD provides no exception and represents, in mammalian cells especially, a paradigmatic case where kinetic effects are intertwined with induction mechanisms and with stability-affecting parameters in a highly sophisticated way. To focus on human erythrocyte G6PD only, relevant examples are the steady and flexible control of catalytic activity by the $NADP^+$/NADPH ratio and the still undefined mechanisms that mediate the characteristic decay of both G6PD activity and protein throughout the life-span of erythrocytes.

The peculiar oligomeric structure of erythrocyte G6PD, mostly regulated in its dimer-monomer equilibrium by the "apoenzyme-bound" or "structural" NADP (1-7), may provide a molecular basis for both properties, i.e. activity and stability. Thus, it is possible, although so far unproven, that in the intracellular conditions that are normally characterized by a low $NADP^+$/NADPH ratio (8, 9), G6PD is working at a minimal fraction of its potential activity as a result of extensive dissociation of catalytically competent dimers to intrinsically sub-active monomers[2]. This

---

[2] G6PD monomers are defined "sub-active" since they are not catalytically competent, yet they regain activity shortly after mixing with $NADP^+$ throughout conventional assays, as demonstrated by stopped-flow experiments (10).

possibility is suggested by direct "in vitro" dissociation of the dimeric enzyme by NADPH concentrations that are inhibitory to G6PD activity (11) and by re-association of the monomers to yield dimers and, in part, tetramers as soon as $NADP^+$ levels are increased (5, 12-14). Also, G6PD monomers might represent more susceptible targets to erythrocyte proteinases than the dimeric and tetrameric species.

At variance with the situation occurring in nucleated and protein-synthesizing cells, in the mature erythrocyte re-aggregation of catalytically sub-active monomers represents the only mechanism allowing this cell to express higher G6PD activity in cases of emergency, typically as a response to an oxidative challenge that will rapidly increase the intracellular $NADP^+$/NADPH ratio. Thus, some structural variants of G6PD having near-normal intracellular activity under steady-state conditions might in principle fail to steadily adjust their intracellular function following an oxidative stress, because of abnormalities in the $NADP^+$-apoenzyme interactions and of looser than normal contacts between constitutive monomers.

Little is known of the opposite step of G6PD turnover, i.e. its degradation processes. This is due to the fact that only recently has a clear picture of the proteolytic activities of human erythrocytes been emerging (15-17). Therefore, mechanisms and pathways of degradation of G6PD B by the highly integrated and complex proteolytic system of the erythrocyte still require elucidation. Accordingly, even less or no information at all is available on proteolysis of mutant G6PD. The state of the art at present allows us to provide only a phenomenological description of levels of G6PD activity in erythrocyte fractions of increasing density and to draw some tentative conclusions as to the patterns of G6PD stability during cell aging. A better knowledge on the basic properties of erythrocyte proteinases and of their possible attack on either normal or mutant G6PD is expected to advance significantly our understanding of G6PD stability in molecular terms. Such goal is strongly suggested by the possibility of influencing by means of specific inhibitors the overall rate of intracellular degradation of unstable mutant G6PD - an approach that would be useful for preventing or attenuating expression of G6PD deficiency at the cellular level.

II. MATERIALS AND METHODS

Blood samples from male subjects were drawn and processed immediately, in order to remove leukocytes and platelets according to the procedure of Beutler *et al.* (18). In addition to the subjects found to have the structural G6PD variants reported previously, i.e. G6PD Napoli (19), G6PD Ferrara II (19) and G6PD Cagliari (20), all other subjects were of Sardinian ancestry. Out of these, some, with normal G6PD activity, had G6PD B. Other asymptomatic subjects, who were born in several towns and villages of Sardinia, had abnormally low levels of G6PD activity and were considered to have the Mediterranean variety (21) of G6PD deficiency. Finally, other severely G6PD-deficient subjects were favic patients during acute hemolytic crisis.

Assays of G6PD activity (22) and purification of mutant G6PD both from "pure" erythrocytes and from "pure" leukocytes (23) were described previously. Entrapment of G6PD purified from leukocytes into G6PD-deficient erythrocytes was obtained according to the procedure of Dale _et al_. (24), as described previously (25). Hexose monophosphate shunt (HMS) activity of erythrocytes (either normal or stimulated with 100 μM methylene blue) was assayed according to Morelli _et al_. (25). Fractionation of erythrocytes according to density (26) was obtained by centrifugation on discontinuous gradients of Stractan II (27), as described (28). Four erythrocyte layers were recovered, which were defined as fractions 1, 2, 3 and 4, respectively, in order of increasing density.

Ca^{2+}-activated neutral proteinase (CANP) was purified from human erythrocytes according to Melloni _et al_. (29). Assays of CANP activity were carried out as described by Melloni _et al_. (30, 31), using acid-denatured globin as substrate. CANP activity in freshly prepared membrane-free hemolysates was determined following removal of the natural inhibitor (32) by DEAE-cellulose chromatography (31). Acidic proteinase activity, as a sum of the three acidic endopeptidases identified previously (15), was assayed both on membrane preparations (15) and on membrane-free hemolysates (16), using acid-denatured globin as substrate. Right-side out membrane vesicles were prepared according to Steck and Kant (33). SDS-PAGE analysis was carried out according to the procedure described by Fairbanks et al. (34).

Experiments on normal erythrocytes treated with 100 µM Ca^{2+} and the ionophore A 23187 were carried out in the conditions reported by Pontremoli et al. (35).

Incubations of purified G6PD B (0.1 µg) with purified CANP (2 µg) and with right-side out white ghosts (250 µg of protein) were carried out for 60 min at 37°C in 50 mM Na borate, pH 7.5, in the experimental conditions specified under "Results".

III. RESULTS

Table I lists the half-lives of three structural G6PD variants (two Class 3 and one Class 2 variants of the usual tabulations) we have recently identified and characterized. It is clear than an enhanced decay of the mutant proteins is involved in the three cases examined; an impaired catalytic efficiency contributes significantly to deficiency associated with G6PD Ferrara II (19), while the functional competence of the two other variants under study is totally unaffected (19, 20).

When the same experimental approach, i.e. fractionation of erythrocytes on gradients of increasing density, was followed with a number of G6PD Mediterranean subjects, the results were quite different. Thus, even the upper fractions in the gradients,

Table I. Patterns of decay of three G6PD variants.

Variant	Intracellular half-life (days)	Activity of 100% reticulocytes (I.U./g Hb)	Ref.
B	63	12.8	20
Napoli	16	9.8	19
Ferrara II	29	10.5	19
Cagliari	10	10.5	20

containing high reticulocyte percentages, showed barely detectable G6PD activity, not remarkably higher than that found in the heaviest cell fractions (23) and in the unfractionated erythrocytes as well (22). These results indicate that in the Mediterranean type of G6PD deficiency the reticulocytes enter circulation with a very low complement of G6PD activity. Since the rate of synthesis of G6PD Mediterranean seems to be unaffected in nucleated cells (36), a markedly increased instability of this mutant protein should characterize the erythroid cells at some as yet unidentified stage of their maturation (23).

In order to better define the relationship between G6PD Mediterranean and the intracellular milieu of the affected erythrocytes, we attempted to enrich them significantly in mutant G6PD up to easily detectable levels of activity and to investigate the intracellular stability accordingly. Several controls were required for this approach to be conveniently pursued. First, the mutant enzyme to be encapsulated within the G6PD-deficient erythrocytes was purified from leukocytes of the same subjects, as the latter cells express much more enzyme protein than the corresponding erythrocytes (23). In addition, although structural comparison was not feasible because of the exceedingly low amount of G6PD in the deficient erythrocytes, evidence for functional identity or close similarity between the leukocyte and the erythrocyte enzymes was obtained (23).

An additional requirement, concerning the correct subcellular localization of the erythrocyte-encapsulated G6PD, had been already verified in a previous study demonstrating full catalytic competence of the B type enzyme, once this is entrapped in the G6PD-deficient erythrocytes (25). Extrapolation of such "intracellular fitness" to the mutant G6PD was achieved by checking its susceptibility to activation by methylene blue: this oxidative stress stimulated the intracellular HMS activity of the G6PD-loaded deficient erythrocytes up to values 14-fold as high as those observed in the same unloaded cells (not shown). Thus, although the final G6PD activity of the loaded cells was distinctively lower compared with normal erythrocytes (because of the limited amount of enzyme that could be purified from the G6PD-deficient leukocytes), the mutant enzyme protein was correctly positioned within the erythrocytes and its activity was high enough to be followed at several time intervals.

Finally, an important requirement was to verify whether the manipulations inherent to the entrapment procedure (basically, transient hypotonic hemolysis, followed by isotonic resealing) would produce substantial alterations in the functional properties of the complex proteolytic machinery of mature erythrocytes (17). Although this includes a number of endopeptidases, dipeptidylaminopeptidases and aminopeptidases (15, 16), a central role is apparently played by a cytosolic, Ca^{2+}-activated neutral proteinase (CANP) (17) and, though to a lesser extent, by three acidic endopeptidases mostly associated with the erythrocyte membrane in normal conditions (15). As a first observation, levels of the (CANP) activity were found to be only moderately reduced following entrapment. Thus, actual values, expressed as I.U./ml packed erythrocytes, were 23±24 before and 19±3 after the hemolysis-resealing procedure, respectively. Available evidence suggests that this slight decrease of CANP activity, far from indicating leakage of enzyme protein throughout the transient hemolysis step, reflects the reported self-inactivation process - based on autoproteolytic mechanisms - that follows preliminary dissociation of the heterodimeric enzyme protein (110 KDa) and the consequent full activation of the 80 KDa subunit to a 75 KDa one (37). Such conversion mechanism, that can be triggered by a variety of signals in reconstructed systems (including fluctuations in Ca^{2+} levels and translocation of CANP to the inner face of erythrocyte membrane - see ref. 17), might be responsible for either increased or decreased CANP activity, depending on the step of this irreversible cycle at which the proteinase is actually assayed. Indeed, SDS-PAGE analysis of the erythrocytes following the entrapment procedure indicated a slight degree of degradation of bands 2.1 and 4.1 (not shown), consistent with moderate intracellular activation of the CANP (35). It is therefore possible to conclude that the entrapment technique per se (24), although resulting in partially decreased intracellular activity of this system by limited triggering of its activation-inactivation cycle, does not "kill" the CANP activity; on the contrary, the intraerythrocytic stability of entrapped proteins might be impaired, especially if these proteins would represent natural targets of CANP itself.

These results indicate that the strategy of investigating the intracellular stability of G6PD variants following their entrapment within erythrocytes, is a workable one and can lead to reliable conclusions. As shown in Table II, G6PD Mediterranean encapsulated in the G6PD-deficient erythrocytes keeps quite stable throughout incubation of the loaded cells, which still are competent to express CANP activity. Accordingly, mutant G6PD proves not to be susceptible to proteolysis in the G6PD Mediterranean mature erythrocytes. Of course, this may not be true for other structural variants whose intracellular

Table II. Stability of G6PD Mediterranean purified from leukocytes and encapsulated in G6PD Mediterranean erythrocytes (from ref. 23).

Erythrocytes	G6PD activity (I.U./g Hb) zero time	15 hours
Native	0.004	0.003
Unloaded[a]	0.004	0.004
G6PD-loaded	0.097	0.102

[a] Processed as for the entrapment technique, i.e. hemolysis and resealing, yet without addition of G6PD.

stability is consistently impaired, like for instance those listed in Table I.

A further check for the refractoriness of G6PD Mediterranean to intracellular proteolysis was recently obtained in a number of Sardinian subjects during acute hemolysis following ingestion of fava beans (Morelli, A., Grasso, M., Meloni, T., Forteleoni, G., and De Flora, A., unpublished data). Evidence has been provided that favism is characterized by severe perturbation of Ca^{2+} homeostasis, with intraerythrocytic Ca^{2+} levels of up to 0.5 mM (38, 39). In these conditions, CANP activity should be almost maximally stimulated to degrade natural protein substrates intracellularly. Indeed, preliminary "in vitro" experiments with Ca^{2+}-loaded normal erythrocytes indicated the following alterations, all consistent for Ca^{2+}-mediated intracellular modulation of CANP activity (Table III): a) significantly decreased levels of CANP activity (the decrease being the result of "suicide" activation through conversion of the 80 KDa subunit to the 75 KDa one - see above); b) degradation of band 4.1 in the SDS-PAGE tracings of membrane proteins and parallel conversion of band 2.1 to band 2.3; c) release into the cytosol of the acidic endopeptidase activity that is normally associated to the membrane. Changes b and c seem to be selectively related to unrestrained intracellular activation of the CANP, following ionophore-mediated penetration of Ca^{2+} within erythrocytes. Parallel assays of cytosolic enzyme activities, including G6PD and 6-PGD, indicated their almost complete stability in conditions of overall stimulation of erythrocyte proteinases.

Table III. Effects of enhanced Ca^{2+} influx on levels and subcellular distribution of proteinase activities in normal erythrocytes.

	Erythrocytes	
	Normal	Ca^{2+}-loaded
CANP (I.U./ml packed cells)	25.2±4.3	10.2±4.8
Acidic endopeptidases (I.U./ml packed cells)		
a) membrane-bound	15.1±3.6	3.2±1.1
b) cytosolic	0.2±0.05	10.5±3.4
Membrane proteins		
a) band 2.1	++	+
b) band 2.3	+	++
c) band 4.1	+++	+
G6PD (I.U./g Hb)	4.1±0.5	4.0±0.5
6-PGD (I.U./g Hb)	3.5±0.4	3.5±0.6

The corresponding picture we observed in the erythrocytes of favic patients, characterized by abnormally elevated Ca^{2+} content, was remarkably complicated by superimposed effects and especially by the severe oxidant damage (CANP is an SH enzyme - see ref. 31) and by the attendant reticulocytosis. However, erythrocytes during the hemolytic crisis seem to be characterized on the whole by activation of their proteolytic systems, including the CANP (found to be variably decreased as compared with controls represented by erythrocytes from asymptomatic G6PD-deficient subjects) and the acidic endopeptidase activities that were abnormally detected at consistently high levels in the cytosol. In addition, taking into account the high number of reticulocytes where these alterations were completely absent, heavily damaged erythrocytes (e.g. filled with Heinz bodies) showed unequivocal selective degradation of bands 2.1 and 4.1, i.e. of two specific membrane targets of fully activated CANP (35). In spite of this evidence of enhanced intracellular proteolysis, careful assay of G6PD activity in erythrocytes from favic patients showed levels comparable with (and in no case lower than) those found in erythrocytes from healthy G6PD-deficient subjects of the same ancestry, whose proteolytic activities were in a resting state.

A final set of experiments were devoted to exploring in a totally reconstructed system the susceptibility of G6PD B to degradation by the purified erythrocyte CANP.

A factor possibly influencing the attack of G6PD by CANP is its state of aggregation. Earlier studies had suggested that in the tetrameric form of G6PD, the apoenzyme-bound NADP is somewhat less accessible than the dimeric species (13). However, under physiological conditions, the balance between the two active forms is largely in favour of the dimers (40). We therefore investigated, following full activation of the CANP by Ca^{2+}, the stability of G6PD B under experimental conditions that are known to shift the equilibrium between active dimers and sub-active monomers. The experimental systems, all supplemented with 1 mM Ca^{2+} to trigger activation of purified CANP, included : a) purified native G6PD B (mostly dimeric); b) G6PD in the presence of 1 mM glucose 6-P (mostly monomeric - see ref. 14); c) dimeric G6PD with white ghosts (as a source of acidic endopeptidases); d) monomeric G6PD with white ghosts. Although some inactivation occurred in the glucose 6-P-containing control systems (i.e., lacking proteinases) because of extensive dissociation of G6PD to monomers (14), no extra-inactivation was observed upon addition of purified CANP and/or white ghosts. On the other hand, dimeric G6PD was completely stable in all systems containing CANP and white ghosts (not shown). These negative results indicate that in no instance is G6PD B susceptible to "in vitro" degradation by the integrated proteolytic systems of mature erythrocytes, not even when it is mostly in a monomeric form.

IV. DISCUSSION

All available evidences support the contention that the proteolytic machinery of circulating erythrocytes is not competent to degrade G6PD B. Both experiments with intact erythrocytes and studies with purified enzymes are in agreement with this conclusion. Accordingly, the view that the physiological function of erythrocyte CANP is restricted to degradation of specific cytoskeletal proteins rather than being the attack of native cytosolic proteins (17) receives further confirmation.

On the other hand, specific point mutations may make G6PD recognizable by CANP, this resulting in lowered intraerythrocytic stability of the corresponding variants. This is certainly not the case in G6PD Mediterranean, whose extensive degradation seems to take place in the erythroid cell and which is conversely resistant to proteolysis within the circulating erythrocytes. Therefore, the mutant G6PD-proteinases relationship is re-established as a key condition for variable expression of G6PD deficiency in specific cell types (41).

Two future trends of research seem to be particularly promising in this context. First, further studies on the properties of proteinases in mature erythrocytes and especially on the physiological role of the acidic endopeptidases (15) and on their functional correlations with the CANP system: it is relevant that some cytosolic proteins, including G6PD B (42), can adhere to the inner face of erythrocyte membrane. Moreover, following enhanced Ca^{2+} influx into the erythrocyte (both as artificially obtained by means of the A 23187 ionophore and in favism), there is substantial release of acidic endopeptidase activity from the membrane to the cytosol. A second important strategy stems from assessment of the primary structure of G6PD, coupled with assignment of conformation, and by sequence analysis of the Gd gene. It should become possible, once these data are available, to relate individual point mutations and aminoacid substitutions with abnormal sites of attack by otherwise silent proteinases and especially by the CANP, whose pleomorphic activation mechanisms (17) make it a reasonable candidate for intracellular removal of mutant gene products.

ACKNOWLEDGMENTS

We are indebted to Dr. E. Melloni for many helpful discussions and to Dr. T. Meloni for kindly providing several blood samples from Sardinian subjects.

REFERENCES

1. Tsutsui, E.A., and Marks, P.A. (1962). Biochem. Biophys. Res. Commun. 8, 338
2. Kirkman, H.N., and Hendrickson, E.M. (1962). J. Biol. Chem. 237, 2371
3. Chung, A.E., and Langdon, R.G. (1963). J. Biol. Chem. 238, 2317
4. Yoshida, A. (1966). J. Biol. Chem. 241, 4966
5. Cohen, P., and Rosemeyer, M.A. (1969). Eur. J. Biochem. 8, 8
6. Bonsignore, A., Lorenzoni, I., Cancedda, R., and De Flora, A. (1970). Biochem. Biophys. Res. Commun. 39, 142
7. De Flora, A., Morelli, A., and Giuliano, F. (1974). Biochem. Biophys. Res. Commun. 59, 406
8. Gaetani, G.F., Parker, J.C., and Kirkman, H.N. (1974). Proc. Natl. Acad. Sci. U.S.A. 71, 3584
9. Kirkman, H.N., Gaetani, G.F., Clemons, E.H., and Mareni, C. (1975). J. Clin. Invest. 55, 875
10. Cancedda, R., Ogunmola, G.B., and Luzzatto, L. (1983). Eur. J. Biochem. 34, 199
11. Luzzatto, L. (1967). Biochim. Biophys. Acta 146, 18
12. Bonsignore, A., Lorenzoni, I., Cancedda, R., Cosulich, M.E., and De Flora, A. (1971). Biochem. Biophys. Res. Commun. 42, 159
13. Bonsignore, A., Cancedda, R., Nicolini, A., Damiani, G., and De Flora, A. (1971). Arch. Biochem. Biophys. 147, 493
14. Wrigley, N.G., Heather, J.V., Bonsignore, A., and De Flora, A. (1972). J. Mol. Biol. 68, 483
15. Pontremoli, S., Salamino, F., Sparatore, B., Melloni, E., Morelli, A., Benatti, U., and De Flora, A. (1979). Biochem. J. 181, 559
16. Pontremoli, S., Melloni, E., Salamino, F., Sparatore, B., Michetti, M., Benatti, U., Morelli, A., and De Flora, A. (1980). Eur. J. Biochem. 110, 421
17. Pontremoli, S., and Melloni, E. (1986). Annu. Rev. Biochem., in the press

18. Beutler, E., West, C., and Blume, K.G. (1976). J. Lab. Clin. Med. 88, 328
19. De Flora, A., Morelli, A., Benatti, U., Giuntini, P., Ferraris, A.M., Galiano, S., Ravazzolo, R., and Gaetani, G.F. (1981). Br. J. Haematol. 48, 417
20. Morelli, A., Benatti, U., Guida, L., and De Flora, A. (1984). Hum. Genet. 66, 62
21. Kirkman, H.N., Schettini, F., and Pickard, B.M. (1964). J. Lab. Clin. Med. 63, 726
22. Morelli, A., Benatti, U., Lenzerini, L., Sparatore, B., Salamino, F., Melloni, E., Michetti, M., Pontremoli, S., and De Flora, A. (1981). Blood 58, 642
23. Morelli, A., Benatti, U., Guida, L., and De Flora, A. (1984). Scand. J. Haematol. 33, 144
24. Dale, G.L. Villacorte, D.G., and Beutler, E. (1977). Biochem. Med. 18, 220
25. Morelli, A., Benatti, U., Salamino, F., Sparatore, B., Michetti, M., Melloni, E., Pontremoli, S., and De Flora, A. (1979). Arch. Biochem. Biophys. 197, 543
26. Piomelli, S., Corash, C.M., Davenport, D.D., Miraglia, J., Amorosi, E.L. (1968). J. Clin. Invest. 47, 940
27. Corash, C.M., Piomelli, S., Chen, H.C., Seaman, C., and Gross, E. (1974). J. Lab. Clin. Med. 84, 147
28. Morelli, A., Benatti, U., Gaetani, G.F., and De Flora, A. (1978). Proc. Natl. Acad. Sci. U.S.A 75, 1979
29. Melloni, E., Salamino, F., Sparatore, B., Michetti, M., and Pontremoli, S. (1984). Biochem. Int. 8, 477
30. Melloni, E., Sparatore, B., Salamino, F., Michetti, M., and Pontremoli, S. (1982). Biochem. Biophys. Res. Commun. 106, 731
31. Melloni, E., Sparatore, B., Salamino, F., Michetti, M., and Pontremoli, S. (1982). Biochem. Biophys. Res. Commun. 107, 1053
32. Murakami, T., Hatanaka, M., and Murachi, T. (1981). J. Biol. Chem. 90, 1809

33. Steck, T.L., and Kant, J.A. (1974). Methods Enzymol. 31, 172
34. Fairbanks, G., Steck, T.L., and Wallach, D.F.H. (1971). Biochemistry 10, 2606
35. Pontremoli, S., Melloni, E., Sparatore, B., Michetti, M., and Horecker, B.L. (1984). Proc. Natl. Acad. Sci. U.S.A. 81, 6714
36. Persico, M., Battistuzzi, G., Mareni, C., Nobile, C., D'Urso, M., Toniolo, D., and Luzzatto, L. (1982). In "Advances in Red Blood Cell Biology" (D.J. Weatherall, G. Fiorelli, S. Gorini, eds.), p. 309, Raven Press, New York
37. Pontremoli, S., Sparatore, B., Melloni, E., Michetti, M., and Horecker, B.L. (1984). Biochem. Biophys. Res. Commun. 123, 331
38. De Flora, A., Benatti, U., Guida, L., Forteleoni, G., and Meloni, T. (1985). Blood 66, 294
39. Turrini, F., Naitana, A., Fischer, T.M., Pescarmona, G.P., and Arese, P. (1985). Blood 66, 302
40. Bonsignore, A., and De Flora, A. (1972). Curr. Topics. Cell. Reg. (B.L. Horecker and E.R. Stadtman, eds.), vol. 6, p. 21, Academic Press, New York
41. Beutler, E. (1983). Proc. Natl. Acad. Sci. U.S.A. 80, 3767
42. Benatti, U., Morelli, A., Frascio, M., Melloni, E., Salamino, F., Sparatore, B., Pontremoli, S., and De Flora, A. (1978). Biochem. Biophys. Res. Commun. 85, 1318

OXIDANT-INDUCED MEMBRANE DAMAGE IN G-6-PD DEFICIENT RED BLOOD CELLS

Gerhard J. Johnson[1], David W. Allen[1,2] and Thomas P. Flynn

Department of Medicine,
Veterans Administration Medical Center
and St. Paul-Ramsey Medical Center,
and The University of Minnesota
Minneapolis, MN, U.S.A.

I. INTRODUCTION

A. Acute Hemolysis In G-6-PD Deficiency

Glucose-6-phosphate dehydrogenase (G-6-PD) deficiency renders the red blood cell susceptible to oxidant-induced injury that leads to hemolysis. The common form of hemolysis seen in patients with G-6-PD deficiency is an episodic event that is the consequence of an acute oxidant stress that exceeds the red cell's defense mechanisms against oxidation. The sequence of events that begins with an acute oxidant stress, precipitated by drugs, infection or fava bean ingestion, progresses to formation of toxic oxygen species, oxidation of glutathione and formation of mixed disulfides and finally results in precipitation of hemoglobin as Heinz bodies is well documented (1). The key abnormality in the G-6-PD deficient red cell's defenses against oxidation is its inability to reduce $NADP^+$ to NADPH at a normal

1. Supported by The Veterans Administration
2. Supported by Grants 2-P01-HL16833-06 and 5HT-32-HL07062 from The National Institutes of Health

rate. The failure of the red cell to maintain adequate NADPH results in deficient reduced glutathione (GSH). The acute hemolytic events that occur in patients with the common G-6-PD variants may be very severe, but they are self-limited (1).

B. Chronic Hemolysis In G-6-PD Deficiency

A number of less common G-6-PD variants are characterized by chronic hemolysis (1). Since the red cells from these patients lack Heinz bodies despite ongoing hemolysis, the mechanism of hemolysis appears to be different from that described for the more common G-6-PD variants. Studies of the mechanism of chronic hemolysis in G-6-PD-deficient human red cells and in experimentally oxidant-stressed red cells have led us to the conclusion that oxidant-induced membrane injury is responsible for hemolysis. This paper will review the data that leads to this conclusion.

Our interest in the pathogenesis of chronic hemolysis in patients with G-6-PD deficiency began with the study of a patient with G-6-PD Long Prairie (2). This mutant G-6-PD was of interest because Dr. Beutler found that it did not have increased sensitivity to inhibition by NADPH (normal K_i for NADPH). Dr. Yoshida had previously postulated that increased susceptibility to inhibition by NADPH explained why some mutants demonstrated chronic hemolysis while others with similarly low G-6-PD activity did not (3).

II. STUDIES OF THE MECHANISM OF CHRONIC HEMOLYSIS

A. Membrane Polypeptide Aggregates

Since G-6-PD Long Prairie obviously was an exception to Dr. Yoshida's hypothesis, and since chronic hemolysis occurred in the patient with this variant G-6-PD without increased Heinz body formation (over that expected postsplenectomy), we asked, "Is there a membrane lesion that could explain hemolysis?". We were encouraged to look for such a lesion by studies of Palek, et al.

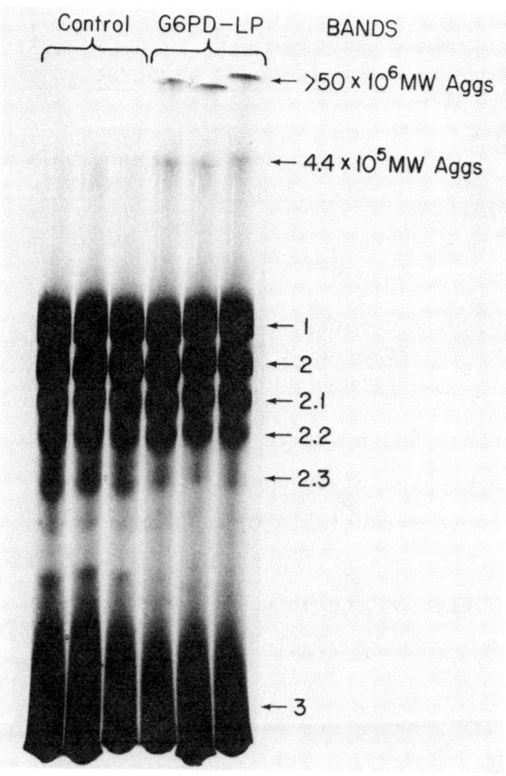

Figure 1. SDS-PAGE of red cell membranes stained with Coomassie Blue. Red cell membranes from normal subjects are shown on the left (control) and from G-6-PD Long Prairie on the right (G-6-PD-LP). Reproduced with permission of The Journal of Laboratory and Clinical Medicine 91:323, 1978.

(4). These investigators observed membrane polypeptide aggregates in the red cells of a patient with G-6-PD Worcester following in vitro incubation. We found that sodium dodecyl sulfate polyacrylamide gel electrophoresis (SDS-PAGE) of red cell membranes revealed similar aggregates in normal membranes following in vitro incubation without glucose until the glutathione concentration had fallen to approximately one-third of normal (5). We also observed the same type of membrane polypeptide aggregates in fresh G-6-PD Long Prairie red cells (Fig. 1). These aggregates were of two types; spectrin dimers (4.4×10^5 daltons) and high molecular weight ($>50 \times 10^6$ daltons)

TABLE I. CLINICAL CHARACTERISTICS OF PATIENTS WITH GLUCOSE-6-PHOSPHATE DEHYDROGENASE (G-6-PD) DEFICIENCY ACCOMPANIED BY CHRONIC HEMOLYSIS

G-6-PD Variant	Age (yr)	Spleen	Heinz Bodies (%)	Blood Hemoglobin (g/dl)	Reticulocyte Count (%)
Unknown	17	Present	0	12.8	7.4
Long Prairie	48	Absent	5	13.9	20.4
"Tomah"	25	Absent	0.7	14.0	17.9
Pea Ridge	43	Present	0	13.2	14.5
Pea Ridge	43	Present	0	12.6	17.4
Minneapolis	18	Present	0	13.8	13.0

aggregates that contained spectrin plus other proteins (5). Both types of aggregates appeared to contain intermolecular disulfide bonds since they were dissociable by mercaptoethanol or dithiothreitol but not by boiling in SDS (5). Of significance is the fact that the high molecular weight aggregates did not contain substantial amounts of globin chains as would be expected if they were associated with Heinz body formation (5).

These results suggested our postulate that membrane abnormalities might be responsible for chronic hemolysis in G-6-PD deficiency was a viable one. However, since the patient with G-6-PD Long Prairie had been splenectomized it was necessary to demonstrate that red cells from non-splenectomized patients with chronic hemolytic G-6-PD variants also contained the same lesions.

With the assistance of Dr. Virgil Fairbanks and Dr. Beatrice Lampkin we had the opportunity to study red cell membranes from five other patients with four different chronic hemolytic G-6-PD variants (6,7). The variants studied and some of their clinical and laboratory characteristics are recorded on Table I. Of note is the fact that only those who had undergone splenectomy had Heinz bodies in their red cells. All six patients including both splenectomized and non-splenectomized individuals were found to have both $4.4x10^5$ and $>50x10^6$ dalton polypeptide aggregates on SDS-PAGE of their red cell membrane proteins (Table II) (6). These aggregates, like those observed in G-6-PD Long Prairie red cells were dissociable with mercaptoethanol or dithiothreitol, and they did not contain significant amounts of globin. Red cell membranes from patients with non-hemolytic G-6-PD variants (probable A-) (Table II) or other hemolytic disorders (5) did not contain aggregates. Others have confirmed our findings of aggregates in G-6-PD deficient red cells from patients with (8) or without (following GSH depletion) chronic hemolysis (9,10).

Since our previous studies of normal red cells incubated without glucose indicated that membrane polypeptide aggregates were related to red cell GSH level, we noted with interest the fact that the red cell GSH in the patients with chronic hemolysis was reduced to approximately 50% of the normal cells and the G-6-PD deficient cells from patients without chronic hemolysis (Table II). When we pooled our data and examined the relationship between red cell GSH and high molecular weight aggregates, we found that the cells from the patients with chronic hemolysis could be located on the curve at approximately the same place as normal red cells incubated without glucose for 24 hr (Fig. 2). Although the in vitro-aged red cell is not an ideal model for the G-6-PD deficient cell, since it becomes depleted of ATP, the fact that aggregates were observed in patients' fresh cells that had normal ATP (5) indicates that ATP depletion is not essential for aggregate formation.

The demonstration of membrane aggregates in the red cells of patients with G-6-PD variants characterized by chronic hemolysis yielded both a "footprint" of oxidative damage and an important clue to the mechanism of hemolysis since it identified a potentially significant membrane lesion. However, the next question to be answered was, "How are membrane protein aggregates related to hemolysis?".

TABLE II. G-6-PD ACTIVITY, GLUTATHIONE (GSH) AND POLYPEPTIDE AGGREGATES IN G-6-PD DEFICIENT AND NORMAL RED CELLS

Subjects	Number	G-6-PD Activity (IU/g Hb)	GSH (μmol/g Hb)	Polypeptide Aggregates (% Membrane Protein)	
				$4.4x10^5$ Da	$>50x10^6$ Da
G-6-PD Variants With Chronic Hemolysis	6	0.75 ± 0.52	3.70 ± 0.80	0.32 ± 0.09	0.30 ± 0.10
G-6-PD Variants With-Out Chronic Hemolysis	4	2.14 ± 1.48	6.88 ± 1.53	0.00	0.06 ± 0.06
Normal	4	12.60 ± 1.12	7.06 ± 0.85	0.00	0.03 ± 0.02

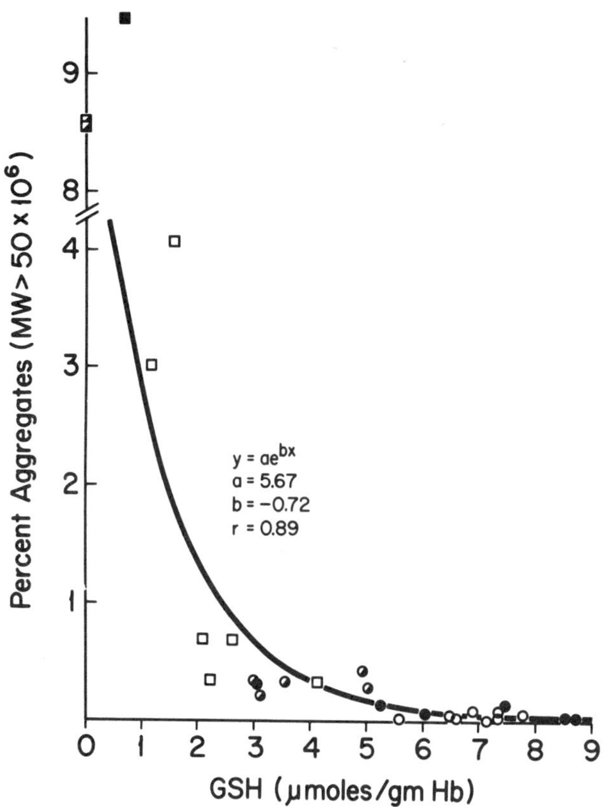

Figure 2. Polypeptide aggregates (>50x10^6 daltons) and glutathione in the following red cells: Fresh normal○; Fresh G-6-PD deficient (without chronic hemolysis)●; Fresh G-6-PD deficient (with chronic hemolysis)◐; 24-hr incubated normal□; 48-hr incubated normal◪; 24-hr incubated G-6-PD deficient (with chronic hemolysis)■. Reproduced with permission of The New England Journal of Medicine 301:525, 1979.

B. Relationship Between Membrane Polypeptide Aggregates and Red Red Cell Deformability

Since the membrane skeleton component, spectrin, was prominently involved in the aggregates, we reasoned that cellular deformability may be diminished in cells with spectrin disulfide bonded to itself and to other membrane proteins. Fischer, et al. (11) had demonstrated that red cell deformability was decreased after in vitro incubation with sulfhydryl reagents that crosslinked spectrin. Our prediction regarding the functional consequences of disulfide bonding of spectrin proved to be correct. Membrane deformability measured in Dr. James White's laboratory by a micropipette aspiration technique was reduced regardless of whether large (2μm) or small (∿1μm) pipettes were used (Table III). Additional evidence in support of the hypothesis that decreased membrane deformability was related to oxidative damage to membrane skeletal components came from our studies of patients with chronic hemolytic disease secondary to the unstable hemoglobin, Hb Köln. Patients with Hb Köln who had been splenectomized had high molecular weight polypeptide aggregates in their red cell membranes (12). These aggregates contained spectrin. Red cells from splenectomized patients that had spectrin-containing aggregates demonstrated reduced micropipette deformability, while red cells from patients with Hb Köln who had not been splenectomized (and whose cells did not contain polypeptide aggregates) had normal deformability (12).

The observations that red cells from patients with chronic hemolytic G-6-PD variants had low GSH and contained disulfide-bonded membrane protein aggregates, and that these cells had decreased deformability, formed the basis for the hypothesis that oxidant-induced membrane damage resulted in cell rigidity that in turn was responsible for premature cellular destruction in the microcirculation, particularly in the spleen. The fact that splenectomy does not substantially reduce the rate of hemolysis in these patients presented a problem for this hypothesis, so we sought further confirmatory evidence.

C. Experimental Induction of Membrane Polypeptide Aggregates

Diamide is a thiol-oxidizing diazine derivative that depletes erythrocyte GSH without forming Heinz bodies (13). Previous studies by Fischer, et al. (11) demonstrated that diamide-incubated red cells had decreased deformability. Since diamide-

TABLE III. MICROPIPETTE DEFORMABILITY OF CHRONIC HEMOLYTIC G-6-PD DEFICIENT AND NORMAL RED CELLS

Subject	Tongue Extension (μm at $-50cm\ H_2O$) 2.0μm Pipette	1.0μm Pipette
Normal	11.2 ± 0.8	
G-6-PD Long Prairie	7.9 ± 0.5[a]	
Normal		14.8 ± 1.3
G-6-PD "Tomah"		8.1 ± 0.7[a]

[a]Significantly different from normal ($p<0.001$)

incubated red cells appeared to closely resemble those obtained from nonsplenectomized patients with chronic hemolytic G-6-PD variants, we studied the effect of diamide on dog red cell membrane proteins, deformability and in vivo survival (14).

Incubation of dog red cells with 0.4mM diamide for 90 min at $37^{\circ}C$ resulted in a marked decrease in GSH without any decrease in ATP (Table IV). SDS-PAGE study of membranes from diamide-treated cells revealed the same type of membrane protein polypeptide aggregates seen in red cells from patients with chronic hemolytic G-6-PD variants (Table V, Fig. 3). The quantity of aggregates was greater in diamide-incubated cells (5-10% membrane protein) than in human cells (<1% membrane protein). Like the chronic hemolytic human G-6-PD deficient cells, diamide-incubated cells demonstrated substantially reduced micropipette deformability (Table VI) and increased viscosity at shear rates from 11.25-90 s^{-1} (14).

D. Membrane Polypeptide Aggregates and Red Cell Survival

When poorly deformable, aggregate-containing, diamide-incubated dog red cells were reinjected into the animals from whom they were obtained, rapid hemolysis followed (Fig. 4). Dog red cells incubated with 0.2mM diamide did not form measurable polypeptide aggregates. They demonstrated decreased micropipette

TABLE IV. GSH AND ATP FOLLOWING DIAMIDE INCUBATION OF DOG RED CELLS

Number Studied	Diamide (mM)	GSH (μmol/g Hb)	ATP (μmol/g Hb)
2	0	7.24	1.28
2	0.2	7.06	1.40
4	0.4	0.22±0.19	1.38±0.11

deformability (Table VI), but their survival in vivo was not different from the control (Fig. 4). Red cell polypeptide aggregates, decreased micropipette deformability and increased viscosity induced by 0.4mM diamide were partially or completely reversible after incubation with 4mM dithiothreitol (Table VII). Strong evidence in favor of an etiologic relationship between

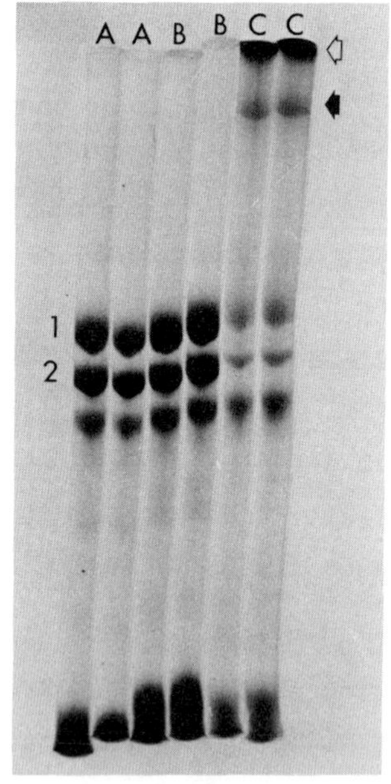

Figure 3. SDS-PAGE of dog red cell membranes stained with Coomassie Blue. Membranes incubated for 90 min. at 37°C with A buffer, B 0.2mM diamide, C 0.4mM diamide. Reproduced with permission of The Journal of Clinical Investigation 66:957, 1980.

TABLE V. MEMBRANE POLYPEPTIDE AGGREGATES FOLLOWING DIAMIDE INCUBATION OF DOG RED CELLS

Number Studied	Diamide (mM)	Polypeptide Aggregates (% membrane protein) 4.4×10^5 Da	$>50 \times 10^6$ Da
2	0	0	0
2	0.2	0	0
4	0.4	6.8 ± 1.8	6.8 ± 2.2

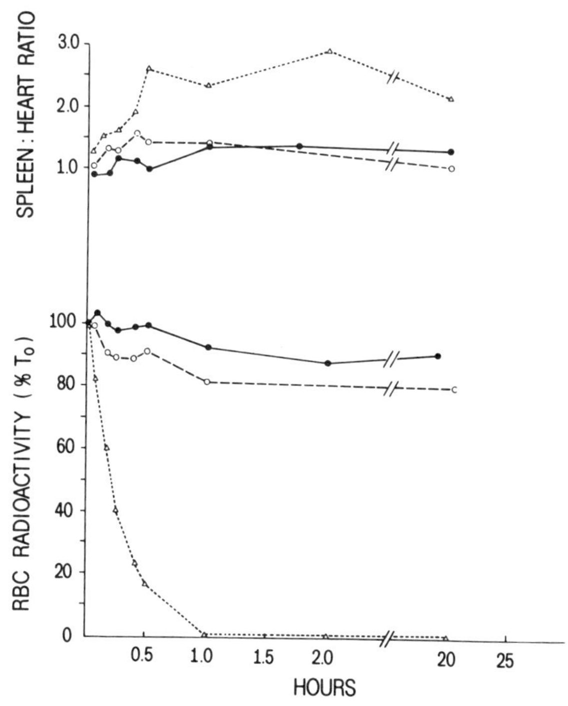

Figure 4. BOTTOM. Dog red cell radioactivity (% initial count) following reinjection of [^{51}Cr]-labeled cells incubated for 90 min. at 37°C with buffer ○; 0.2mM diamide ●; or 0.4mM diamide △. TOP. Splenic sequestration of [^{51}Cr] (ratio spleen: heart counts) observed during study in bottom panel. Reproduced with permission of the Journal of Clinical Investigation 66:959, 1980.

TABLE VI. MICROPIPETTE DEFORMABILITY FOLLOWING DIAMIDE INCUBATION OF DOG RED CELLS

Diamide	Tongue Extention (μm at $-25 cm\ H_2O$)	(μm at $-50 cm\ H_2O$)
(mM)	1.8 - 2.0μm Pipette	1.0μm Pipette
0	13.2 ± 0.3	
0.2	10.0 ± 0.6[a]	
0.4	9.6 ± 0.5[a]	
0		3.2 ± 0.4
0.4		2.1 ± 0.4[a]

[a]Significantly different from preincubation values ($p<0.001$)

sulfhydryl oxidation, decreased deformability and hemolysis was provided by the observation that red cell survival was markedly improved when 0.4mM diamide-incubated cells were treated with 4mM dithiothreitol prior to injection (Table VII).

E. Structure of Membrane Polypeptide Aggregates

The molecular weight of the 4.4×10^5 dalton aggregates and the fact that they were dissociable by sulfhydryl reagents strongly suggested they were composed of spectrin dimers.

TABLE VII. MEMBRANE POLYPEPTIDE AGGREGATES, MICROPIPETTE DEFORMABILITY AND SURVIVAL OF DIAMIDE-INCUBATED DOG RED CELLS ± DITHIOTHREITOL (DTT)

	Aggregates (% membrane protein)		Tongue Extension	Red Cell Survival[a]
Incubation	4.4×10^5 Da	$>50 \times 10^6$ Da	(μm)[b]	(%)
Buffer	0	0	11.8 ± 0.3	82
0.4mM Diamide	4.0	5.2	7.9 ± 1.9	11
0.4mM Diamide +DTT	0.4	0	11.2 ± 0.9	62

a % of [^{51}Cr] counts at T_0 remaining at 60 min.
b 1.8μm pipette; $-25 cm\ H_2O$.

Further evidence in support of this interpretation came from two dimensional gel electrophoresis. This procedure clearly demonstrated bands 1 and 2 (spectrin) were the only constituents of the lower molecular weight aggregates.

Additional studies directed toward a definition of the components of the high molecular weight aggregates revealed evidence of several membrane constituents plus additional cytoplasmic proteins. SDS-PAGE of the >50x10^6 dalton aggregates dissociated by mercaptoethanol revealed evidence of bands 1 and 2 (spectrin), band 2.1 (ankyrin) and a significant quantity of unidentified protein between bands 4.2 and 5 (Fig.

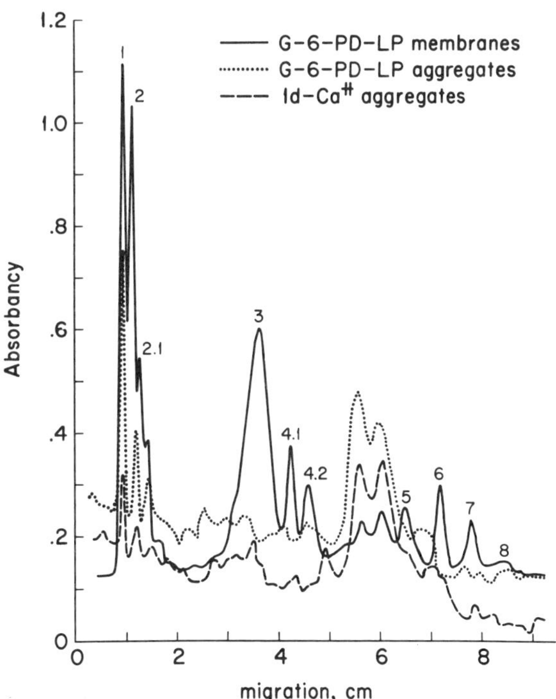

Figure 5. Scans of SDS-PAGE of red cell membrane polypeptide aggregates (>50x10^6 daltons) dissociated by mercaptoethanol from normal red cells incubated for 24 hr without glucose (---) and fresh G-6-PD Long Prairie red cells (....). A scan of whole G-6-PD Long Prairie red cell membranes (——) is included for comparison with the aggregates. Reproduced with permission of The Journal of Laboratory and Clinical Medicine 91:325, 1978.

5). Sucrose density gradient analysis of red cell membranes from a variety of hemolytic disorders, including chronic hemolytic G-6-PD variants (15) revealed evidence of increased membrane density due to adsorbed cytoplasmic protein (Fig. 6). Membranes from diamide-incubated cells have even greater density (Fig. 6). Therefore, SDS-PAGE and sucrose density gradient analysis indicates binding of cytoplasmic proteins (largely nonglobin) to the membrane.

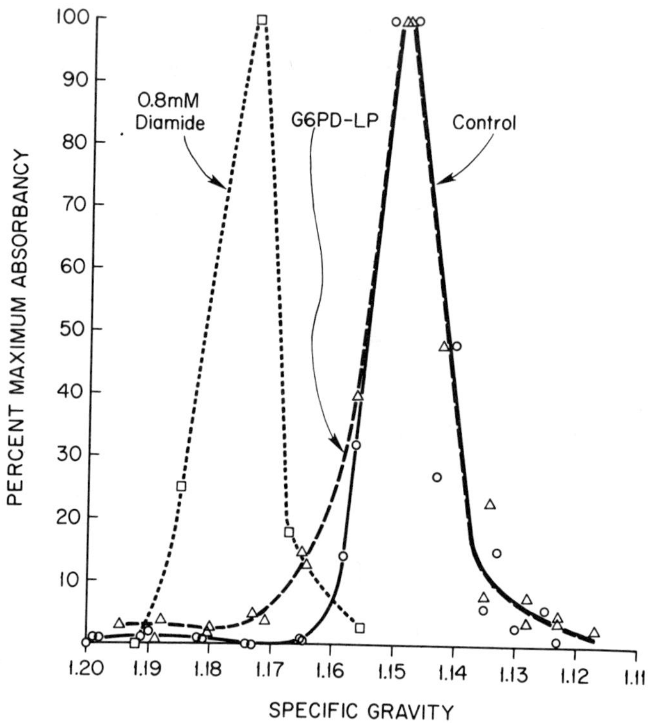

Figure 6. Sucrose density analysis of red cell membranes: G-6-PD Long Prairie (---); normal subject incubated with 0.8mM diamide (x90 min, 37°C) (....); control (____). Note increased membrane density (>1.165) in some G-6-PD Long Prairie cells and nearly all diamide-incubated cells. Reproduced with permission of Alan R. Liss, Inc., New York from Membranes and Genetic Diseases. Sheppard, J.R., Anderson, V.E. and Eaton, J.W., Eds. 1982, p. 38.

TABLE VIII. SULFHYDRYL GROUPS LABELED WITH [^{14}C]-IODOACETAMIDE IN G-6-PD LONG PRAIRIE AND NORMAL RED CELLS

	Ratio G-6-PD Long Prairie:Normal								
Band	1	2	2.1	2.2,2.3	3	4.1	4.2	5	6
Directly Reactive	1.1	1.2	1.1	1.0	1.0	1.1	1.2	0.9	0.8
Disulfide	0.7	0.9	_1.7_	_1.9_	1.1	1.1	0.9	1.1	_1.4_

To discern whether the intermolecular bonds resulting in membrane aggregates were the only consequences of sulfhydryl oxidation, we measured reactive and disulfide sulfhydryl groups using [^{14}C]-iodoacetamide (16). When the distribution of disulfide bonds in G-6-PD Long Prairie membranes was compared to that in normal membranes, an abnormal increase in bands 2.1, 2.2 and 2.3 was observed (16) (Table VIII). Since this protein component, ankyrin, is not a prominent aggregate component, it is likely that these bonds are intramolecular (Fig. 7). Note the difference between the distribution of total disulfides in membrane polypeptides (Fig. 7) and the disulfides involved in intermolecular bonds as approximated by the composition of high molecular weight aggregates (Fig. 5). The oxidation of ankyrin may impair this molecule's key role of providing a site of attachment of the membrane skeleton to the intrinsic membrane protein (17).

III. PATHOPHYSIOLOGIC IMPLICATIONS

The results of the experimental studies with diamide provided strong support for the hypothesis that the polypeptide aggregates we identified in the membranes of red cells from patients with chronic hemolytic G-6-PD variants are the consequence of oxidant-induced membrane injury. The reversibility of these membrane

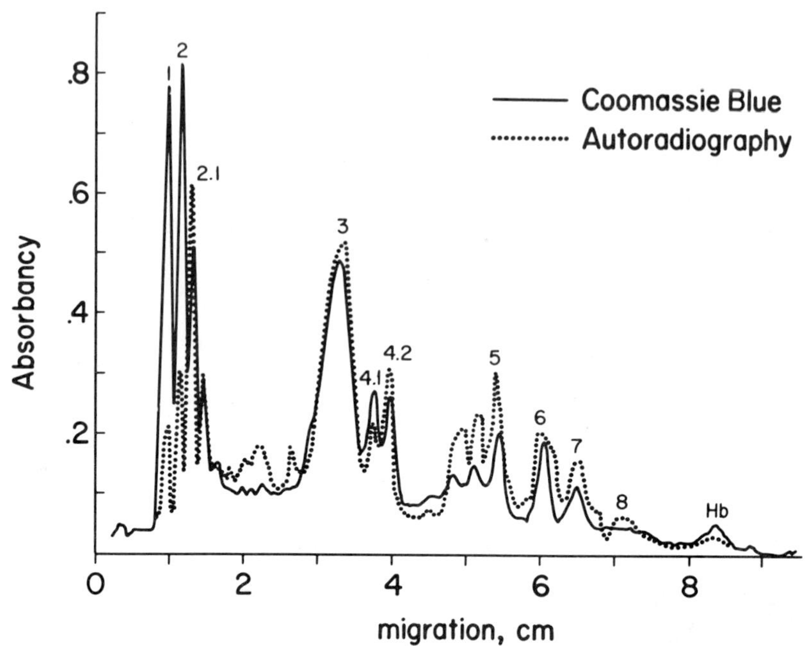

Figure 7. Reducible disulfides in membrane proteins from the red cells of a patient with G-6-PD Tomah. Disulfides were analyzed by blocking reactive sulfhydryl groups with iodoacetamide, followed by reduction with mercaptoethanol and exposure to [^{14}C]-iodoacetamide. The autoradiogram (.....) is compared with a scan (550nm) of the Coomassie blue-stained gel (_____). Reproduced from Erythrocyte Membranes. 2. Recent Clinical and Experimental Advances. Kruckeberg, W.C., Eaton, J.W. and Brewer, G.J., Eds., 1981, p. 238, with permission of Alan R. Liss, Inc., New York.

abnormalities with sulfhydryl reagents indicates that the primary pathophysiologic process is the formation of intermolecular disulfide bonds by toxic oxygen radicals generated within the red cell. Whether the intermolecular disulfide bonds of spectrin-containing aggregates result in the observed impaired deformability, or this results from other oxidant damage, such as intramolecular disulfide bonds in ankyrin, is not clear. It is quite possible that both lesions contribute to decreased deformability. Our current concept of the oxidant damage to the membranes of red cells from patients with chronic hemolytic G-6-PD variants is presented in Fig. 8. Note the presence of both intra-and intermolecular disulfide bond formation.

Although we have demonstrated a plausible relationship between membrane polypeptide aggregates and red cell deformability, it is not so clear that decreased deformability is the primary mechanism responsible for hemolysis. Decreased red cell deformability is considered to be an important mechanism leading to hemolysis in hereditary spherocytosis (18). Splenectomy cures hemolysis in this disorder. However, there are several reasons to question whether decreased cell deformability is primarily responsible for hemolysis in G-6-PD deficiency. First, splenectomy does not substantially reduce hemolysis in chronic hemolytic G-6-PD variants. Second, patients with Hb Köln who have been splenectomized have decreased red cell membrane deformability, but they hemolyze at approximately the same rate as patients who have not been splenectomized and whose red cell deformability is normal (12). Third, treatment with pentoxifylline, an agent that increases red cell deformability (see below), does not alter the rate of hemolysis in G-6-PD Long Prairie. Fourth, dog red cells incubated with 0.2mM diamide have impaired micropipette deformability, but their survival does not differ from the control (14). Therefore it is appropriate to question whether decreased red cell deformability is primarily responsible for hemolysis in G-6-PD deficiency. It is also appropriate to ask, "What is the mechanism responsible for final clearance of oxidant-damaged red cells from the circulation?".

This question has not been satisfactorily answered in chronic hemolysis secondary to G-6-PD deficiency nor in most other non-immune chronic hemolytic diseases. Several potential mechanisms have been proposed (19-21). Many of these postulates involve modification of membrane carbohydrate by removal of

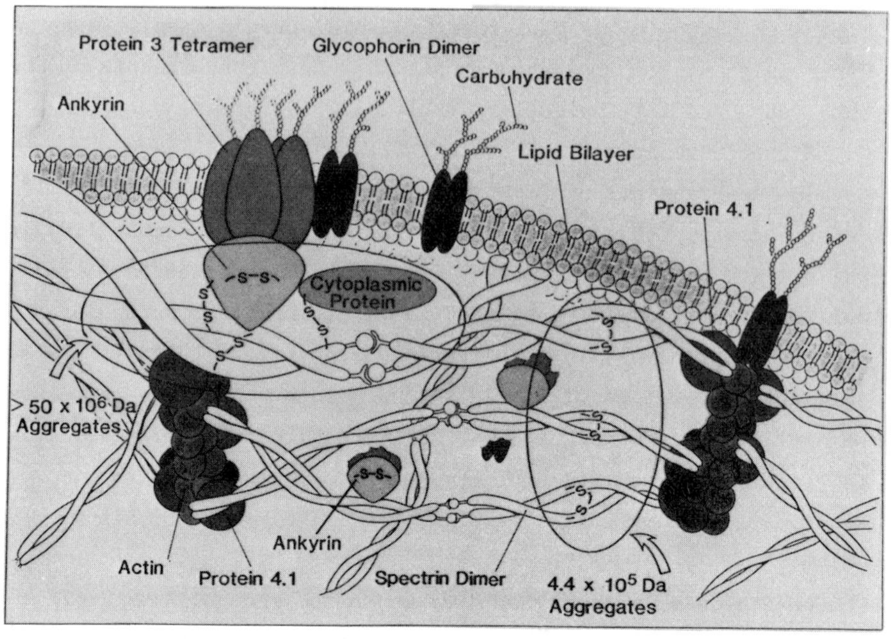

Figure 8. Proposed structure of membrane polypeptide aggregates in red cells from patients with chronic hemolytic G-6-PD variants. The 4.4×10^5 Da aggregates are composed of disulfide-bonded spectrin dimers. The $>50 \times 10^6$ Da aggregates include spectrin, ankyrin and cytoplasmic protein linked by disulfide bonds. Intramolecular disulfide bonds in ankyrin are also shown. Adapted from Hospital Practice, October, 1984, p.82.

sialic acid or structural alteration of membrane protein or phospholipid. Alteration of the red cell surface may directly or indirectly, via binding of autologous antibodies to previously masked antigens, lead to macrophage recognition and removal by the reticuloendothelial system.

In collaboration with Dr. Marguerite Kay we evaluated G-6-PD Long Prairie red cells for evidence of an immune mechanism of cell destruction. Dr. Kay found no increase in cell-associated immunoglobin in the young or middle-aged red cell population obtained from this patient. This suggests that the mechanism of final clearance of damaged red cells in this chronic hemolytic

G-6-PD variant is not mediated by the appearance of a senescent antigen. Therefore, to date the primary functional consequence of oxidant-induced membrane damage identified in the chronic hemolytic G-6-PD variants we have studied is altered cell deformability. Although it is possible that this results in more prolonged contact of damaged red cells with macrophages in the reticuloendothelial system which in turn may facilitate their removal, it is likely that other consequences of oxidant-induced membrane injury are responsible for macrophage recognition and removal. This is consistent with our earlier hypothesis that membrane polypeptide aggregates are a "footprint" of oxidant damage (6), rather than a precise definition of the mechanism of hemolysis. One possible mechanism that warrants exploration is loss of membrane phospholipid asymmetry. Haest, et al (22) demonstrated that diamide incubation of red cells resulted in movement of phosphatidylserine and phosphatidylethanolamine from their normal position in the inner lipid layer of the membrane to the outer layer. In view of the recent observation by Schroit, et al (19) that red cells with phosphatidylserine in their surface membranes are rapidly cleared from the peripheral blood and accumulate in the liver and spleen, it would be of interest to evaluate this mechanism of oxidant-damaged cell clearance.

IV. THERAPEUTIC IMPLICATIONS

A. Antioxidants

Since chronic hemolysis in G-6-PD deficiency appears to be the consequence of oxidant-induced membrane injury we, like others, postulated that antioxidant drugs might reduce hemolysis. We evaluated one anti-oxidant, vitamin E, in our G-6-PD deficient patients with chronic hemolysis. While our study was in progress, two reports of beneficial effects of vitamin E on oxidant-induced hemolysis appeared (23,24). In contrast to the

TABLE IX. EFFECT OF VITAMIN E THERAPY ON ALPHA TOCOPHEROL LEVELS IN G-6-PD DEFICIENCY

G-6-PD Variants	Vitamin E Therapy	Alpha Tocopherol Plasma (mg/dl)	Platelets (μg/mg protein)	RBC (μg/ml RBC)
Long Prairie	No	1.402	0.28	1.08
	Yes	4.024	0.50	2.83
Minneapolis	No	0.794	<0.02	1.17
	Yes	1.762	0.44	1.62
"Tomah"	No	0.946	0.06	-
	Yes	1.096	0.13	-

results of these studies, however, we found no significant benefit could be ascribed to vitamin E in patients with high grade hemolysis secondary to G-6-PD deficiency (7). Despite substantial increases in cellular vitamin E (Table IX), the reticulocyte count remained elevated, the [^{51}Cr]-red cell survival remained shortened and membrane polypeptide aggregates were unchanged (Table X). Our results were confirmed by a subsequent study by Corash, et al (25).

B. Agents That Increase Membrane Deformability

More recently we have initiated an evaluation of another potentially beneficial therapeutic modality, pentoxifylline. Pentoxifylline (1-[5-oxohexyl]-3,7-dimethylxanthine) is a drug that has recently been approved for use in the U.S. as a treatment for symptoms of occlusive peripheral vascular disease (26). It has been shown to normalize the decreased filterability of red cells seen in occlusive vascular disease and to increase the distance patients with occlusive peripheral vascular disease can walk before they have to stop because of pain (27). It has been postulated that the increase in exercise tolerance seen in patients treated with pentoxifylline is attributable to a decrease in blood viscosity secondary to improvement in red cell

TABLE X. EFFECT OF VITAMIN E THERAPY ON CHRONIC HEMOLYSIS AND RED CELL POLYPEPTIDE AGGREGATES IN G-6-PD DEFICIENCY

					Polypeptide Aggregates (% Membrane Protein)	
G-6-PD Variant	Vitamin E Therapy	Hb (g/dl)	Reticulocytes (%)	[^{51}Cr] Half Life (days)	4.4×10^5 Da	$>50 \times 10^6$ Da
Long Prairie	No	13.7	14	7.7	0.4	0.7
	Yes	14.3	19	6.7	0.3	-
Minneapolis	No	13.8	13	7.3	0.4	0.14
	Yes	13.6	10	10.0	0.4	0.15
"Tomah"	No	12.5	45	5.8	0.4	0.45
	Yes	13.4	42	6.9	0.4	0.39

deformability (26,27). Since the drug reportedly corrects decreased filterability that may be due to abnormal red cell deformability, we evaluated its effect in one patient with chronic hemolysis due to G-6-PD Long Prairie. Treatment with pentoxifylline 400mg three times per day for 6 weeks resulted in no change in the patient's blood hemoglobin concentration or reticulocyte count. However, Dr. Clark Smith found micropipette deformability (extentional static rigidity measured with 1μm pipettes), that was abnormal prior to therapy, to be equal to the normal control when studied following treatment with pentoxifylline. The disparity between the reticulocyte count and the change in red cell micropipette deformability again raises the question, "What specific characteristic of the oxidant-damaged red cell is responsible for its final clearance from the circulation?". As implied above, the answer to this question must await further studies of the mechanism responsible for the final clearance of red cells from the circulation.

REFERENCES

1. Beutler, E. (1983). Glucose-6-phosphate dehydrogenase deficiency. In Williams, W.J., Beutler, E., Erslev, A.J., and Lichtman, M.A. Eds. Hematology, 3rd Edition, p. 561-574 McGraw-Hill Book Co., New York.

2. Johnson, G.J., Kaplan, M.E., and Beutler, E. (1977). G-6-PD Long Prairie: A new glucose-6-phosphate dehydrogenase mutant exhibiting normal sensitivity to inhibition by NADPH and accompanied by nonspherocytic hemolytic anemia. Blood 49:247-251.

3. Yoshida, A. (1973). Haemolytic anemia and G-6-PD deficiency. Science 179:532-537.

4. Palek, J., Liu, S.C., and Snyder, L.M. (1978). Metabolic dependence of protein arrangement in human erythrocyte membranes. I. Analysis of spectrin-rich complexes in ATP-depleted red cells. Blood 51:385-395.

5. Allen, D.W., Johnson, G.J., Cadman, S., and Kaplan, M.E. (1978). Membrane polypeptide aggregates in glucose-6-phosphate dehydrogenase-deficient and in vitro aged red blood cells. J. Lab. Clin. Med. 91:321-327.

6. Johnson, G.J., Allen, D.W., Cadman, S., Fairbanks, V.F., White, J.G., Lampkin, B.C., and Kaplan, M.E. (1979). Red-cell-membrane polypeptide aggregates in glucose-6-phosphate dehydrogenase mutants with chronic hemolytic disease. A clue to the mechanism of hemolysis. New Eng. J. Med. 301:522-527.

7. Johnson, G.J., Vatassery, G.T., Finkel, B., and Allen, D.W. (1983). High-dose vitamin E does not decrease the rate of chronic hemolysis in glucose-6-phosphate dehydrogenase deficiency. New Eng. J. Med. 308:1014-1017.

8. Sheetz, M. (1980). Glucose-6-Phosphate Dehydrogenase Deficiency. In Genetic disorders of glutathione and sulfur amino-acid metabolism:new biochemical insights and therapeutic approaches. Ann. Int. Med. 93:330-346.

9. Coetzer, T., Zail, S. (1980). Membrane protein complexes in GSH-depleted red cells. Blood 56:159-167.

10. Kosower, N.S., Zipser, Y., and Faltin, Z. (1982). Membrane thiol-disulfide status in glucose-6-phosphate dehydrogenase deficient red cells. Relationship to cellular glutathione. Biochim. Biophys. Acta 691:345-352.

11. Fischer, T.M., Haest, C.W.M., Stöhr, M., Kamp, D., and Deuticke, B. (1978). Selective alteration of erythrocyte deformability by SH-reagents. Evidence for an involvement of spectrin in membrane shear elasticity. Biochim. Biophys. Acta 510:270-282.

12. Flynn, T.P., Allen, D.W., Johnson, G.J., and White, J.G. (1983). Oxidant damage of the lipids and proteins of the erythrocyte membranes in unstable hemoglobin disease. J. Clin. Invest 71:1215-1223.

13. Kosower, N.S., Kosower, E.M., Wertheim, B., and Correa, W.S. (1969). Diamide, a new reagent for the intracellular oxidation of glutathione to the disulfide. Biochem. Biophys. Res. Commun. 37:593-596.

14. Johnson, G.J., Allen, D.W., Flynn, T.P., Finkel, B., and White, J.G. (1980). Decreased survival in vivo of diamide-incubated dog erythrocytes. A model of oxidant-induced hemolysis. J. Clin. Invest. 66:955-961.

15. Flynn, T.P., Johnson, G.J., and Allen, D.W. (1981). Sucrose density gradient analysis of erythrocyte membranes in hemolytic anemias. Blood 57:59-65.

16. Flynn, T.P., Johnson, G.J., and Allen, D.W. (1981). Mechanisms of decreased erythrocyte deformability and survival in glucose-6-phosphate dehydrogenase mutants. In Erythrocyte Membranes 2: Recent Clinical and Experimental Advances, p. 231-245, Alan R. Liss, New York.

17. Marchesi, V.T. (1983). Review: The red cell membrane skeleton: Recent Progress. Blood 61:1-11.

18. Jandl, J.H., and Cooper, R.A. (1983). Hereditary Spherocytosis. In Williams, W.J., Beutler, E., Erslev, A.J., and Lichtman, M.A. Eds. Hematology, 3rd Edition, p. 547-553, McGraw-Hill Book Co., New York.

19. Schroit, A.J., Madsen, J.W., and Tanaka, Y. (1985). In vivo recognition and clearance of red blood cells containing phosphatidylserine in their plasma membranes. J. Biol. Chem. 260:5131-5138.

20. Kay, M.M.B. (1985). Senescent cell differentiation antigen. In Eaton, J.W., Konzen, D.K., and White, J.G. Eds. Cellular and molecular aspects of aging. The red cell as a model, p. 251-262, Alan R. Liss, Inc., New York.

21. Aminoff, D. (1985). Senescence and sequestration of RBC from circulation. In Eaton, J.W., Konzen, D.K., and White, J.G. Eds. Cellular and molecular aspects of aging. The red cell as a model, p. 279-300, Alan R. Liss, Inc., New York.

22. Haest, C.W.M., Plasa, G., Kamp, D., and Deuticke, B. (1978). Spectrin as a stabilizer of the phospholipid asymmetry in the human erythrocyte membrane. Biochim. Biophys. Acta 509:21-32.

23. Spielberg, S.P., Boxer, L.A., Corash, L.M., and Schulman, J.D. (1979). Improved erythrocyte survival with high-dose vitamin E in chronic hemolyzing G-6-PD and glutathione synthetase deficiencies. Ann. Int. Med. 90:53-54.

24. Corash, L., Spielberg, S., Bartsocas, C., Boxer L., Steinherz, R., Sheetz, M., Egan, M., Schlessleman, J., and Schulman, J.D. (1980). Reduced chronic hemolysis during high-dose vitamin E administration in Mediterranean-type glucose-6-phosphate dehydrogenase deficiency. N. Eng. J. Med. 303:416-420.

25. Corash, L.M., Sheetz, M., Bieri, J.G., Bartsocas, C., Moses, S., Bashan, N., and Schulman, J.D. (1982). Chronic hemolytic anemia due to glucose-6-phosphate dehydrogenase deficiency or glutathione synthetase deficiency: The role of vitamin E in its treatment. Ann. N.Y. Acad. Sci. 393: 348-360.

26. Spittell, J.A., Jr. (1985). Pentoxifylline and intermittent claudication. Ann. Int. Med. 102:126-127.

27. Porter, J.M., Cutler, B.S., Lee, B.Y., Reich, T., Reichle, F.A., Scogin, J.T., and Strandness, D.E. (1982). Pentoxifylline efficacy in the treatment of intermittent claudication: Multicenter controlled double-blind trial with objective assessment of chronic occlusive arterial disease patients. Am. Heart J. 104:66-72.

G6PD VARIATION

ORIGIN OF G6PD POLYMORPHISM: MALARIA AND G6PD DEFICIENCY

Lucio Luzzatto, Stella O'Brien, Essien Usanga and Wanchai Wanachiwanawin

Department of Haematology, Royal Postgraduate Medical School, London, U.K.

The notion that malaria has played a major role in the micro-evolution of the human species was first formulated by J.B.S. Haldane (1949) with respect to thalassaemia. As soon as the world wide distribution of G6PD deficiency began to emerge, A.C. Allison (1960) and A.G. Motulsky (1960) clearly identified the possibility that this genetic system was also a case of polymorphism balanced by malaria selection. Since then, for a quarter of a century, evidence for this concept has continued to accumulate. It seems, therefore, somewhat incongruous that it is still referred to as the "malaria hypothesis". Whether this phrase persists because of conviction or because of habit is perhaps a matter of relevance to psychology and linguistics rather more than to biology. The overwhelming evidence in support of malaria selection has been reviewed elsewhere (Luzzatto, 1979; Miller, 1985; Luzzatto & Battistuzzi, 1985), and therefore it will not be analyzed in detail in this brief paper. We shall accept the fact that certain _Gd_ genotypes do confer resistance against Plasmodium falciparum, and we shall concentrate on the mechanism whereby this resistance operates.

Rationale for Biological Selection by Malaria

In considering how malaria may have increased the frequency of a Gd^- mutant gene in a particular population, several principles must be borne in mind. First, we will give attention only to one species of the parasite, namely P. falciparum, since it is by far the most likely, and perhaps the only likely candidate, by virtue of the high rate of mortality caused by this parasite. This is in contrast to the relatively low mortality, in spite of high morbidity, associated with infection by the other human Plasmodia (see Bruce-Chwatt, 1985). Second, for biological selection to take place only partial protection by the gene concerned is required. Indeed, death from malaria is associated with high levels of parasitaemia. In endemic areas repeated infections lead to the development of acquired immunity within the first decade of life (see Fig. 1). From then on immune phenomena limit the parasitaemia arising from subsequent attacks to levels that are not life-threatening (indeed, in endemic

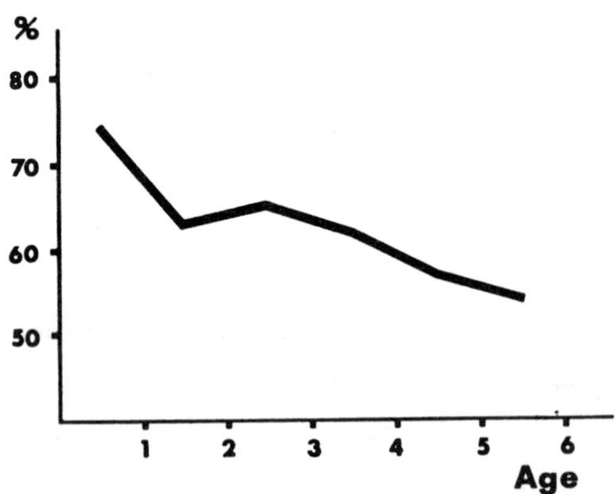

Fig. 1. P. falciparum malaria morbidity rate as a function of age in Nigerian children (from Bienzle et al., 1981).

areas asymptomatic parasitaemia is commonplace in adults). Therefore, any gene that, early in life, can reduce the multiplication of the parasite in the blood, without necessarily suppressing it completely, has the potential to allow host survival while giving a chance for acquired immunity to set in. Third, we must consider which genotypes are protective. Mathematical models and simulations (Livingstone, 1965; Cavalli-Sforza & Bodmer, 1971) indicate that for a two-allele X-linked system equilibrium can be reached only with certain sets of relationships amongst fitness value. Of the two models in Table I that are compatible with stable balanced polymorphism, model II seems unlikely, since it implies selection operating in opposite directions in males and in females. Model IV, instead, postulates heterozygote advantage in the females without prejudice to whether, in males, G6PD deficient individuals have the same or a lower fitness than G6PD normal individuals. Of course it is always possible that in many populations G6PD polymorphism is not at equilibrium. However, if it is, heterozygote advantage is, *a priori*, a requirement for equilibrium to be attained, regardless of the fitness of males, who cannot be heterozygous.

Studies in Cultures of *P. falciparum*

Because of the well characterized biochemical alterations in G6PD deficient red cells (Kirkman et al., 1980; Beutler, 1978), it seemed quite reasonable that they might be less hospitable hosts for the malaria parasite. However, if this alone was the basis for resistance, one would expect hemizygotes to be more protected than heterozygotes, contrary to the prediction of the model just outlined, and contrary to findings in the field (Bienzle et al., 1972). Protection of heterozygotes suggested that it might result not merely from the abnormal biochemistry of G6PD deficient red cells, but from their coexistence with G6PD normal red cells in a mosaic situation.

Table 1. MALARIA SELECTION AND POLYMORPHISM AT THE G6PD LOCUS

Possible situations	Fitness ranking of various genotypes		Possibility of balanced polymorphism
	Males	Females	
I	$Gd^+ \gtreqless Gd^-$	$Gd^+/Gd^+ > Gd^+/Gd^- > Gd^-/Gd^-$	No
II	$Gd^+ > Gd^-$	$Gd^+/Gd^+ < Gd^+/Gd^- < Gd^-/Gd^-$	Yes
III	$Gd^+ < Gd^-$	$Gd^+/Gd^+ < Gd^+/Gd^- > Gd^-/Gd^-$	No
IV	$Gd^+ \geq Gd^-$	$Gd^+/Gd^+ < Gd^+/Gd^- > Gd^-/Gd^-$	Yes

Based in part on a table in section 4.10 of Cavalli-Sforza and Bodmer (1971). The fitness relationships apply to an environment with heavy malaria endemicity. In the absence of malaria the Gd^- allele will be selected against because of the adverse clinical effects it has *per se*, and no stable polymorphism will be expected at all.

The advent of P. falciparum cultures (Trager & Jensen, 1976) made experimental tests of the resistance mechanism possible. When parasites grown in normal red cells were transferred to G6PD deficient cells it was found indeed that their growth was substantially impaired (Friedman, 1979; Luzzatto, 1981, Roth et al., 1983). However, repeated passages through G6PD deficient cells under appropriate conditions resulted in gradual restoration of growth to virtually normal rate (see Fig. 2; and Luzzatto et al., 1983; O. Sodeinde, personal communication).

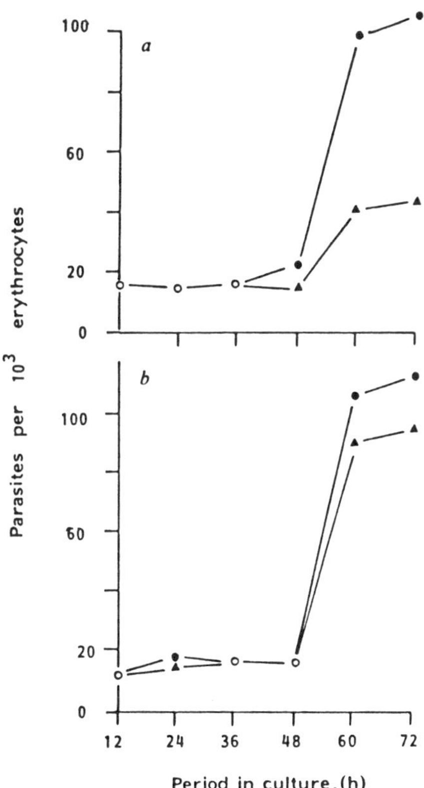

Fig. 2. Growth of P. falciparum in vitro in G6PD(+) (●—●) and in G6PD(-) (▲—▲) red cells. (a) Upon first passage from G6PD(+) to G6PD(-). (b) After four cycles in G6PD(-) red cells. From Usanga & Luzzatto (1985).

These findings were consistent with the previously formulated hypothesis (Luzzatto, 1974) that lack of resistance in G6PD deficient hemizygotes resulted from adaptation of the parasite to the initially unfavourable host cell environment.

Mechanism of Adaptation

Metabolic changes in response to the environment have been little studied in intracellular parasites, but they are characteristic of unicellular organisms, such as bacteria and yeasts. In bacteria, amongst the best known cases are the induction of enzymes required for the biosynthesis of a particular amino acid when this is lacking in the culture medium, and their repression when the same amino acid is abundantly available. By analogy, it seemed at least possible that *P. falciparum* might respond to a host cell environment in which G6PD activity is limiting by inducing its own G6PD. If this is the case, one could expect that the reverse regulatory phenomenon also occurs, whereby parasite G6PD synthesis is repressed when the same enzyme is not limiting, as in G6PD-normal host cells. Recent experimental evidence supports this idea. G6PD activity in extracts of *P. falciparum*-infected G6PD deficient red cells is higher than in sham cultures (Fig. 3). Electrophoretic analysis revealed that the extra enzyme activity in these extracts had the same (B-like) mobility, irrespective of whether the host cells were G6PD(-) A^- or G6PD(-) Mediterranean. No similar parasite-coded G6PD could be detected in extracts of G6PD(+) parasitized red cells (Usanga & Luzzatto, 1985).

Cloning of G6PD Gene from Malaria Parasites

In order to confirm that G6PD can be synthesized by *P. falciparum*, and in order to investigate the molecular mechanism whereby the parasite can regulate G6PD synthesis in response to the host cell environment, we thought that isolating the corresponding gene would be the most direct approach. A genomic DNA library of *P. falciparum* in λ phage was kindly supplied by

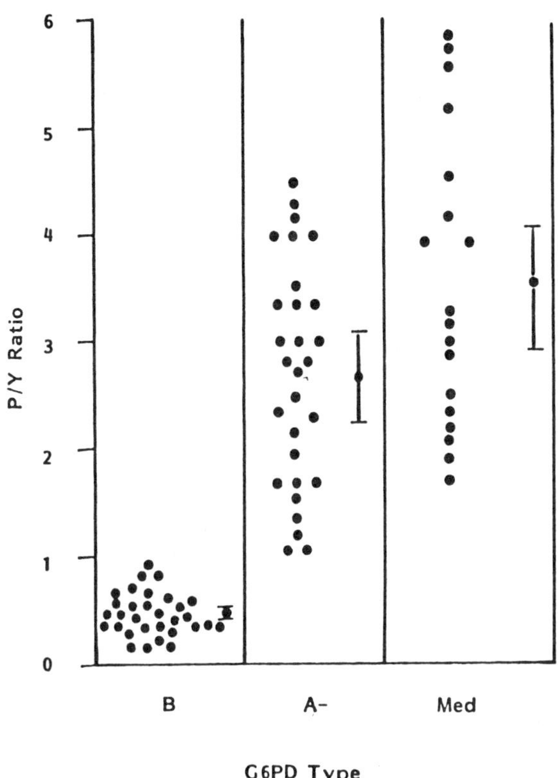

Fig. 3. G6PD activity of human red cells upon infection with P. falciparum. Each dot is one sample, with parasite rate between 5 and 12%. P/Y on the ordinate is the ratio between the enzyme activity of the parasitized red cells and of a comparable number of "young" red cells (see Usanga & Luzzatto, 1985). It is seen that the ratio is less than 1 with normal red cells, but always above 1 with G6PD deficient red cells, indicating net synthesis of enzyme by the parasite.

J. Scaife. About 10^5 plaques were screened (see Table II) by using a human G6PD-specific cDNA probe containing about 800 nucleotides of the coding sequence (see Persico et al., 1985). Hybridization was carried out by standard techniques (Maniatis et al., 1982), and washing of filters was carried out

Table II. ISOLATION OF G6PD GENE FROM PLASMODIUM FALCIPARUM

Library	Vector	Restriction enzyme	Range of insert size (kb)	Number of inserts corresponding to one genome	Number of plaques screened	Positive plaques in successive rounds of hybridization		
						I	II	III
RIK1	λNM 1149	Eco R 1	0 - 11	$\cong 2 \times 10^3$	9×10^4	2	2	1
HPF 8	λNM 788	Hind III	2 - 12	$\cong 1.8 \times 10^3$	8×10^4	5	1	0

The first library is from Langsley et al. (1983); the second library is from Goman et al. (1982).

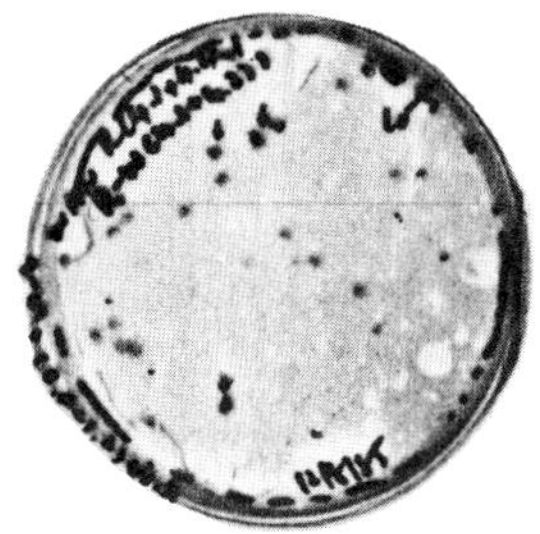

Fig. 4. Purification of recombinant λ phage containing G6PD gene sequence from *P. falciparum*. Filter hybridization was carried out with a human G6PD cDNA probe. Both positive and negative plaques are seen.

at relatively low stringency (6 x SSC at 65°C). Two plaques gave a signal, and after two successive rounds of purification (see Fig. 4) one of the two phages was confirmed as positive. In order to ensure that the DNA insert was of parasite origin, and not due to contamination with human DNA, the insert was excised from the phage by digestion with the restriction enzyme *Eco* RI. After electrophoresis and Southern blotting on duplicate filters, one filter was washed at high stringency and the other at low stringency. A positive signal was obtained only at low stringency, whereas a control human DNA insert showed no difference between the two filters. In addition, a preliminary restriction map of the phage insert shows no resemblance to that of the human genomic *Gd* gene (data to be published). Thus, the *P. falciparum* genome contains a sequence homologous to that of human *Gd*. This sequence is likely to be part of the parasite's own *Gd* gene, and therefore the data suggest significant evolutionary conservation, in keeping with the existence of an almost identical peptide in yeast and human G6PD (Persico et al., 1985). Further characterization of the *P. falciparum* *Gd* gene is in progress. Availability of a parasite-specific probe for this gene should make it possible to confirm the occurrence of induction and repression of enzyme biosynthesis and to investigate its mechanism.

Parasite Adaptation and Host Resistance

We can now ask how the data obtained *in vitro* can help to interpret how different *Gd* genotypes are associated with high or low susceptibility to malaria *in vivo*. We may refer to the G6PD normal genotype (Gd^+ in males and Gd^+/Gd^+ in females) as having high susceptibility. In G6PD deficient individuals (Gd^- males and Gd^-/Gd^- females) we infer from the adaptation observed in cultures that parasites will be able to attain a similar rate of multiplication, with a delay of only a few cycles. If it were possible to measure the time interval from parasite inoculation to life-threatening parasitaemia this ought to be prolonged by some days. From the practical point of view this delay will be slight, and therefore the susceptibility to malaria mortality of the G6PD deficient genotypes will be similar to that of the G6PD normal genotypes, whether hemizygous males or homozygous females. However, in heterozygous females parasite adaptation will be persistently frustrated by the coexistence of G6PD normal and G6PD deficient red cells (see Fig. 5). As a result, the susceptibility to malaria mortality of the Gd^+/Gd^- genotype is reduced, in accord with model IV in Table I. This is perhaps the first clear example in the human species in which X-linked mosaicism, rather than heterozygosity *per se*, displays a significant biological advantage with respect to selection by an environmental agent.

ACKNOWLEDGMENTS

This work has received financial support through an MRC Research Group. EU was a Research Training Fellow supported by the WHO/UNDP/World Bank Tropical Disease Research Programme, on leave of absence from the University of Ibadan, Ibadan, Nigeria. WW is on leave of absence from Mahidol University, Bangkok, Thailand.

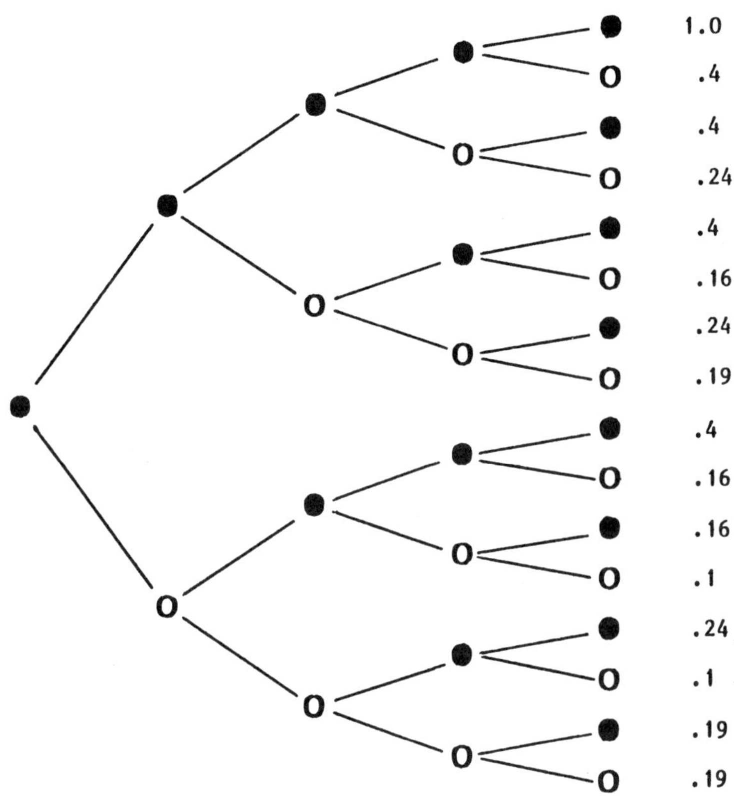

Fig. 5. Diagram of parasite multiplication in a $\underline{Gd}^+/\underline{Gd}^-$ heterozygous female mosaic. The full circles represent G6PD(+) red cells and the empty circles G6PD(-) red cells. Each connecting line represents one schizogonic cycle. It is assumed that the ratio between the two cell types is 1, and therefore that at each cycle a parasite has an even probability to infect a normal or a deficient cell. The growth rate in a passage from G6PD(+) to G6PD(+) is assigned a reference value of 1.0 (in fact, the multiplication factor is about 5 *in vitro* and probably higher *in vivo*). The growth rate in a *first* passage from G6PD(+) to G6PD(-) is assigned a relative value of 0.4, based on the data in Fig. 2 (see also Luzzatto et al., 1985). Since full adaptation takes about 4 cycles, we assign relative values of 0.6, 0.8 and 1.0 to the second, third and fourth passage respectively from G6PD(-) to G6PD(-). Passage from G6PD(-) to G6PD(+) is assigned a relative value of 1.0. After such a passage, it is assumed that the adaptation process is interrupted, and therefore

subsequent passage to G6PD(-) starts again from a relative value of 0.4. The figures in the right hand column give the overall relative growth rate values for each of the 16 four-cycle paths outlined. The average of the sixteen figures is 0.29. This figure which must be regarded as merely an illustrative one because of the many assumptions involved, signifies a potential reduction in parasitaemia of the order of 70%, sufficient to bring it below the life-threatening threshold.

REFERENCES

Allison, AC. (1960). Nature, 186, 531.

Beutler, E. (1978). "Hemolytic Anemia in Disorders of Red Cell Metabolism." Plenum Medical Book Co., New York.

Bienzle, U., Ayeni, O., Lucas, A.O., Luzzatto, L. (1972). Lancet, i, 107.

Bienzle, U., Guggenmoos-Holzmann, I., Luzzatto, L. (1981). Int. J. Epidemiol., 10, 9.

Bruce-Chwatt, L.J. (1985). "Essential Malariology." 2nd Edition. Heinemann Medical Books Ltd., London.

Cavalli-Sforza, L.L., Bodmer, W.F. (1971). "The Genetics of Human Populations." Freeman, San Francisco.

Friedman, M.J. (1979). Nature, 280, 245.

Goman, M., Langsley, G., Hyde, J.E., Yankovsky, N., Werner Zolg, J., Scaife, J.S. (1982). Molec. Biochem. Parasitol., 5, 391.

Haldane, J.B.S. (1949). Ric. Sci. 19 (Suppl. 1), 3.

Kirkman, H.N., Wilson, W.G., Clemons, E.H. (1980). J. Lab. Clin. Med., 95, 877.

Langsley, G., Hyde, J.E., Goman, M., Scaife, J.G. (1983). Nucleic Acid Res., 11, 8703.

Livingstone, F.B. (1964). Am. J. Human Genet., 16, 435.

Luzzatto, L. (1974). Bull. Wld Hlth Org., 50, 195.

Luzzatto, L. (1979). Blood, 54, 961.

Luzzatto, L. (1981). In "Modern Genetic Concepts and Techniques in the Study of Parasites" (F. Michal, ed.), p. 257. Tropical Diseases Research Series No. 4, Schwabe & Co., Basel.

Luzzatto, L., Battistuzzi, G. (1985). Advances Hum. Genet., 14, 217.

Luzzatto, L., Usanga, E.A., Modiano, G. (1985). In "Ecology and Genetics of Host-Parasite Interactions" (D. Rollinson and R.M. Anderson, eds), p. 205. Academic Press, London.

Luzzatto, L., Sodeinde, O., Martini, G. (1983). In "Malaria and the Red Cell", Ciba Foundation Symposium 94, p. 159, Pitman, London.

Maniatis, T., Fritsch, E.F., Sambrook, J. (1982). "Molecular Cloning: A Laboratory Manual," Cold Spring Harbor Laboratory, USA.
Miller, L.H. (1984). In "Textbook on Malariology" (I. McGregor and W. Wernsdorfer, eds).
Motulsky, A.G. (1960). Hum. Biol. 32, 28.
Persico, M.G., Viglietto, G., Martini, G., Toniolo, D., Paonessa, G., Moscatelli, C., Dono, R., D'Urso, M., Vulliamy, T., Luzzatto, L. (198), this Symposium.
Roth, E.F.Jr, Raventos-Saurez, C., Rinaldi, A., Nagel, R.L. (1983). Proc. Natl. Acad. Sci. USA 80, 298.
Trager, W., Jensen, J.B. (1976). Science, 193, 673.
Usanga, E.A., Luzzatto, L. (1985). Nature, 313, 793.

G6PD VARIANTS IN SOUTHERN ASIAN POPULATIONS

Vicharn Panich

Department of Pathology, Faculty of Medicine
Prince of Songkla University, Songkhla, Thailand

Glucose-6-phosphate dehydrogenase (G6PD) deficiency is common in Southern Asia. Very high incidence of 50% or over (among males) has been reported from some population groups of the Middle East and of Papua New Guinea (1-4). This review will be confined to areas below the latitude of 30° N. Countries having area both above and below the latitude are included. Altogether 78 G6PD variants, some of which are not fully characterized, have been reported from this area and are summarized alphabetically in Table I.

Table I. G6PD variants in Southern Asia.

Variant	Properties[a]	Population origin
Class 1		
Ashdod	1,6,14,23,26	N.African Jew (5)
Bangkok	2,5,9,14,22,25	Thai (6,7)
Bat-Yam	2,4,14,23,25	Iraqi Jew (5)
Boston	2,4,7,14,19,24,26	Polish Jew (8)
Hong Kong	2,4,8,14,22,27	Chinese (9)

Table I. (Continued)

Variant	Properties[a]	Population origin
Linda Vista	3,4,9,10,14,19,23,25	Laotian (10)
Panama	1,4,7,14,19,23,26	Sephardic Jew (11)
Ramat-Gan[b]	1,4,14,23,25	Iraqi Jew (5)
Class 2		
Amboin	3,4,9,14,19,21,22,26	Papua New Guinea (12)
Angoram	1,4,8,14,19,21,22,27	Papua New Guinea (12)
Bagdad[c]	2,5,9,14,23	Iraqi Jew (13)
Birmingham[d]	4,9,11,13,19,23,27	Pakistani (14)
Bnei Brak[e]	2,5,9,14,27	Iraqi Jew (15)
Bogia	1,5,9,14,18,21,22,27	Papua New Guinea (16)
Bukitu[f]	6,9,14,18,21,22,26	Papua New Guinea (16)
Campbellpore	2,4,14,23,26	Pakistani (17)
Canton	3,4,7,14,16,19,20,23,26	Chinese (18-22), Thai (19,23,24) Laotian (25), Vietnamese (26).
Castilla-like	3,4,9,14,19,21,22,26	Papua New Guinea (12)
Chainat	1,5,9,13,15,19,20,22,28	Thai (24)
Goodenough[f]	4,9,14,19,21,23,26	Papua New Guinea (12)
Haad-Yai	3,6,14,16,19,20,22,26	Chinese in Thailand (22)
Hong Kong Pokfulam	1,5,8,13,15,19,20,22,27	Chinese (originally called Panay-like) (19-22), Thai (24)
Hualien	3,4,14,23,25	Taiwanese (27)
Hualien-Chi	3,4,14,23,27	Taiwanese (27)
Indonesia	2,4,9,23,28	Indonesian (28)
Intanon	1,5,7,13,15,17,20,23,27	Karen in Thailand (29)
Jammu[g]	5,9,13,18,22,27	Kashmir Indidan (30)
Kaluan	1,6,9,13,17,21,22,26	Papua New Guinea (12,16)
Kar Kar	1,4,7,14,19,21,22,27	Papua New Guinea (12)
Kilgore	2,4,8,14,19,23,27	Hungarian Jew (31)
Kirovograd[h]	4,7,24,27	Ashkenazi Jew in USSR (32,33)
Lifta	1,4,14,23,25	Iraqi Jew (5)
Long Xuyen	3,4,14,16,19,20,23,26	South Vietnamese (26)
Madang	1,4,8,14,18,21,22,26	Papua New Guinea (12)
Mainoki	1,4,8,13,19,21,22,26	Papua New Guinea (12)
Manus	1,4,7,14,19,21,23,26	Papua New Guinea (12)
Markham	3,4,7,12,14,16,19,21,23,26	Papua New Guinea (3,4,12,16,34)

Table I. (Continued)

Variant	Properties[a]	Population origin
Mediterran-like	2,4,7,14,16,19,20,23,26	Papua New Guinea (3,4), Southern Thai, (24,25) Northern Indian, Pakistani (36)
N-Sawan	3,4,7,13,15,19,20,23,26	Thai (24)
Padrew	3,4,9,14,19,21,23,26	Thai (24)
Palakau[i]	5,9,14,17,21,22,26	Papua New Guinea (12)
Panay	1,4,8,13,23,26	Philipino (37)
Popondetta	1,4,7,14,19,21,23,26	Papua New Guinea (12)
Salata	1,6,9,14,18,21,22,27	Papua New Guinea (16)
Siriraj	2,4,14,16,19,20,23,26	Thai (23,38)
Swit[j]	4,7,14,19,21,23,26	Papua New Guinea (12)
'Taiwan-Ami5'	3,4,14,23	Taiwanese (27)
'Taiwan-Ami6'	3,4,14,23,25	Taiwanese (27)
Taiwan-Hakka	3,4,14,19,23,26	Hakka Chinese (19)
Teheran	2,5,13,23,27	Iranian (27)
Union	3,4,7,12,14,16, 19,20,23,25	Philipino (34,39), Thai (24,35,40), Laotian (25), Papua New Guinea (4)
West Bengal	1,4,9,13,22,27	Indian (36)
Wewak	1,4,7,14,19,21,22,26	Papua New Guinea (12)
Zhitomir	1,4,7,12,14,19,23,26	Ashkenazi Jew in USSR (33)
Class 3		
A(-)-like	3,5,8,11,13,15,18,20, 22,27	Thai (24), Laotian (25), Chinese from Cambodia (41)
Anant	2,4,14,16,19,20,22,27	Thai (23,24)
Kalyan	1,4,14,16,19,22,26	Koli (West India) (42)
Kan	3,14,16,19,21,25	Thai (23)
Kerala	1,4,7,14,16,23,27	Indian from Kerala (36)
Lizu-Baisha	5,8,14,22,27	Chinese (43)
Mahidol (B(-)Chinese)	2,4,8,13,15,18,20,22,28	Thai (44), Mon (24), Malay (24) Laotian (25), Cambodian (24,45), Vietnamese (26), Chinese (20-22)
Taipei-Hakka (Dhon)	3,4,8,13,15,19,20,22,26	Hakka Chinese (19), Chinese, (20,22), Thai (24,35)
Tel-Hashomer	1,4,13,23,27	Sephardic Jew (46,47)

Table 1. (Continued)

Variant	Properties[a]	Population origin
Tenganan	1,4,9,14,19,21,22,26	E.Balinese (Indonesia) (48)
Vientiane	1,4,9,12,14,16,19,22,28	Laotian (49)
Yangoru[k]	1,4,9,14,19,21,26	Papua New Guinea (12)
Class 4		
?A	not characterized	Saudi Arabia (50,51)
'Andra Pradesh'	3	Indian (42)
Ayutthaya	1,4,7,14,16,19,20,23,27	Central Thai (52)
Bali	3,5,9,13,17,20,22,26	E.Balinese (Indonesia) (48)
Chao Phya	3,5,9,13,15,18,20,22,27	Thai (52)
'Khartoum'	1,13	Saudi Arabia (51,53)
Porbandar	1,4,7,14,19,22,27	Indian (54)
S-Sarkorn	3,6,9,13,15,17,20,22,27	Thai(52)
RBC enzyme activity unknown		
'B(-) intermediate'	2	Saudi Arabia (50)
'Corinth-like'	14,18,20	Papua New Guinea (4)
'Cutch'	1	Indian (42)
'Munum'	13,19,21	Papua New Guinea (4)

[a] 1 = slow, 2 = normal, 3 = fast electrophoretic mobility; 4 = low, 5 = normal, 6 = high Km G6P; 7 = low, 8 = normal, 9 = high Km NADP; 10 = low, 11 = normal, 12 = high Ki NADPH; 13 = normal, 14 = high 2deoxyG6P utilization; 15 = normal, 16 = high galactose-6-phosphate utilization; 17 = low, 18 = normal, 19 = high deaminoNADP utilization; 20 = normal, 21 = high NAD utilization; 22 = normal, 23 = biphasic, 24 = peak pH curve; 25 = severely reduced, 26 = reduced, 27 = normal, 28 = increased stability to heat.

[b] some cases do not have CNHD.

[c] increased activity by heat.

[d] fast electrophoretic mobility in phosphate buffer, pH 7.0, but normal mobility in EBT buffer pH 8.6 & Tris HCl, pH 8.8.

[e] triphasic pH curve (6.0,7.5,10.5).

[f] fast electrophoretic mobility in EBT buffer, pH 8.6, but normal mobility in phosphate buffer, pH 7.0.

[g] fast electrophoretic mobility in phosphate buffer, pH 7.0, but slow mobility in EBT buffer, pH 8.0 and Tris HCl buffer, pH 8.8.

[h] slow electrophoretic mobility in phosphate buffer, pH 7.0, but normal mobility in EBT buffer, pH 8.0 and Tris HCl buffer, pH 8.8.

[i] slow electrophoretic mobility in phosphate buffer, pH 7.0 and Tris HCl buffer, pH 8.8, but normal mobility in EBT buffer, pH 8.0.

[j] fast electrophoretic mobility in EBT buffer, pH 8.0, but slow mobility in phosphate buffer, pH 7.0 and Tris HCl buffer, pH 8.8.

[k] Triphasic pH curve (5.5, 7.0, 9.0).

Quotes used with name to indicate incomplete characterization or uniqueness not established.

INCIDENCE OF G6PD DEFICIENCY AND DESCRIPTION OF VARIANTS BY POPULATION GROUPS

Australia

Because G6PD is an X-linked intermediate condition, the incidence mentioned in this paper is the incidence in males. No G6PD deficiency was found among 435 male Australian aborigines (55) and there have been no reported G6PD variants from this population group.

Bangladesh

Six out of 150 Bengali Moslems (4%) were G6PD deficient (56), but there has been no report of G6PD variant in this country.

Cambodia

The incidence of G6PD deficiency among Cambodians is 14-15% (45,57). The lower incidence of 7% among Cambodian refugees arriving in the USA might be due to gene dilution by the Chinese who populate every city in Southeast Asia (58). Sixteen G6PD-deficient Cambodian males have been characterized and all possessed G6PD Mahidol (22,45). Gene frequency of G6PD Mahidol of 0.14-0.15 in Cambodian is the highest thus far reported.

China

In Chinese, G6PD deficiency shows highest incidence of 4.5-6.8% among the Hakka Chinese (59). In Kwangtung Province the incidence of 5.5% was reported, and that in Fukien Provice was 3% (60,61). Generally, the incidence of G6PD in Southern Chinese is 3-6%, but non-existent in Northern Chinese (22,62-67).

Nine G6PD variants have been reported from Chinese, the most common of which is G6PD Canton. McCurdy et al reported that all 5 G6PD-deficient Chinese males living in USA and Canada studied had G6PD Canton and later reported that 5 out of 6 G6PD-deficient Cantonese in Hong Kong had this variant (18,19). Among 41 G6PD-deficient Chinese originated from Southern China living in Hong Kong, 25 had G6PD Canton (20). In Thailand, 8 cases of G6PD Canton were found among 25 G6PD-deficient Chinese (22).

G6PD Mahidol or B(-) Chinese is the second most common variant among the Chinese. It was found in 11 out of 41 G6PD-deficient Chinese in Hong Kong and 6 out of 25 G6PD-deficient Chinese in Thailand (20,22).

G6PD Taipei-Hakka (Dhon), originally reported in 3 out of 6 G6PD-deficient Hakka Chinese in Taiwan, was reported in 1 out of 41 G6PD-deficient Chinese in Hong Kong and 8 out of 25 G6PD-deficient Chinese in Thailand (19,20,22).

The variant originally called G6PD Panay-like by McCurdy et al (19), and called G6PD Hong Kong by Chan & Todd (20), Chan, Todd & Lai (21), Panich (35,68,69), Panich & Na-Nakorn (24), Panich, Na-Nakorn & Wasi (22) should be called G6PD Hong Kong Pokfulam after Beutler & Yoshida and Chan, Todd & Tso (70,71). This variant never causes CNHD and shows slow electrophoretic mobility. It has been found in 5 out of 41 G6PD-deficient Chinese in Hong Kong and in 1 out of 25 G6PD-deficient Chinese in Thailand(20,22).

G6PD Haad-Yai was found in 2 out of 25 G6PD-deficient Chinese in Thailand (22). G6PD Taiwan-Hakka was found in 3 out of 6 Hakka Chinese in Taiwan (19). G6PD A(-)-like was found in 2 brothers whose mother was Chinese from Cambodia, now living in France (41). G6PD Lizu-Baisha was reported from China (43). G6PD Hong Kong was the only variant causing CNHD reported in Chinese (9).

Table II. G6PD Variants in Southern Asia classified by erythrocyte enzyme activity and electrophoretic mobility

Class[a]	Total No.	Electrophoretic Mobility			
		Slow	Normal	Fast	Undetermined
1	8(10.3%)	3	4	1	0
2	46(59.0%)	19[b]	8	19[c]	0
3	12(15.4%)	6	2	3	1
4	8(10.3%)	3	0	5	0
Undetermined	4(5.1%)	1	1	0	2
Total	78	32(41.0%)	15(19.2%)	28(35.9%)	3(3.8%)

[a] class 1 associated with congenital nonspherocytic hemolytic disease (CNHD), class 2 rbc enzyme activity 0-9.9%, class 3 rbc enzyme activity 10-59.9%, class 4 rbc enzyme activity 60-150% of normal.

[b] G6PD Kirovograd shows slow electrophoretic mobility in phosphate buffer, pH 7.0, but normal mobility in EBT buffer, pH 8.0 and Tris HCl buffer, pH 8.8. G6PD Palakau shows slow mobility in phosphate buffer pH 7.0 and Tris HCl buffer, pH 8.8, but normal mobility in EBT buffer, pH 8.0.

[c] Three variants had fast and normal electrophoretic mobility. G6PDs Bukitu and Goodenough show fast mobility in EBT buffer, but normal mobility in phosphate buffer. G6PD Birmingham shows fast mobility in phosphate buffer, but normal mobility in EBT and Tris HCl buffers. Two variants show fast and slow mobility. G6PD Jammu shows fast mobility in phosphate buffer but slow mobility in EBT and Tris HCl buffers. G6PD Swit has fast mobility in EBT buffer, but slow mobility in phosphate and Tris HCl buffers.

India

The incidence of G6PD deficiency in India is very variable between regions and ethnic groups. The highest incidence of 17-19% was reported among the Parsees in Bombay, Western India (72,73). High incidence of 17.3% was reported from the Santhals of West Bengal, 11.6-12.2% in the Koya Doras (Dravidian) in Andra Pradesh, and 9.0-12.5% in the tribal people in the Nilgiri Hills, Southern India (73-76). The incidence among the Koya Doras was, however, very variable between different groups, ranging from zero to 11.6% and average 3.6% (75). In Delhi, Uttar Pradesh, the incidence was reported to be 2.8% and 2.1%, but later the high incidence of 10.1% was reported (77-79). In Assam, the incidence was 7.0%, 5.4% and 4.3% in Khasi, Ahom and Assamese respectively (80). Other high incidence of 7.7% was found in Calcutta Muslims, 4.1% in Hindus, 6.7% in Bhils of Madhya Pradesh, 6.6% in tribal groups in Maharashtra, 7.7% in non-tribal groups of Orissa, and 4.4% in non-tribal Andhas group of Visakhapatnam town, Andra Pradesh (73,81-84). The incidence of 7.4% in mixed group in Bombay has been reported, but was only 0.4% in another report (85,86). The incidence among Bombay Muslims was between 0-2% (87). The incidence of 1-4% was reported in tribal groups other than Koya Doras of Thallavaram, and Hindus of Polavaram, Andra Pradesh, Hindus of Tirupati, Andra Pradesh, Bengalee of Orissa, Raj Gond (Dravidian), Naikpod (Dravidian) and Lambadi (Aryan) of Andra Pradesh, Dravidian of Kerala, Hindus of Maharashtra, Hindus and Muslims of Madhya Pradesh, Hindus, Sindhis and Sikhs of Punjab (73-75,82,88-90). The Kadar (tribal group) of Kerala, the Pardhan (Aryan) of Andra Pradesh and the Toda tribe of Nilgiri Hills were reported to have no G6PD deficiency (75,76,91).

Ten variants, some of which are not fully characterized have been reported from Indian population. G6PD West Bengal was found in an Indian from West Bengal (Northwest), G6PD Mediterranean was found in an Indian from Punjab (Northeast), and G6PD Kerala was found in 2 brothers from Kerala, Southwest India (36). The Porbandar variant was found in an Indian family in Johannesburg, South Africa (54). G6PD Kalyan was found in a Koli family, a tribal group of Western India (42). G6PD Jammu was reported from a Kashmir Indian (30). G6PDs 'Andra Pradesh' and 'Cutch' were mentioned by Ishwad and Naik but have not yet been fully characterized (42). Judging from historical population movement from Europe to India and from the finding of other 2 cases of G6PD Mediterranean in Pakistanis (36), G6PD Mediterranean should be common in India, at least in the Northern part of the country. One male having fast electrophoretic G6PD variant was found among 49 Jalari males, a fishing caste population in the coastal villages of Kothajalaripet and Kothaveedhi of Visakhapatnam, Andra Pradesh, was assigned to have G6PD A. One female out of 51 from the same population was reported to be heterozygous for G6PD AB. Another female was assigned the genotype of BC, indicating the occurrence of an electrophoretically slow variant with normal erythrocyte enzyme activity in the population (92). This slow variant is not included in Tables I and II. There was no electrophoretic G6PD variant among 85 Telugu Brahmin boys, also of Visakahapatnam, Andra Pradesh (92). The frequency of G6PD ?A in the Jalari of Andra Pradesh should be around 2%. Frequency of other G6PD variants found in India could not be estimated due to a lack of systematic characterization studies.

Indonesia

The highest incidence of G6PD deficiency thus far reported from Indonesia was 10.8% among the isolated population of Tenganan Pageringsingan, East Bali (93). The incidence among

East Balinese was 2% (93). In Djarkarta the incidence was 1.1% in adult males, and 3.2% in newborns (94,95). In Ambon the incidence among the islanders was 2-3% (96,97).

There have been 3 G6PD-deficient variants and one normal variant reported from Indonesia. G6PD Indonesia was found in 3 males and thus should be a common variant in this population. A variant very similar to G6PD Taipei-Hakka (or G6PD Dhon) was found in an Indonesian woman with Chinese ancestry by the same group of investigators (28). G6PD Tenganan was characterized from an East Balinese man (48). G6PD Bali, the only polymorphic variant with normal erythrocyte enzyme activity in Southeast Asian population, has been found in 2% of East Balinese men (48).

It has been mentioned that G6PD Indonesia has very similar properties to G6PD Mahidol which is very widely destributed in Southeast Asia (24,68,69).

Iran

In Iran the highest frequency of G6PD deficiency of 20.3% was reported from the Mamassani, Southwest Iran (98). The general incidence among Iranians was 9.8% (99,100). Among the Armenians from Soviet Armenia living in New Julfa Isfahan, the incidence was 0.7% (101).

The only well-characterized variant among Iranians was G6PD Teheran (27). Daneshbod found that among 26 G6PD-deficient males in Tehran 18 had normal enzyme electrophoretic mobility and assumed that they should have G6PD Mediterranean (102). This author also mentioned other fast (105-110%) and slow (85-93%) electrophoretically moving variants, but no detailed characterization was performed. These 2 variants are not included in Tables I and II.

Iraq

The incidence of G6PD deficiency ranged from 6-13% in different population groups of Iraq (103,104). No G6PD variant has been reported.

Israel

A very high G6PD deficiency incidence of more than 50% has been reported from some groups of Kurdish and Iraqi Jew(1). The incidence among Jewish population is, however, very variable from very low to very high. The high incidence of 10-70% was reported in Iranian Jews, Afghanistan Jews, Iraqui Jews, Kurdish Jews, and Caucasus Jews (1,105,106). The low incidence of 0.4-2.2% was reported in Sephardic and Ashkenazi Jews (1,105). No case of G6PD deficiency was found in 46 Jews from Bukhara, USSR, 52 Jews from Djerba, 208 Falasha Jews from Ethiopia, 201 Sammaritan, and 23 Jews from Syria and Lebanon (1,105).

Thirteen G6PD variants, six of which were associated with CNHD, have been reported from Jewish people. G6PDs Ramat-Gan, Bat-Yam, and Mediterranean were each reported in one Iraqi Jewish male with very mild CNHD and G6PD Ashdod was reported in a North African Jewish male with very mild CNHD (5). Another Ashkenazi Jewish male with very mild CNHD also had G6PD Mediterranean (5). Other two variants associated with CNHD were G6PDs Panama and Boston found in Sephardic and Polish Jews respectively (8,11). G6PD Bagdad was found in one Iraqi Jewish family in the Netherlands (13). G6PDs Lifta and Bnei-Brak were each found in one Iraqi Jewish male with massive hemolysis associated with viral hepatitis (5,15). G6PD Kilgore was described in one Jewish family from Hungary living in USA (31). The Tel-Hashomer variant was found in Sephardic Jews from Tunis (46,47). G6PDs Kirovograd and Zhitomir was each reported from one Ashkenazi Jewish male in USSR (32,33).

Jordan and Kuwait

The incidence of G6PD deficiency among Jordanian blood donors in Amman was 1.7% (107). The incidence in Kuwaiti was 18.9% (108). Red cell G6PD electrophoretic study of 543 Jordanian blood donors showed that 2.4% had fast mobility with normal enzyme level and was assumed to have the A variant. All the 9 G6PD-deficient subjects had enzyme with normal electrophoretic mobility, probably having G6PD Mediterranean (107). No detailed characterization study has been reported from these two population groups.

Laos

The incidence of G6PD deficiency was 13% among Laotian in Laos and 11.5% among Laotians in Thailand (25,109). Among Laotian refugees arriving in USA the incidence was 9.4% (58).

Six G6PD variants have been reported from this population group. The most common variant should be G6PD Mahidol which was found in 3 out of 8 G6PD deficient Laotians characterized (25). Sicard, Kahn & Labie found that most of G6PD deficiency in Laos had normal electrophoretic mobility which is compatible with G6PD Mahidol (109). G6PD Union, originally described from Filipinos, was found in 3 out of 8 G6PD-deficient Laotians(25,39). G6PDs Canton and A(-)-like (N-Pathom) was each described in one Laotian (25). Sicard, Kahn & Labie found that some G6PD deficiency in Laos had electrophoretic mobility between A and B variants which should be compatible with the Union and Canton variants, and some had mobility similar to A variant which should be compatible with the Canton and A(-)-like variants (109). G6PD Vientiane was reported in a Laotian male in France (49). The only variant associated with CNHD in Laotian was G6PD Linda Vista (10).

Malaysia

The population of Malaysia is, as in many countries, very heterogeneous. Among the Malay ethnic group the frequency of G6PD deficiency ranged from 2% among the peninsular population, 5.6-6.3% in Brunei (now an independent country), 4.1-9.1% in Sabah, to 11.6% in Sarawak (67,110,111). The incidence among the Chinese was 2.3-2.4% in the peninsular Malaysia, and was 5.8% in Sabah (67,110,111). G6PD deficiency in Indians was 1.7-3.3% (110,112). The Aboriginal Malay showed the incidence of 16.1-17.0%, while the Senoi (a tribal group) had higher incidence of 11.8-23.1% (113).

In Sarawak, the Sea Dyak and the Iban had the incidence of 1.3% and 5.4% respectively (111).

In Sabah, the Murut, Kadezan, Bajau, and Bisaya had the G6PD deficiency incidence of 16.7-24.2%, 7.1-12.1%, 3.4-6.6%, and 6.7-16.7% respectively (67,111).

No study of G6PD variants has been reported in population groups of Malaysia. One Thai Muslim of Southern Thailand, an ethnic group related to the Malay, was found to have G6PD Mahidol (24).

Micronesia

The incidence of G6PD deficiency in Caroline Islands was 8.6% among the Angaur, 5.9% among the Ifalik, 8.3% among the Koror, but no G6PD deficiency was found among 117 Ulithi males (114). In Guam, only 1 case of G6PD deficiency was found in 246 (0.4%) Chamorros males (115). No G6PD characterization study has been reported in these population groups.

New Guinea

The population of New Guinea consists of many isolated inbred groups. The highest incidence of G6PD deficiency in this area was 60% (21 out of 35 males) among the people of Butibam village near the town of Lae at the mouth of the Markham River (3). The 50% incidence was later reported from a larger sample from the same area including people from Butibam as well as five other villages (4). High incidence of 20-30% was reported from the Bukawa, Wampur, and Melanesian at Labubutu area, all living in Markham River Valley (116,117). The other groups living in the Markham River Valley with around 10% incidence are the Papuan at Mamamban and the Melanesian at Wankum, Kaiapit, and Sukurum-Dumlinan (117). The other groups living in this area but having low incidence of 1.5-3% are the Melanesian at Gnarowein and Tsile-Tsile, the Papuan at Binumarein and Tumbuna and the Mumeng (116,117). In the area of Sepik River Valley, high incidence of 16.7 (3/18 males) was found in the Abelam at Neligum, but low incidence of 3.2% and 0.9% was found in the Abelam at Kuminibus and the Sause respectively (117). In the Eastern and Western Highlands G6PD deficiency incidence was very low or practically none (116,118,119).

In New Britain most of the isolated group studied show the incidence of around 10 to over 20%. The only two groups with lower incidence were the Taulil with 4.0% and the Tolai with 0.4% incidence (116). In the Manus Island, the Usiai showed a high incidence of 15% (3 out of 20) but the Manus showed only 5.1% deficiency (120).

In West Irian, the highest G6PD deficiency incidence of 16.7% was reported among the Papuan at Asmat (116). The Papuan at Merauke and another group of Papuan had the incidence of 8 % (116,121). The population of Biak had 4.4 % G6PD deficiency incidence (121). The low incidence of 0-2.1% was found among the Dani Papuan and the Paluaka Papuan (121,122).

Among the studies of mixed population groups in Papua New Guinea, the students from various regions showed the incidence of 3%, and soldiers showed the result of 5.8% (16,123). Another group of males from various parts of Papua New Guinea had the incidence of 7.2% (12).

Twenty-two G6PD variants have been reported from Papua New Guinea. The one attaining the highest frequency was G6PD Mediterranean-like, being found in 18 out of 35 males (51.4%) of Butibam village (3). The number of G6PD deficient subjects in the group was 21. In another study by Yoshida et al, this variant was found among deficient males as follows, 43 out of 51 in Butibam village, 8 out of 10 in Yanga village, 1 out of 4 in Yalu village, and all 16 of Kamkumung village (4). All of the 7 G6PD-deficient cases in Manus Island studied by Malcolm et al had normal electrophoretic mobility and probably had this variant (120). G6PD Markham was found, in Butibam village, in 3 out of 21 and 7 out of 51 G6PD deficient males (3,4). The variant was found in 3 out of 9 G6PD deficient males of Wagan village and in one male each from Madang District and Goodenough Island (4,16). It was also found in 2 men from West New Britain (12).

G6PD Corinth-like was found in 2 out of 10 G6PD-deficient males of Yanga village and in one out of 10 G6PD-deficient males of Munum village (4). G6PD 'Munum' was found in 3 out of 4 and 9 out of 10 G6PD-deficient males of Yalu and Munum village respectively (4). G6PD Salata was found in 1 male each from East Sepik District, Central District and Manus Island (12,16). G6PD Kaluan was found in 2 unrelated men from New Ireland and G6PD Goodenough was found in 6 unrelated males from Goodenough Island (12,16). The 7 variants mentioned above are found in more than one independent studies and should be considered common G6PD variants in Papua New Guinea.

The fifteen variants described only once are G6PDs Union-like found in 1 out of 51 G6PD-deficient males in Butibam village, Bogia found in Madang District, Bukitu from East Sepik District, Wewak, Amboin, Angoram, Yangoru and Castilla-like from East Sepik

Region, Swit and Madang from Madang District, Kar Kar from Kar Kar Island, Madang District, Manus from Manus Island, Palakau from New Ireland, Mainoki from North Solomon Islands, and Popondetta from Northern District (4,12,16).

Pakistan

The incidence of G6PD deficiency among Pakistani was 2.3-8% (17,124-126). Three variants have been reported from this population group. G6PD Campbellpore was found in 3 men from Campbellpore (17). The Mediterranean variant was found in 2 Pakistani males living in USA (36). G6PD Birmingham was described in one Pakistani family living in USA (14).

The Philippines

The highest incidence of G6PD deficiency among the Filipinos was found to be 25% in the Panay and Negros Island (127). The incidence was around 10% in North Luzon and in Cebu (127,128). In Manila the incidence was 2.5-6.7% (128-130). In Mindanao and Leyte the incidence was 6.7% and 3.2% respectively (127). In South Luzon the incidence was low, being 1.5% (128).

G6PD Union was described in two families of Naguilian Town, La Union Province (39). G6PD Panay was found in a Philippino female native of Panay island (37). Only these two variants have been reported from this population group.

Saudi Arabia

The highest incidence of G6PD deficiency in Saudi Arabia was around 65% in children of Safwah and Al Ajam of Qatif Oasis. The incidence was 53.3% in children Mansura, Al Hasa Oasis (2). Adults of Qatif Oasis had the frequency of 34.2%, and 24% of children in Qarah of Al Hasa Oasis had G6PD deficiency (2). All data mentioned above were from Shia Muslim. The incidence in local population of Dhahran was 18.2% (50). In Al-Hafouf and Jaizan the incidence was 14.1% and 15.0% respectively (131). The 19.9% incidence was found in 607 Saudis attending the outpatient clinics in Hafouf and Damman, in the Eastern Province (132). Adult Shia and Sunni Muslims of Al Hasa Oasis had the incidence of 4.7%. In other Eastern Provinces the Sunni adults had the incidence of 2.9% (2). In Western Saudi Arabia, the highest and lowest incidence of 8.5% and 1.7% was found in the Mowallad and Harbi tribe respectively. Among the Ghamid, Sahafi, Zahran and Mograbi tribes the incidence was 5-7.5% (51).

There has been no report of detailed G6PD characterization in Saudi Arabians. The following assignment of variants is from electrophoretic studies only. Among males from oasis Shia and nonoasis Sunni in Dhahran 13% showed erythrocyte G6PD deficiency with normal electrophoretic mobility and was assumed to have G6PD Mediterranean, and 1.9% showed normal enzyme activity with fast mobility and was assumed to have the A variant (50). By the same inference, G6PD A was found in 1.7% of the Harbi, Sahafi and Zahran tribes, 1.6% of the Ghamid tribe, 8.5% of the Mowallad tribe, and 15.1% of the Mograbi tribe, all living in Western Saudi Arabia (51). The A variant was found in Al-Hafouf and Jaizan at the frequency of 15.8% and 5.7% respectively (131). G6PD deficiency with fast electrophoretic mobility, found in 0.8% of oasis Shia and nonoasis Sunni, Dhahran, was assumed to have G6PD A(-) (50). Persons with normal erythrocyte G6PD activity and slow electrophoretic mobility were assumed to have G6PD 'Khartoum', an incompletely characterized variant. The variant was found in 16.1% of the Ghamid, 3.3% of the Sahafi and the Zahran, and 2.1% of the Mowallad tribe, all of Western Saudi Arabia (51). Another variant with moderate enzyme activity and normal electrophoretic mobility, called B(-), was found in 11% of oasis Shia and nonoasis Sunni males in Dhahran (50).

Singapore

The three main ethnic groups of Singapore are Chinese, Malay, and Indian. Among the Chinese the incidence of G6PD deficiency was report to be around 3%. The reported incidence was 2-3.5% among the Malay and 0.4-4.7% among the Indian (62,65,133,134). No characterization study has been reported from the citizens of Singapore.

Sri Lanka (Ceylon)

The highest incidence of G6PD deficiency in Sri Lanka was 20.9% among Tihara Balankulana Singhalese. The incidence in other groups ranged from 1.2-7.1% (135,136). There has been no report of G6PD characterization among this population group.

Taiwan (Aborigines)

The Ami is the aborigines group of Taiwan with the highest incidence of G6PD deficiency of 3.5% (137). Other tribes showed low frequency of 0.3% (59,137). Four variants, G6PDs Hualien-Chi, Hualien, 'Taiwan-Ami 5', and 'Taiwan-Ami 6', have been reported but no data of variant frequency is available (27).

Thailand

The average incidence of G6PD deficiency among Thai people was 12% (63,138). The incidence in Northern, Eastern and Northeastern Thai was 13-16% (63,138-140). Among the Thai people in the central area of the country the incidence was 10-14% (63,138). It was reported that the incidence in Southern Thailand was low, being 2.8% (138). But the study in a larger population of the South showed the incidence of 9.3% (141). In the earlier study of G6PD deficiency in the South, blood samples were taken from people of one village of Nakhon-Srithummaraj

Province, the people might be Thai Muslim, a different ethnic group from the Thai. It has been shown that Thai Muslims of Southern Thailand have a low incidence of 2.8% (142). In Sichang Island, with no malaria for very long time, the incidence was 3.4% (143). No G6PD deficiency was found among the Miao, a tribal group of Northern Thailand (63).

Among other ethnic groups populating Thailand, the incidence was around 12% in Mon, Thai Yong, Thai Lue, and Karen, all of Northern part of the country. The highest incidence of 20% was found among Thai Ya. Kmer and Suay of Surin Province, Northeast had 16.7% incidence (63). Pootai, an ethnic group in Sakon Nakhon Province (Northeast), had the incidence of 9.7% (144). Low incidence of 1.9-3.8% was found in other tribal groups; Lahu, Lissu, Lawa, all of Northern Provinces, and So of Sakon Nakhon Province (63,144).

Seventeen G6PD variants have been reported from the Thai population. The most common one is G6PD Mahidol found in about 70% of G6PD-deficient Thai (24,44). G6PD Canton, the most common variant among the Chinese, was first reported in a Thai boy by McCurdy et al and was later found in 15 out of 137 unrelated Thai males with G6PD deficiency (19,24). Among these 137 unrelated G6PD-deficient Thais 15 had G6PD Taipei-Hakka (Dhon), 10 had G6PD Union, 3 had G6PD A(-)-like (N-Pathom), 3 had G6PD Hong Kong Pokfulam, 2 had G6PD Mediterranean-like (Songkhla), and G6PD Siriraj, Chainat, Kan, Anant, N-Sawan, and Padrew was each found in one person (24). G6PD Bangkok was the only variant associated with CNHD reported from Thailand (6,7). Among 1157 G6PD-normal male blood donors we found 4 persons with electrophoretic variants. Among these four people, G6PD Ayutthaya was identified in two and G6PD S-Sarkorn and Chao Phya was each found in one donor (52). It should be mentioned that G6PD Chao Phya had very similar properties to the A variant.

Vietnam

The Southern Vietnamese had a G6PD deficiency incidence of 1.4-4.1% (57,100,145). The incidence was 1.3% among the Vietnamese refugees in Thailand and 3.6% among the Vietnamese refugees arriving in USA (26,58). In other ethnic groups in Vietnam, the highest incidence of 15.3% was found in the Kmer (Cambodian), and the second highest was 9.1% among the Cham. The other groups with decreasing frequency were the Stieng, Rhade, Montagnard, and Sedang with the incidence of 5.4, 2.3, 1.6, and 0.4% respectively (57,145).

Three variants were reported from six unrelated G6PD-deficient Vietnamese refugees in Songkhla, Thailand. G6PDs Mahidol, Canton, and Long Xuyen were found in 3, 2 and 1 respectively (26).

Miscellaneous

In Fiji, the Melanesians had very low incidence of G6PD deficiency, being found in only one male out of 913 tested (146). Among the Polynesians of Cook Islands no G6PD-deficient case was found in 110 men (147). One G6PD-deficient boy was found among 540 New Zealand Maoris, a Polynesian group. No G6PD deficiency was found among 86 male Samoans, 7 male Tongans, 34 male Niue Islanders, and 10 male Tokelau Islanders (147). The Samoans, another group of Polynesian, in New Zealand were found to have a 2.3% incidence of G6PD deficiency (148). In the United Arab Emirates, 100 men of all tribes were tested, 10% were found to be G6PD-deficient (149).

No characterization study has been reported among these population groups.

COMMON G6PD VARIANTS IN SOUTHERN ASIAN POPULATIONS

There are 17 variants with reports from more than one countries; G6PDs Mahidol, Canton, Taipei-Hakka (Dhon), Hong Kong Pokfulam, Union, Panay, Mediterranean, A(-), ?A, Corinth, Tel-Hashomer, West Bengal, Castilla, Kilgore, Boston, Ramat-Gan and 'Khartoum'. Distribution of these variants are shown in Figure 1. The last 13 variants have also been reported from population groups outside Southern Asia.

G6PD Union was originally reported from Filipinos and was later found in Greece (39,150). The enzyme in Greeks was called U-M (Union-Markham) because NAD utilization was not tested to differentiate between these two very similar variants. Five per

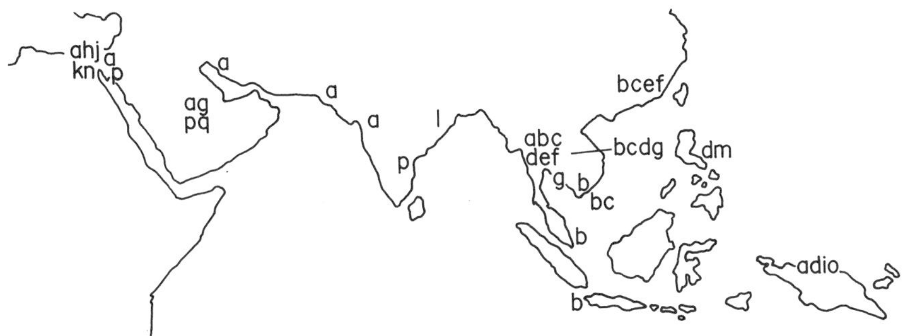

FIGURE 1. *Geographical distribution of G6PD variants reported from Southern Asia having two or more than two population origins. Symbol for each variant is a = Mediterranean, b = Mahidol, c = Canton, d = Union, e = Taipei-Hakka, f = Hong Kong Pokfulam, g = A(-), h = Tel-Hashomer, i = Castilla, j = Kilgore, k = Boston, l = West Bengal, m = Panay, n = Ramat-Gan, o = Corinth, p = ?A, q = 'Khartoum'.*

cent of Greeks with severe erythrocyte G6PD deficiency had this variant (151). It was later decribed in Thai, Laotian, Papua New Guinean, and Spanish (4,25,40,152).

G6PD Panay, earlier reported from a Filipino, was also found outside Asia in Bulgaria (37,153,154). G6PD Mediterranean, a very common variant in the Mediterranean area and the vicinity, is also found in many population groups of Southern Asia as shown in Figure 2 but in a much lower frequency except in some area of Papua New Guinea where it was found in 51.4% of men at the Butibam

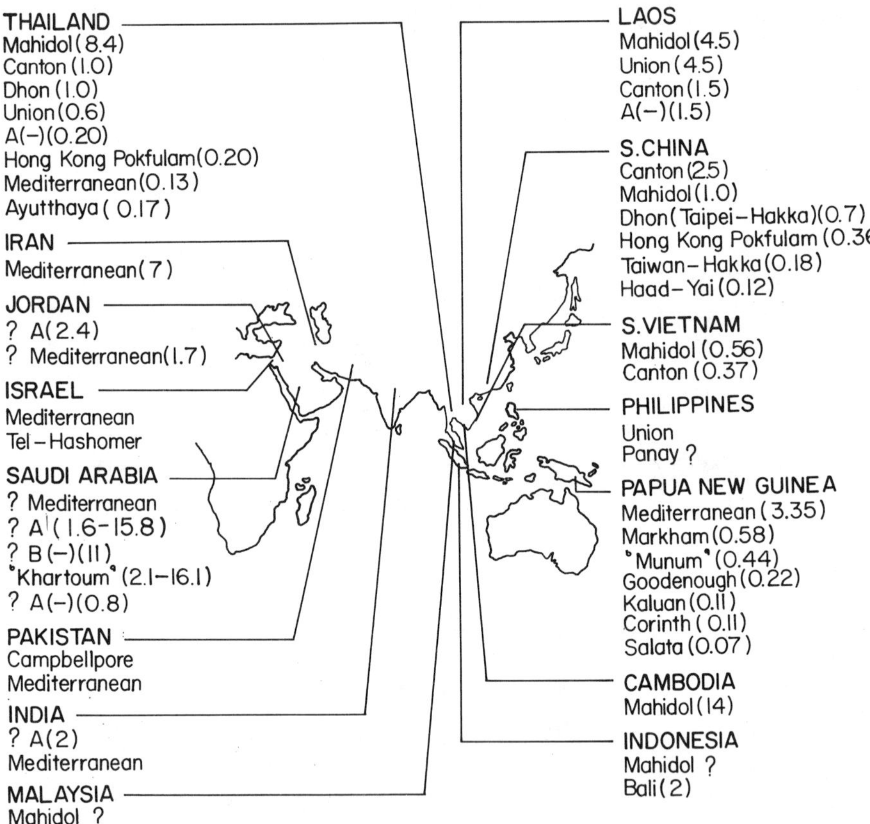

FIGURE 2. *G6PD variants found in 2 families or more in Southern Asia. Number in parenthesis indicates estimated frequency(%) in the population.*

village studied (3,4). It was also found in Japan (155). G6PDs A and A(-) are well known to be very common in African populations and the surrounding areas. It can be questioned whether the A and A(-) variants identified only by electrophoretic mobility reported in Saudi Arabia are really the same variants as those in African populations (50,51). The A(-) variant found in Thailand was compared to the African A(-) side by side in the same laboratory and no difference was found by the WHO characterization criteria (24,156). G6PD Corinth was reported from Greece, the Mediterranean area, Southeast Asia, and from Papua New Guinea (4,27). It has also been reported from Romania and Bulgaria (153,154,157). The Tel-Hashomer variant was reported originally in a Tunisian Jew and later in 3 Egyptians (46,47,158). G6PD Castilla was first reported in a Spaniard and was later found in Papua, New Guinea (12,159). G6PD Kilgore was oiriginally reported in a Hungarian Jew living in USA, and was found also in 4 Bulgarians (31,153,154). G6PD Boston was earlier found in a Polish Jew with CNHD, but later found in 3 Bulgarians without CNHD (8,153,154). G6PD Ramat-Gan, originally reported from an Iraqi Jew with mild CNHD, was later reported in 6 Egyptians without CNHD but one of whom had favism (5,158). G6PD 'Khartoum', an incompletely characterized variant with normal erythrocyte enzyme activity and slow electrophoretic mobility, has been reported from Sudan and Saudi Arabia (51,53,160). In Saudi Arabia, G6PD 'Khartoum' was very common among the Ghamid tribe(16.1%) and common among the Mowallad tribe(2.1%), the Zahran and Sahafi tribes(3.3%), but was not found in the Harbi and Mograbi tribes (51).

Among the 4 variants distributed in more than one countries in Southern Asia, G6PD Mahidol is the variant attaining the highest frequency, being 14% in Cambodia and 8.4% in Thailand. This variant was first described in Thailand and was independently described among the Southern Chinese under the name of B(-) Chinese (20,21,44). It has been additionally found in

Mon, an ethnic group in Southeast Asia, Cambodian, Laotian, Vietnamese, Chinese in Thailand, and in Thai Muslim, an ethnic group related to the Malay (22,23,25,26,45). It was also found in Japanese (161,162). G6PD Ogori and another variant (not named) described in Japanese also had very similar properties to G6PD Mahidol (155,163). The differences between G6PD Indonesia and G6PD Mahidol are so slight that one could consider them to be due to inter-laboratory variation and assume these two variants are the same. Considering the historical connection between Indochina and Indonesia, the assumption could be very likely. G6PD Canton, originally found in the Chinese, has been found in Thai, Laotian, Vietnamese and Chinese in Thailand (18,19,22-26). G6PD Taipei-Hakka, firstly identified in the Chinese, was independently reported in Thai under the name of G6PD Dhon and later found to be the same variant (19,35). It has been found in Chinese, Thai, and Chinese in Thailand (20,22-24,35). G6PD Hong Kong Pokfulam has been reported in Chinese and Thai (19-22,24). It was earlier called G6PD Panay-like because of its resemblance to G6PD Panay. It also bears very similar properties to G6PD West Bengal. At this stage, no one can be sure whether these 3 variants are actually the same or different ones.

In addition to the 17 variants found in more than one countries, there are additional 10 common or probably common (found in more than 1 families) variants each reported from only one country or in one specific area. They are G6PDs Campbellpore, Markham, 'Munum', Salata, Kaluan, Goodenough, Bali, Ayutthaya, Haad-Yai and Taiwan-Hakka.

There are 13 variants with normal erythrocyte enzyme activity in 3, showing polymorphism (gene frequency > 1%) in Southern Asia. The detailed distribution and frequency of these polymorphic variants are shown in Table III. It should be stressed that this is the minimum estimation due to limited information in many areas.

It is well known to investigators in the G6PD field that variant identification following the characterization methods recommended by WHO expert committee can lead to misassignment of different variants to cases actually containing identical enzyme even in good quality laboratories due to intra- and inter-laboratory variation (156). A good example is the eventual finding that G6PDs Chicago and Cornell are actually the same

Table III. Distribution and frequency of 13 polymorphic G6PD variants in Southern Asia.

Variant	Population groups with polymorphism
Mediterranean	Papua New Guinea Butibam Village 36.4%, 51.4%; Yanga Village 13.1%, Wagan village 7%, Kamkumung Village 23.9%, Yalu Village 9.1%; Iran 7%; Saudi Arabia Dhahran 13%; ? Israel; Jordan 1.7%; ? N.India; ? Pakistan
A(-)	Laos ? 1.5%; ? Saudi Arabia
Mahidol	Cambodia 14%; Thailand 8.4%; Laos 4.5%; S.China 1%
Canton	S.China 2.5%; Laos 1.5%; Thailand 1%
Dhon (Taipei-Hakka)	Thailand 1%
Union	Laos 4.5%; ? Philippines
Markham	Papua New Guinea Butibam Village 5.9%, 8.6%; Wagan Village 3.5%
'Munum'	Papua New Guinea Munum Village 22%, Yalu Village 7%
Corinth-like	Papua New Guinea Yanga Village 3.3%; Munum Village 1.4%
Taiwan-Hakka	Taiwan Hakka Chinese 2.7%
? A	Saudi Arabia Dhahran 1.9%, Mograbi tribe 15.1%, Mowallad tribe 8.5%, Harbi tribe 1.7%, Sahafi and Zahran tribes 1.7%, Ghamid tribe 1.6%, Al-Hafouf 15.8%, Jaizan 5.7%; Jordan 2.4%; India Jalari (Andhra Pradesh) 2%
'Khartoum'	Saudi Arabia Ghamid tribe 16.1%, Sahafi and Zahran tribes 3.3%, Mowallad tribe 2.1%
Bali	Indonesia East Balinese 2%

variant reported from related subjects (164). On the other hand, a reported variant could actually be several variants with similar properties. One example is the "Mediterranean group" in Papua New Guinea which has been shown to be comprised of G6PDs Mediterranean, Corinth-like, 'Munum', Goodenough, Swit, Kalapau, and others (3,4,12). These limitations should be kept in mind when considering G6PD variants mentioned in this article.

Among the 13 polymorphic G6PD variants in Southern Asia, G6PD Mediterranean is the one attaining the highest frequency in a specific area, being 51.4 and 36.4% in Butibam village, Papua New Guinea(3,4). G6PDs Mahidol and Canton are probably the two variants most widely distributed in the Far East.

CLINICAL PICTURES OF THE VARIANTS

In general, clinical manifestations of classes 2 and 3 G6PD variants are acute hemolysis associated with infection, certain kinds of drugs and chemicals, favism, and severe neonatal jaundice (165,166). But certain variants can be more susceptible to acute hemolysis induced by environmental agent(s) or produce a more severe clinical pictures as has been well documented for G6PD Mediterranean when compared to G6PD A(-). In comparing clinical pictures of G6PD variants in different populations it should be kept in mind that the difference could also be caused by other genetic and environmental factors.

In Southern Asia, acute hemolysis due to drugs and/or infection in G6PD deficiency is well documented (5,15,38,45,69, 103,109,112,167-184). In some cases the hemolytic episodes were very severe leading to acute renal failure (38,167-169,176-178). But in most of the clinical reports, no identification of G6PD variants was performed. Correlation of clinical pictures and G6PD variants could thus be done with the constraint of limited information. G6PD Mahidol, the most common variant in Indochina, when challenged with a daily dose of 45 mg primaquine the percentage of hemolysis was found to be 0-18% of circulating red cells. The hemolytic process was found to be self-limited (185). The Mahidol variant is in class 3 with erythrocyte enzyme activity just above 10% and other properties very similar to G6PD B except in vivo lability. This variant is thus simlar to G6PD A(-). Leukocyte and platelet enzyme activity were found to be slightly lower in G6PD Mahidol when compared to G6PD A(-) (21,186). As a group, patients with G6PD Mahidol having acute hemolytic crises showed less severe clinical pictures than those with G6PDs Union and Canton, variants in class 2. But if considered as an individual patient, clinical pictures in acute hemolysis due to G6PD Mahidol cannot be differentiated from those in G6PD Union and Canton (167,168). G6PDs Union and Canton, despite their lower erythrocyte G6PD activity, have more favorable enzyme kinetics with lower Km G6P and Km NADP. Acute hemolysis in G6PD Mahidol can be very severe leading to acute renal failure and even death (109,167,168). The largest study of drug effects upon survival of G6PD deficient red cells in Asia is reported by Chan et al who showed that red cells with G6PDs Mahidol, Canton, and Hong Kong Pokfulam did not differ in the hemolytic susceptibility to drugs (71). It should be mentioned that, when studying ^{51}Cr-labelled G6PD-deficient red cell survival in G6PD-normal recipient volunteers, there is a wide variation in lifespan shortening of G6PD-deficient red cells from the same individual challenged by the same dose of the same drug. We have observed in one instance when ^{51}Cr-labelled red cells with G6PD Mahidol from an individual were transfused into two G6PD-normal male volunteers. Primaquine was given at the dose of 45 mg/d to one, and 15 mg/d to another. It was found that there was more destruction of ^{51}Cr-labelled G6PD-deficient red cells in the recipient taking 15 mg/d primaquine. The two volunteers were healthy throughout the experiment and did not receive any agent that may cause hemolysis in G6PD deficiency (187). This indicates that severity of hemolysis in G6PD deficiency does not depend only on the type of the G6PD variant and dose of the hemolytic agent but also on other factors possibly absorption and metabolism of the agent.

Severe neonatal jaundice leading to kernicterus in some cases is a very important clinical effect of G6PD deficiency. Flatz et al concluded from a population study that kernicterus was the major negative balancing factor of G6PD deficiency, at least in Thailand (188). Significantly increased incidence of neonatal jaundice was found in G6PD-deficient newborns in Israel, Hong Kong, Taiwan, Thailand, Singapore, New Guinea, and Chinese in Hawaii (62,64,189-203). It is not known whether different G6PD variants would cause different incidence and/or severity of neonatal jaundice. In Thailand, with G6PD Mahidol as a predomination variant, it was estimated that 5% of G6PD-deficient and 1% of G6PD-normal newborns had severe neonatal jaundice that might lead to kernicterus (188,194). In Singapore, 22% of G6PD-deficient and 10% of normal Chinese full-term newborns in a controlled hospital enviroment had a bilirubin higher than 15 mg/dl (200). These G6PD-deficient Chinese newborns should have G6PD Canton as the predominating variant. Significantly higher reticulocyte counts were found in cord blood of G6PD-deficient Chinese and Malay newborns, but in G6PD-deficient Thai newborns the reticulocyte count was normal (194,198). There are at least 2 predisposing factors for severe hyperbilirubinemia of the newborn in G6PD-deficient infants, the increased hemolysis and defective bilirubin conjugation mechanism. It has been shown in Italian G6PD-deficient neonates that salicylamide glucuronide formation is lower, indicating defective hepatic conjugation (204). In Singapore, serial determination of serum bilirubin in G6PD-normal full-term newborns resulted in much different peak bilirubin levels between oriental neonates and neonates of British parents. The Chinese, Malay and Indian newborns showed peak serum bilirubin of 10-11 mg/dl, while that of the British newborns was around 4 mg/dl. The former groups of newborns altained peak serum bilirubin on the 5th- 6th day of life, while the British did on the third day (205). These babies were kept during the study period under the same hospital environment. The result of this study indicates that there should be some genetic factor(s) determining the difference of neonatal bilirubin levels of these racial groups. The discrepancy of incidence of severe neonatal jaundice between American and African Negro newborns with G6PD deficiency is well documented (206). It was rare in American Negro but common in African Negro although most of these babies should have the same G6PD variant, the A(-). The difference could be due to enviromental factor(s) and/or protective gene(s) from American whites which have been mixed into the American black population. A different incidence of neonatal jaundice in populations of the same ethnic group with probably the same G6PD variant was also reported from Greeks of Lesbos Island and Rhodes Island (207). Concerning G6PD variants and neonatal jaundice, it could be concluded that, although class 1 variants are usually associated with neonatal jaundice, differences in G6PD variants of classes 2 and 3 play a minor role in the variability of severity and incidence of neonatal jaundice. Other genetic factor(s) and environmental factor(s) play at least an equal role.

In Southern Asia, favism has been reported in the Middle East and in the Chinese (112,181,182,208,209). The variants found in Southern Asia reported to be associated with favism are G6PDs Corinth, Kilgore, Mediterranean, Ramat-Gan, and Union (152-154, 158). Cases of favism associated with the above mentioned variants were, however, found in population groups outside Asia. Since favism was reported to be common among the Chinese, it could be speculated that G6PD Canton might be another variant associated with favism. There has been no evidence that favism is variant-specific and it is well-known that, another genetic factor is interacting with G6PD deficiency in causing favism (208).

In Sardinia, it has been reported that G6PD deficiency is associated with pre-senile cataracts (210,211). But, in Thailand the association was not found (212). It is not known whether the difference in G6PD variant is responsible for the difference or not. In Sardinia there are 3 common variants with very low erythrocyte enzyme activity (class 2), G6PDs Mediterranean, Sassari, and Cagliari, and one class 3 variant, G6PD Seattle-like (213,214). In Thailand it has been mentioned earlier that 70% of G6PD deficiency are associated with G6PD Mahidol which is in class 3. Some variants in class 1 were reported to be associated with cataracts (215,216), suggesting that very low enzyme activity in red cells and lens might be a predisposing factor of cataracts. G6PD Camperdown with 15% red cell enzyme activity (class 3) was found in an Australian boy with Maltese maternal grandfather. The boy had bilateral lamellar cataracts since he was 4 years old (217).

Two cases of Gilbert's syndrome were reported in association with G6PD Taipei-Hakka (Dhon), a class 3 variant (218).

It is well documented that very low or complete G6PD deficiency in white cells is associated with impaired bactericidal activity (219-223). It was shown that G6PD deficiency in general might predispose to certain infections (224,225).

CONCLUSION

There are 78 reported G6PD variants in Southern Asia. Eight variants are associated with CNHD. Sixty-six variants are in classes 2 and 3. Of the 78 variants, 30 are considered common ones, being found in more than one families. At least 13 variants are polymorphic in Southern asian populations, 3 of these variants have normal red cell enzyme activity. G6PD Mediterranean is the variant having the highest frequency, being

36.4-51.4% in Butibam Village, Papua New Guinea. This variant is widely distributed in Southern Asia, possibly with high frequency in the Middle East and northern part of the India subcontinent, and found sporadically in the Far East. Other very common variants in the Middle East are G6PD 'Khartoum' and probably G6PDs A and A(-). The most common variant among the southern Chinese is G6PD Canton which is also common in Indochina. G6PD Mahidol is the most common variant in Indochina. It is also common in southern Chinese and probably is common in Indonesia. G6PD Union is as widely distributed as G6PD Mediterranean, being found in the Far East as well as in the Mediterranean area. Three other very commn variants in Papua New Guinea are G6PDs Markham, 'Munum', and Corinth-like. The only polymorphic variant with normal red cell enzyme activity in the Far East is G6PD Bali found in East Balinese.

G6PD-deficient variants are associated with acute hemolysis induced by drugs, chemicals, infections, and with severe neonatal jaundice. Clinical pictures of acute hemolysis are variable, the severity of which shows little difference between common variants. Incidence of severe neonatal jaundice associated with G6PD deficiency is also variable among different population groups. There is no clear evidence that difference in G6PD variants is the important cause of this variability. Contribution of other genetic factor(s) and enviromental factor(s) to the severity of neonatal jaundice is evident.

REFERENCE

1. Sheba C. Glucose-6-phosphate dehydrogenase (G6PD) deficiency as a possible marker for studying ethnic origin and migration. East Afr Med J 1962;39:261-263. Cited in Livingstone FB. Abnormal Hemoglobins in Human Populations. Aldine Publishing Company, Chicago, 1967;225.
2. Gelpi AP. Glucose-6-phosphate dehydrogenase deficiency in Saudi Arabia : a survey. Blood 1965;25:486-493.

3. Kirkman HN, Kidson C, Kennedy M. Variants of human glucose-6-phosphate dehydrogenase. Studies of samples from New Guinea. In : Beutler E (Ed.) Hereditary Disorders of Erythrocyte Metabolism. Grune & Stratton, New York, 1968;126-145.
4. Yoshida A, Giblett ER, Malcolm RA. Heterogeneous distribution of G6PD variants in the Markham Valley of New Guinea. Ann Hum Genet 1973;37:145-150.
5. Ramot B, Ben-Bassat I, Shchory M. New glucose-6-phosphate dehydrogenase variants observed in Israel and their association with congenital nonspherocytic hemolytic disease. J Lab Clin Med 1969;74:895-901.
6. Talalak P, Beutler E. G6PD Bangkok : A new variant found in congenital nonspherocytic hemolytic disease (CNHD). Blood 1969;33:772-776.
7. Talalak P. G6PD Bangkok : Clinical follow up and family study. J Med Assoc Thailand 1983;66:429-434.
8. Necheles TF, Snyder LM, Strauss W. Glucose-6-phosphate dehydrogenase Boston. A new variant associated with congenital nonspherocytic hemolytic disease. Humangenetik 1971;13:218-221.
9. Wong PWK, Shih L-Y, Hsia DY-Y, Tsao YC. Characterization of glucose-6-phosphate dehydrogenase among Chinese. Nature 1965;208:1323-1324.
10. Yoshida A, Beutler E. G6PD variants : another up-date Ann Hum Genet 1983;47:25-38.
11. Beutler E, Matsumoto F, Daiber A. Nonspherocytic hemolytic anemia due to G6PD Panama. IRCS 1974;2:1389.
12. Chockkalingam K, Board PG, Nurse GT. Glucose-6-phosphate dehydrogenase in Papua New Guinea. The description of 13 new variants. Hum Genet 1982;60:189-192.
13. Geerdink RA, Horst R, Staal GEJ. An Iraqi Jewish family with a new red blood cell glucose-6-phosphate dehydrogenase variant (Gd Bagdad) and kernicterus. Israel J Med Sci 1973;9:1040-1043.
14. Prchal JT, Crist WM, Malluah A, Vitek A, Tauxe WN, Carroll AJ. A new glucose-6-phosphate dehydrogenase deficient variant in a patient with Chediak-Higashi syndrome. Blood 1980;56:476-480.
15. Sidi Y, Aderka D, Brok-Simoni F, Benjamin D, Ramot B, Pinkhas J. Viral hepatitis with extreme hyperbilirubinemia, massive hemolysis and encephalopathy in a patient with a new G6PD variant. Isr J Med Sci 1980;16:130-133.
16. Chockkalingam K, Board PG. Further evidence for heterogeneity of glucose-6-phosphate dehydrogenase deficiency in Papua New Guinea. Hum Genet 1980;56:209-212.
17. McCurdy PR, Mahmood L. Red cell glucose-6-phosphate dehydrogenase deficiency in Pakistan. J Lab Clin Med 1970;76:943-948.

18. McCurdy PR, Kirkman HN, Naiman JL, Jim RTS, Pickard BM. A Chinese variant of glucose-6-phosphate dehydrogenase. J Lab Clin Med 1966;67:374-385.
19. McCurdy PR, Blackwell RQ, Todd D, Tso SC, Tuchinda S. Further studies on glucose-6-phosphate dehydrogenase deficiency in Chinese subjects. J Lab Clin Med 1970;75:788-797.
20. Chan TK, Todd D. Characteristics and distribution of glucose-6-phosphate dehydrogenase-deficient variants in South China. Am J Hum Genet 1972;24:475-484.
21. Chan TK, Todd D, Lai MCS. Glucose-6-phosphate dehydrogenase : identity of erythrocyte and leukocyte enzyme with report of a new variant in Chinese. Biochem Genet 1972;6:119-124.
22. Panich V, Na-Nakorn S, Wasi P. G6PD variants in Chinese in Thailand. Southeast Asian J Trop Med Pub Hlth 1980; 11:250-255.
23. Panich V, Sungnate T. Characterization of glucose-6-phosphate dehydrogenase variants in Thailand. The occurrence of 6 variants among 50 G6PD deficient Thai. Humangenetik 1973;18:39-46.
24. Panich V, Na-Nakorn S. G6PD variants in Thailand. J Med Assoc Thailand 1980;63:537-543.
25. Panich V. G6PD variants in Laotians. Hum Hered 1974; 24:285-290.
26. Panich V, Bumrungtrakul P, Jitjai C, Kamolmatayakul S, Khoprasert B, Klaisuwan C, Kongmuang U, Maneechai P, Pornpatkul M, Ruengrairatanaroje P, Surapruek P, Viriyayudhakorn S. Glucose-6-phosphate dehydrogenase deficiency in South Vietnamese. Hum Hered 1981;30:361-364.
27. Yoshida A, Beutler E, Motulsky AG. Human glucose-6-phosphate dehydrogenase variants. Bull Wld Hlth Org 1971;45:243-253.
28. Kirkman HN, Lie Injo LE. Variants of glucose-6-phosphate dehydrogenase in Indonesia. Nature 1969;221:959-960.
29. Panich V. G6PD Intanon. A new glucose-6-phosphate dehydrogenase variant. Humangenetik 1974;21:203-205.
30. Beutler E. Glucose-6-phosphate dehydrogenase deficiency. A new Indian variant G6PD Jammu. In : Sen NN, Basu KK (Eds.). Trends in Haematology. Calcutta India 1975;279-283.
31. Alperin JB, Mills GC. New variants of glucose-6-phosphate dehydrogenase (G6PD). Clin Res 1972;20:76.
32. Shatskaya TL, Krasnopolskaya KD, Annenkov GA. A description of new mutant forms of erythrocyte glucose-6-phosphate dehydrogenase isolated at territory of the Soviet Union. Genetika 1975;11(12):116-122.
33. Shatskaya TL, Krasnopolskaya KD, Idelson LJ. Mutant forms of erythrocyte glucose-6-phosphate dehydrogenase in Ashkenazai. Description of two new variants : G6PD

Kirovograd and G6PD Zhitomir. Humangenetik 1976;33: 175-178.
34. Yoshida A. Hemolytic anemia and G6PD deficiency. Science 1973;179:532-537.
35. Panich V. G6PD characterization in Thailand. Abstract. Thirteenth International Congress of Genetics, Berkeley, California. Genetics 1973;74(suppl):208.
36. Azevedo E, Kirkman HN, Morrow AC, Motulsky AG. Variants of red cell glucose-6-phosphate dehydrogenase among Asiatic Indians. Ann Hum Genet 1968;31:373-379.
37. Fernandez MN, Fairbanks VF. Glucose-6-phosphate dehydrogenase deficiency in the Philippines : Report of a new variant : G6PD Panay. Mayo Clin Proc 1968;43: 645-660.
38. Panich V, Sungnate T, Na-Nakorn S. Acute intravascular hemolysis and renal failure in a new glucose-6-phosphate dehydrogenase variant : G6PD Siriraj. J Med Assoc Thailand 1972;55:726-731.
39. Yoshida A, Baur EW, Motulsky AG. A Philippino glucose-6-phosphate dehydrogenase variant (G6PD Union) with enzyme deficiency and altered substrate specificity. Blood 1970;35:506-513.
40. Panich V. The occurrence of G6PD Union in Thailand. Humangenetik 1973;17:169-171.
41. Viallard JL, Cottreau D, Kahn A, Dastugue B. G6PD deficiency with Gd(-)A like variant in a Chinese family from Cambodia. Hum Genet 1979;51:213-215.
42. Ishwad CS, Naik SN. A new glucose-6-phosphate dehydrogenase variant (G6PD Kalyan) found in a Koli family. Hum Genet 1984;66:171-175.
43. Chuanshu D, Yankang X, Lin W, Xiaoyum H. Studies on erythrocyte glucose-6-phosphate dehydrogenase variants in Chinese I. Gd(-) Lizu-Baisha. Acta Acad Medi Zhong Shan 1981;2:649-658. Cited in Yoshida A, Beutler E. G6PD variants : another up-date. Ann Hum Genet 1983; 47:25-38.
44. Panich V, Sungnate T, Wasi P, Na-Nakorn S. G6PD Mahidol : The most common glucose-6-phosphate dehydrogenase variant in Thailand. J Med Assoc Thailand 1972;55:576-585.
45. Everett WD, Yoshida A, Pearlman E. Hemoglobin E and glucose-6-phosphate dehydrogenase deficiency in the Kmer Air Force (Cambodia). Am J Trop Med Hyg 1977;26:679-601.
46. Ramot B, Brok F. A new glucose-6-phosphate dehydrogenase mutant (Tel-Hashomer mutant). Ann Hum Genet 1964;28: 167-172.
47. Kirkman HN, Ramot B, Lee JT. Altered aggregational properties in a genetic variant of human glucose-6-phosphate dehydrogenase. Biochem Genet 1969;3:137-150.

48. Chockkalingam K, Board PG, Breguet G. Glucose-6-phosphate dehydrogenase variants of Bali Island (Indonesia). Hum Genet 1982;60:60-62.
49. Kahn A, North ML, Cottreau D, Giron G, Lang JM, Oberling F. G6PD Vientiane : A new glucose-6-phosphate dehydrogenase variant with increased stability. Hum Genet 1978;43:85-89.
50. Gelpi AP, King MC. New data on glucose-6-phosphate dehydrogenase deficiency in Saudi Arabia. Hum Hered 1977;27:285-291.
51. Bayoumi RA, Omer A, Samuel APW, Saha N, Sebai ZA, Sabaa HMA. Haemoglobin and erythrocytic glucose-6-phosphate dehydrogenase variants among selected tribes in Western Saudi Arabia. Trop Geogr Med 1979;31:245-252.
52. Panich V. Glucose-6-phosphate dehydrogenase in Thailand. The occurrence of three electrophoretic variants among 1157 nondeficient males. Hum Genet 1980;53:227-228.
53. Saha N, Samuel APW, Omer A, Ahmed MA, Hussein AA, Gaddoura EN. A study of some genetic characteristics of the population of the Sudan. Ann Hum Biol 1978;5:569-575.
54. Cayanis E, Lane AB, Jenkins T, Nurse GT, Balinsky D. Glucose-6-phosphate dehydrogenase Porbandar : A new slow variant with slightly reduced activity in a South African family of Indian descent. Biochem Genet 1977;15: 765-773.
55. Budtz-Olsen O. Absence of red cell enzyme deficiency in Australian Aborigines. Nature 1961;192:765.
56. Papiha SS, Roberts DF, Ali SGM, Islam MM. Some hereditary blood factors of the Bengali Muslim of Bangladesh (red cell enzymes, haemoglobins, and serum proteins). Humangenetik 1975;28:285-293.
57. Bowman JE, Carson PE, Frischer H, Powell PD, Colwell EJ, Legters LJ, Cottingham AJ, Boone SC, Hiser WW. Hemoglobin and red cell enzyme variation in some populations in the Republic of Vietnam with comments on the malaria hypothesis. Am J Phys Anthropol 1971;34: 313-324.
58. Schwartz IK, Chin W, Newman J, Roberts JM. Glucose-6-phosphate dehydrogenase deficiency in Southeast Asian refugees entering the United States. Am J Trop Med Hyg 1984;33:182-184.
59. Lee T-C, Shin L-Y, Huang P-C, Lin C-C, Blackwell B-N, Blackwell RQ, Hsia DY-Y. Glucose-6-phosphate dehydrogenase deficiency in Taiwan. Am J Hum Genet 1963;15:126-132.
60. Chan TK, Todd D, Wong CC. Erythrocyte glucose-6-phosphate dehydrogenase deficiency in Chinese. Br Med J 1964;2:102.
61. Motulsky AG, Lee T-C, Frazer GR. Glucose-6-phosphate dehydrogenase (G6PD) deficiency, thalassaemia, and

abnormal haemoglobins in Taiwan. J Med Genet 1965;2:18-20.
62. Smith GD, Vella F. Erythrocyte enzyme deficiency in unexplained kernicterus. Lancet 1960;i:1133-1134.
63. Flatz G, Sringam S. Glucose-6-phosphate dehydrogenase deficiency in in different ethnic groups in Thailand. Ann Hum Genet 1964;27:315-318.
64. Lu T-C, Wei H, Blackwell RQ. Increased incedence of severe hyperbilirubinemia among newborn Chinese infants with G6PD deficiency. Pediatrics 1966;37:994-999.
65. Saha N, Banerjee B. Erythrocyte G6PD deficiency among Chinese and Malays of Singapore. Trop Geogr Med 1971; 23:141-144.
66. Yue PCK, Strickland M. Glucose-6-phosphate dehydrogenase deficiency and neonatal jaundice in Chinese male infants in Hong Kong. Lancet 1965;1:350-351.
67. Fong T. Prevalence of erythrocyte G6PD deficiency in Sabah. Mod Med Asia 1977;13:14-16.
68. Panich V. Glucose-6-phosphate dehydrogenase deficiency : Tropical Asia. Clin Haematol 1981;10:800-814.
69. Panich V. Glucose-6-phosphate dehydrogenase deficiency : Genetic heterogeneity in Asia. In : Weatherall DJ, Fiorelli G, Gorini S (Eds.). Advances in Red Blood Cell Biology. Raven Press, New York, 1982;329-338.
70. Beutler E, Yoshida A. Human glucose-6-phosphate dehydrogenase variants : a supplementary tabulation. Ann Hum Genet 1973;37:151-155.
71. Chan TK, Todd D, Tso SC. Drug-induced haemolysis in glucose-6-phosphate dehydrogenase deficiency. Br Med J 1976;2:1227-1229.
72. Baxi AJ, Balakrishnan V, Undevia JV, Sanghvi LD. Glucose-6-phosphate dehydrogenase deficiency in the Parsees community, Bombay. Indian J Med Sci 1963; 17:493-500.
73. Kate SL, Mukherjee BN, Malhotra KC, Phadke MA, Mutalik GS, Sainani GS. Red cell glucose-6-phosphate dehydrogenase deficiency and haemoglobin variants among ten endogamous groups of Maharashtra and West Bengal. Hum Gent 1978;44:339-343.
74. Meera Khan P. Glucose-6-phosphate dehydrogenase deficiency in an Indian rural area. J Genet 1964;59:14-18.
75. Rao PR, Goud JD. Sickle cell haemoglobin and glucose-6-phosphate dehydrogenase deficiency in tribal populations of Andhra Pradesh. Indian J Med Res 1979;70:807-813.
76. Saha N, Kirk RL, Shanbhag S, Joshi SR, Bhatia HM. Population genetic studies in Kerala and the Nilgiris (South West India). Hum Hered 1976;26:175-197.
77. Khanduja PC, Agarwal KN, Julka S, Bhargava SK, Taneja PN. Incidence of glucose-6-phosphate dehydrogenase deficiency and some observations on patients with haemoglobinuria. Indian J Paediatr 1966;33:341-345.

78. Gupta S, Ghai OP, Chandra RK. Glucose-6-phosphate dehydrogenase deficiency in the newborn and its relation to serum bilirubin. Indian J Paediatr 1970;37:169-176.
79. Bhatia R, Gupta M, Ichhpujani RL, Khanna KK, Chowdhuri ANR. Glucose-6-phosphate dehydrogenase deficiency---relionship with caste and ABO blood groups. J Com Dis 1984;16:240-243.
80. Flatz G, Chakravartti MR, Das BM, Delbruk, H. Genetic survey in the population of Assam. I. ABO blood groups, glucose-6-phosphate dehydrogenase and hemoglobin type. Hum Hered 1972;22:323-330.
81. Ghosh U, Banerjee T, Banerjee PK, Saha N. Distribution of haemoglobin and glucose-6-phosphate dehydrogenase phenotypes among different caste groups of Bengal. Hum Hered 1981;31:199-121.
82. Papiha SS, Roberts DF, Mukerjee DP, Singh SD, Malhotra M. A genetic survey in the Bhil Tribe of Madhya Pradesh, Central India. Am J Phys Anthropol 1978; 49:179-185.
83. Misra RC, Misra D. Erythrocyte glucose-6-phosphate dehydrogenase deficiency among non-tribal population of Orissa. Trop Geogr Med 1977;29:233-236.
84. Veerraju P, Busi BR, Anand NE. Red cell glucose-6-phosphate dehydrogenase deficiency among Andhas. Am J Phys Anthropol 1978;49:143-144.
85. Baxi AJ, Balakrishnan V, Sanghvi LD. Deficiency of glucose-6-phosphate dehydrogenase --- observations on a sample from Bombay. Curr Sci 1961;30:16-17.
86. DaCosta H, Pattani J, Mehandle K, Desai M, Merchant SM. Glucose-6-phosphate dehydrogenase defect in Indian children. A preliminary report. Indian Paediatr 1967; 4:163-168.
87. Hakim SMA, Baxi AJ, Balakrishnan V, Kulkarni KV, Rao SS, Jhala HJ. Glucose-6-phosphate dehydrogenase deficiency and colour-vision studies in Indian Muslims. Humangenetik 1972;15:90-92.
88. Sethuraman M, Ramana Rao KV. A survey of glucose-6-phosphate dehydrogenase deficiency and sickle cell trait on a local population of Tirupati. Indian J Exp Biol 1978;16:1098-1099.
89. Chatterjea JB. Haemoglobinopathies, glucose-6-phosphate dehydrogenase deficiency and allied problems in the Indian subcontinent. Bull Wld Hlth Org 1966;35:837-856.
90. Saha N, Banerjee B. Incidence of erythrocyte glucose-6-phosphate dehydrogenase deficiency among different ethnic groups of India. Hum Hered 1971;21:78-82.
91. Saha N, Kirk RL, Shanbhag S, Joshi SH, Bhatia HM. Genetic studies among the Kadar of Kerala. Hum Hered 1974;24:198-218.

92. Naidu JM, Mohrenweiser HW, Neel JV. A sero-biochemical genetic study of Jalari and Brahmin caste populations of Andra Pradesh, India. Hum Hered 1985;35:148-156.
93. Breguet G, Ney R, Kirk RL, Blake NM. Genetic servey of an isolated community in Bali, Indonesia. II. Hemoglobin types and red cell isozymes. Hum Hered 1982;32:308-317.
94. Lie Injo LE, Poey-Oey HG. Glucose-6-phosphate dehydrogenase deficiency in Indonesia. Nature 1964; 204:88-89.
95. Suradi R, Monitja HE, Munthe BG, Suparno. Glucose-6-phosphate dehydrogenase deficiency in the Dr. Cipto Mangunkusumo General Hospital. Paediatr Indonesiana 1979;19:30-40.
96. Van Rood JJ, Van de Beek J, Gevers-Potgieser HH, Eernisse JG, Veeger W, Nieweg HO. Enzymedficientie bijtwee patienten met nietsferocytaire hemolytische anemie. Nederlandsch Tijdschrift voor Geneeskunde 1962;106:1217-1220. Cited in Livingstone FB. Abnormal Hemoglobins in Human Populations. Aldine Publishing Company, Chicago 1967;180.
97. Oort M. Red cell glucose-6-phosphate dehydrogenase (G6PD) deficiency. Unpublished thesis, Central Laboratory of the Blood Transfusion Service of the Dutch Red Cross, 1964. Cited in Livingstone FB. Abnormal Hemoglobins in Human Populations. Aldine Publishing Company, Chicago, 1967;180.
98. Walker DG, Bowman JE. In vitro effect of Vicia faba extracts upon reduced glutathione of erythrocytes. Proc Soc Exp Biol Med 1960;103:476-477.
99. Walker DG, Bowman JE. Glutathione stability of the erythrocytes in Iranians. Nature 1959;184:1325.
100. Frischer H, Bowman JE, Carson PE, Rieckman KH, Willerson D Jr, Colwell EJ. Erythrocyte glutathione reductase, glucose-6-phosphate dehydrogenase, and 6-phosphogluconic dehydrogenase deficiencies in populations of the United States, South Vietnam, Iran, and Ethiopia. J Lab Clin Med 1973;81:603-612.
101. Bowman JE, Walker DG. Virtual absence of glutathione instability of the erythrocytes among Armenians in Iran. Nature 1961;191:221-222.
102. Daneshbod G. Erythrocyte glucose-6-phosphate dehydrogenase in Tehran. Acta Haematol 1975;53:152-157.
103. Amin-Zaki L, El-din ST, Kubba K. Glucose-6-phosphate dehydrogenase deficiency among ethnic groups in Iraq. Bull Wld Hlth Org 1972;47:1-5.
104. Hamamy HA, Saeed TKh. Glucose-6-phosphate dehydrogenase deficiency in Iraq. Hum Genet 1981;58:434-435.
105. Zaidman JL, Leiba H, Scharf S, Steinman I. Red cell glucose-6-phosphate dehydrogenase deficiency in ethnic groups in Israel. Clin Genet 1976;9:131-133.

106. Cohen T, Simhai B, Steinberg AG, Levene C. Genetic polymorphisms among Iranian Jews in Israel. Am J Med Genet 1981;8:181-190.
107. Banerjee B, Saha N, Daoud ZF Khalaf FH, Qudah H. A genetic study of the Jordanians. Hum Hered 1981; 31:65-69.
108. Shaker Y, Onsi A, Aziz R. The frequency of glucose-6-phosphate dehydrogenase deficiency in the newborns and adults in Kuwait. Am J Hum Genet 1966;18:609-613.
109. Sicard D, Kaplan JC, Labie D. Haemoglobinopathies and G6PD deficiency in Laos. Lancet 1978;ii:571-572.
110. Lie Injo LE, Ti TS. Glucose-6-phosphate dehydrogenase deficiency in Malayans. Trans Roy Soc Trop Med Hyg 1964;58:500-502.
111. Lie Injo LE, Chin J, Ti TS. Glucose-6-phosphate dehydrogenase deficiency in Brunei, Sabah, and Sarawak. Ann Hum Genet 1964;28:173-176.
112. Vella F. Favism in Asia. Med J Aust 1959;2:196-197.
113. Lie Injo LE, Chin J. Abnormal haemoglobin and glucose-6-phosphate dehydrogenase deficiency in Malayan Aborigines. Nature 1964;204:291-292.
114. Kidson C, Gajdusek DC. Glucose-6-phosphate dehydrogenase deficiency in Micronesian peoples. Aust J Sci 1962; 25:61-62.
115. Plato CC, Cruz MT, Kurland LT. Frequency of glucose-6-phosphate dehydrogenase deficiency, red-green color blindness and Xg^a blood group among Chamorros. Nature 1964;202:728.
116. Gorman JG, Kidson C. Distribution pattern of an inherited trait, red cell enzyme deficiency in New Guinea and New Britain. Am J Phys Anthropol 1962; 20:347-356.
117. Giles E, Curtain CC, Baumgarten A. 1966. Cited in Livingstone FB. Abnormal Hemoglobins in Human Populations. Aldine Publishing Company, Chicago, 1967; 170-171.
118. Mourant AE, Tills D, Kopec AC, Warlow A, Teesdale P, Booth PB, Hornabrook RW. Red cell antigen, serum protein and red cell enzyme polymorphisms in Eastern Highlanders of New Guinea. Hum Hered 1982;32:374-384.
119. Mourant AF, Tills D, Kopec AC, Warlow A, Teesdale P, Booth PB, Hornabrook RW. Red cell antigen, serum protein, and red cell enzyme polymorphisms in inhabitants of the Jimmi Valley, Western Highlands, New Guinea. Hum Genet 1981;59:77-80.
120. Malcolm LA, Woodfield DG, Blake NM,Kirk RL, McDermid EM. The distribution of blood, serum protein and enzyme groups on Manus Island (Admiralty Islands, New Guinea). Hum Hered 1972;22:305-322.

121. Prins HK, Loos JA, Meuwissen JHETh. Glucose-6-phosphate dehydrogenase deficiency in West New Guinea. Trop Geogr Med 1963;15:361-370.
122. Kidson C, Gajdusek DC. Congenital defects of the central nervous system associated with hyperendemic goiter in a neolithic highland society of Netherlands New Guinea. II. Glucose-6-phosphate dehydrogenase in the Mulia population. Pediatrics 1962;29:364-368.
123. Young GP, Smith MB, Woodfield DG. Glucose-6-phosphate dehydrogenase deficiency in Papua New Guninea using a simple methylene blue reduction test. Med J Aust 1974; 1:876-878.
124. Ronald AR, Underwood BA, Woodward TE. Glucose-6-phosphate dehydrogenase deficiency in Pakistani males. Trans Roy Soc Trop Med Hyg 1968;62:531-533.
125. Stern MA, Kynoch PA, Lehmann H. Beta-thalassaemia, glucose-6-phosphate dehydrogenase deficiency, and haemoglobin D-Punjab in Pathans. Lancet 1968;i:1284-1285.
126. Hashmi JA, Farzana F, Ahmed M. Abnormal hemoglobins, thalassemia trait & G6PD deficiency in young Pakistani males. JPMA 1976;26:2-4.
127. Blackwell RQ, Paraan AA, Huang JT-H, Yen L, Chien L-C. Incidence of G6PD deficiency and hemoglobin H among Filipinos. Vox Sang 1968;15:65-69.
128. Motulsky AG, Stransky E, Fraser GR. Glucose-6-phosphate dehydrogenase (G6PD) deficiency, thalassaemia, and abnormal haemoglobins in the Philippines. J Med Genet 1964;1:102-106.
129. Chan AC, Jongco A, Mendoza RR. G6PD deficiency as a cause of jaundice in the newborn period (preliminary report). Phili J Pediat 1965;14:405-410. Cited in Livingstone FB. Abnormal Hemoglobins in Human Populations. Aldine Publishing Company, Chicago, 1967;178.
130. Bayani-Sioson PS. Some biochemical polymorphic traits (Filipino population) 2. Glucose-6-phosphate dehydrogenase deficiency and electrophoretic variants. Acta Med Philipina 1969;5:156-163.
131. El-Hazmi MAF, Warsy AS. Aspects of sickle cell gene in Saudi Arabia --- interaction with glucose-6-phosphate dehydrogenase deficiency. Hum Genet 1984;68:320-323.
132. Warsy AS. Frequency of glucose-6-phosphate dehydrogenase deficiency in sickle-cell disease. Hum Hered 1985;35: 143-147.
133. Wong HB. Genetic diagnosis in obstetrics/gynaecology. J Singapore Paediatr Soc 1984;26:15-22.
134. Saha N, Toh CCS, Ghosh MB. Genetic association in myocardial infarction. Ethnicity; ABO, Rh, Le^a, Xg^a blood groups; G6PD deficiency; and abnormal haemoglobins. J Med Genet 1973;10:340-345.

135. Abeyaratne KP, Premawansa S, Rajapakse L, Roberts DF, Papiha SS. A survey of glucose-6-phosphate dehydrogenase deficiency in the north central province of Śri Lanka (formerly Ceylon). Am J Phys Anthropol 1976;44:135-138.
136. Nagaratnam N, Leelawathie PK, Weerasinghe WMT. Enzyme glucose-6-phosphate dehydrogenase (G6PD) deficiency among Singhalese in Ceylon as revealed by the methemoglobin reduction test. Indian J Med Res 1969; 57:569-572.
137. Blackwell RQ, Blackwell B-N, Yen L, Lee H-F. Low incidence of erythrocyte G6PD deficiency in Aborigines of Taiwan. Vox Sang 1969;17:310-313.
138. Tuchinda S, Rucknagel DL, Na-Nakorn S, Wasi P. The Thai variant and the distribution of alleles of 6-phosphogluconate dehydrogenase and the distribution of glucose-6-phosphate dehydrogenase deficiency in Thailand. Biochem Genet 1968;2:253-264.
139. Wasi P, Na-Nakorn S, Suingdumrong A. Studies of the distribution of Hb E, thalassaemias and glucose-6-phosphate dehydrogenase deficiency in north-eastern Thailand. Nature 1967;214:501-502.
140. Kruatrachue M, Charoenlarp P, Chongsuphajaisiddhi T, Harinasuta C. Erythrocyte glucose-6-phosphate dehydrogenase and malaria in Thailand. Lancet 1962;2:1183-1186.
141. Panich V, Kuptamethi S, Pornpatkul M, Surapruek P, Kamolmatayakul S, Bumrungtrakul P, Viriyayudhakorn S, Ruengrairatanaroje P. Glucose-6-phosphate dehydrogenase deficiency in Southern Thailand. Songklanakarin J 1979;1:52-58.
142. Panich V. G6PD deficiency among Thai Moslems of Southern Thailand. Unpublished data 1981.
143. Panich V, Wasi P, Na-Nakorn S. Hematological survey of Sichang Island, Cholburi Province, Eastern Thailand. Unpublished data 1970.
144. Prasongwatana V, Sriboonlue P, Mularee N, Prasongwatana J, Sanchaisuriya P, Moonamard S, Saowakontha S. G6PD deficieny in the Pootai and the So communities in Northeast Thailand. Southeast Asian J Trop Med Pub Hlth 1984;15:112-114.
145. Youel DB, Strickland GT, Binh BA, Clarkson R, Blackwell RQ. Low incidence of erythrocyte G6PD deficiency in Vietnamese and Montagnards of South Vietnam. Vox Sang 1971;20:555-558.
146. Buchanan JG, Wilson FS, Nixon AD. Survey for glucose-6-phosphate dehydrogenase deficiency in Fiji. Amer J Hum Genet 1973;25:36-41.
147. Nixon AD, Buchanan JG. Survey for erythrocyte glucose-6-phosphate dehydrogenase deficiency in

Polynesians. Am J Hum Genet 1969;21:305-309.
148. Booth PB, Fasagali JL, Kirk RL, Blake NM. HLA types, blood groups, serum protein and red cell enzyme types among Samoans in New Zealand. Hum Hered 1977; 27:412-423.
149. Kamel K, Chandy R, Mousa H, Yunis D. Blood groups and types, hemoglobin variants, and G6PD deficiency among Abu Dhabians in the United Arab Emirates. Am J Phys Anthropol 1980;52:481-484.
150. Stamatoyannopoulos G, Voigtlander V, Kotsakis P, Akrivakis A. Genetic diversity of the "Mediterranean" glucose-6-phosphate dehydrogenase phenotype. J Clin Invest 1971;50:1253-1261.
151. Stamatoyannopoulos G, Kotsakis P, Voigtlander V, Akrivakis A, Motulsky AG. Electrophoretic diversity of glucose-6-phosphate dehydrogenase among Greeks. Am J Hum Genet 1970;22:587-596.
152. Vives Corrons JL, Pujades A. Heterogeneity of "Mediterranean type" glucose-6-phosphate dehydrogenase (G6PD) deficiency in Spain and description of two new variants associated with favism. Hum Genet 1982; 60:216-221.
153. Shatskaya TL, Krasnopolskaya KD, Tzoneva M, Mavrudieva M, Toncheva D. Variants of erythrocyte glucose-6-phosphate dehydrogenase (G6PD) in Bulgarian populations. Hum Genet 1980;54:115-117.
154. Toncheva D, Tzoneva M. Gnetic polymorphism of G6PD in a Bulgarian population. Hum Genet 1984;67:340-342.
155. Miwa S, Nakashima K, Ono J, Fujii H, Suzuki E. Three glucose-6-phosphate dehydrogenase variants found in Japan. Hum Genet 1977;36:327-334.
156. World Health Organization. Standardization of Procedures for the study of glucose-6-phosphate dehydrogenase. WHO Tech Rep Ser, No.366;1967.
157. McCurdy PR, Schneer JH, Hansen IM. Red cell glucose-6-phosphate dehydrogenase variants in Rumania. Rev Eur Etudes Clin Biol 1972;17:66-69.
158. McCurdy PR, Kamel K, Selim O. Heterogeneity of red cell glucose-6-phosphate dehydrogenase (G6PD) deficiency in Egypt. J Lab Clin Med 1974;84:673-680.
159. Lisker R, Briceno RP, Zavala C, Navarrete JI, Wessels M, Yoshida A. A glucose-6-phosphate dehydrogenase Gd(-) Castilla variant characterized by mild deficiency associated with drug induced hemolytic anemia. J Lab Clin Med 1977;90:754-9.
160. Samuel AP, Saha N, Omer A, Hoffbrand AV. Quantitative expression of G6PD activity of different phenotypes of G6PD and haemoglobin in a Sudanese population. Hum Hered 1981;31:110-115.

161. Fujii H, Nakashima K, Miwa S. Incidence and characteristics of G6PD deficiency in Japan. Jap J Hum Genet 1978;23:271-272.
162. Miwa S. Glucose-6-phosphate dehydrogenase variants in Japan. Hemoglobin 1980;4:781-787.
163. Kinugasa A, Kusunoki T, Iwashima A. Deficiency of glucose-6-phosphate dehydrogenase found in a case of hepatic-1, 6-diphosphatase deficiency. Pediatr Res 1979;13:1361-1364.
164. Fairbanks VF, Nepo AG, Beutler E, Dickson ER, Honig G. Glucose-6-phosphate dehydrogenase variants : Reexamination of G6PD Chicago and Cornell and a new variant (G6PD Pea Ridge) resembling G6PD Chicago. Blood 1980;55:216-220.
165. Beutler E. Glucose-6-phosphate dehydrogenase deficiency. In : Stanbury JB, Wyngaarden JB, Fredrickson DS, Goldstein JL, Brown MS (EDs.). The Metabolic Basis of Inherited Disease, 5th ed. McGraw-Hill, New York, 1983; 1629-1653.
166. World Health Organization. Treatment of Haemoglobinopathies and allied disorders. WHO Tech Rep Ser, No. 509,1972.
167. Panich V. Clinical pictures of acute hemolysis in G6PD Mahidol : Report on 10 patients. J Med Assoc Thailand 1973;56:476-484.
168. Panich V, Na-Nakorn S. Acute hemolysis in G6PD Union(Thai). Report on four cases. J Med Assoc Thailand 1973;55:726-731.
169. Na-Nakorn S, Wasi P, Panich V, Sookanek M, Pornpakul M, Pootrakul P. Acute haemólysis in glucose-6-phosphate dehydrogenase deficiency. A study in 20 patients. Siriraj Hosp Gaz 22:1117-1137.
170. Na-Nakorn S, Panich V. Viral hepatitis and acute hemolysis in glucose-6-phosphate dehydrogenase deficiency. Siriraj Hosp Gaz 1971;23:109-119.
171. Lexomboon U, Unkurapiana N. Co-trimoxazole in the treatment of typhoid fever in children with glucose-6-phosphate dehydrogenase deficiency. Southeast Asian J Trop Med Pub Hlth 1978;9:576-580.
172. Puavilai S, Chutha S, Polnikorn N, Timpatanapong P, Tasanapradit P, Churawichitratana S, Boonthanom A, Wongwaisayawan H. Incidence of anemia in leprosy patients treated with dapsone. J Med Assoc Thailand 1984;67:404-407.
173. Tishler M, Abramov M. Phenazopyridine-induced hemolytic anemia in a patient with G6PD deficiency. Acta Haemat 1983;70:208-209.
174. Chatterji SC, Das PK. Chloramphenicol induced haemolytic anaemia due to enzymatic deficiency of erythrocytes. J Indian Med Assoc 1963;40:172-176.

175. Kothari UR, Solanki SV, Oza JJ, Doshi KJ, Anadkat NC, Michta KK. Acute reversible renal failure and Stevens-Johnson syndrome in a patient having glucose-6-phosphate dehydrogenase deficiency. J Assoc Phys India 1977;25:299-302.
176. Choudhry VP, Madan N, Sood SK, Ghai OP. Chloroquine induced haemolysis and acute renal failure in subjects with G6PD deficiency. Trop Geogr Med 1978;30:331-335.
177. Manoharlal SA, Naqvi J. Acute renal failure in glucose-6-phosphate dehydrogenase (G6PD) deficiency. JPMA 1983;33:192-196.
178. Verma M, Aggarwal A. Glucose-6-phosphate dehydrogenase deficiency with methemoglobinemia. Indian Pediatr 1977;14:831-836.
179. Choudhry VP, Madan N, Sood SK. Intravascular haemolysis and renal insufficiency in children with glucose-6-phosphate dehydrogenase deficiency, following antimalarial therapy. Indian J Med Res 1980;71:561-566.
180. Aung-Than-Batu, U Hla-Pe, Thien-Than. Primaquine induced haemolysis in G6PD deficient Burmese. Trans Roy Soc Trop Med Hyg 1970;64:785-786.
181. Lie Injo LE, Pillay RP, Virik HK. Haemolysis due to glucose-6-phosphate dehydrogenase deficiency in Malaya. Trans Roy Soc Trop Med Hyg 1966;60:262-266.
182. Wong HB. Favism in Singapore. J Singapore Paediatr Soc 1972;14:17-25.
183. Wong HB. Acute haemolysis in Southeast Asia. Med Progr 1977;4:12-14.
184. Ryan BPK, Parsons IC. Glucose-6-phosphate dehydrogenase activity in anaemic Papuans. Med J Aust 1962;2:502-506.
185. Charoenlarp P, Areekul S, Harinasuta T, Sirivorasarn P. The haemolytic effect of a single dose of 45 mg of primaquine in G6PD deficient Thais. J Med Assoc Thailand 1972;55:631-638.
186. Kahn A. G6PD variants. Hum Genet 1978; Suppl 1:37-44.
187. Pootrakul P, Manochiewpinij S, Panich V. Unpublished data, 1980.
188. Flatz G, Sringam S, Komkris V. Negative balancing factors for glucose-6-phosphate dehydrogenase polymorphism in Thailand. Acta Genet 1963;13:316-327.
189. Milbauer B, Peled N, Svirsky S. Neonatal hyperbilirubinemia and glucose-6-phosphate dehydrogenase deficiency. Isr J Med Sci 1973;9:1547-1552.
190. Li AMC, Chau AS. Jaundice in Chinese infants. Far East Med J 1967;3:43-48.
191. Lai HC, Lai MPY, Leung KSN. Glucose-6-phosphate dehydrogenase deficiency in Chinese. J Clin Pathol 1968;21:44-47.
192. Yeung CY, Lai HC, Sin WK, Leung NK. Fluorescent spot test for screening erythrocyte glucose-6-phosphate

dehydrogenase deficiency in newborn babies. J Pediatr 1970;76:931-934.
193. Flatz G, Sringam S, Komkris V. Neonatal Jaundice in glucose-6-phosphate dehydrogenase deficiency. Lancet 1963;i:1382-1383.
194. Flatz G, Sringam S, Premyothin C, Penbharkkul S, Ketusingh R, Chulajata R. Glucose-6-phosphate dehydrogenase deficiency and neonatal jaundice. Arch Dis Childh 1963;38:566-570.
195. Phornphutkul C, Whitaker JA, Worathumrong N. Severe hyperbilirubinemia in Thai newborns in association with erythrocyte G6PD deficiency. Clin Pediatr 1969;8:275-278.
196. Angsusingha P, Tanphaichitr V, Suvatte V, Mahasandana C, Tuchinda S. Hyperbilirubinemia in the newborn. Siriraj Hosp Gaz 1980;32:202-214.
197. Weatherall DJ. Enzyme deficiency in haemolytic disease of the newborn. Lancet 1960;ii:835-837.
198. Brown WR, Wong HB. Hyperbilirubinemia and kernicterus in glucose-6-phosphate dehydrogenase-deficient infants in Singapore. Pediatrics 1968;41:1055-1062.
199. Tan KL. Phototherapy for neonatal jaundice in erythrocyte glucose-6-phosphate dehydrogenase-deficient infants. Pediatrics 1977;59(suppl.):1023-1026.
200. Tan KL. Glucose-6-phosphate dehydrogenase status and neonatal jaundice. Arch Dis Childh 1981;56:874-877.
201. Wong HB. Singapore kernicterus. Singapore Med J 1980; 21:556-557.
202. Woodfield DG, Biddulph J. Neonatal jaundice and glucose-6-phosphate dehydrogenase deficiency in Papua New Guinea. Med J Aust 1975;1:443-446.
203. Jim RTS, Chu FK. Hyperbilirubinemia due to glucose-6-phosphate dehydrogenase deficiency in a newborn Chinese infant. Pediatrics. 1963;31:1046-1049.
204. Meloni T, Costa S, Corti R, Cutillo S. Salicylamide glucuronide formation in newborn babies with G6PD deficiency. Biol Neonate 1978;33:189-192.
205. Brown WR, Wong HB. Ethnic group differences in plasma bilirubin level of full-term healthy Singapore newborn. Pediatrics 1965;36:745-751.
206. Beutler E. Hemolytic anemia and disorders of red cell metabolism. Plenum Medical Book Company, New York, 1978;94-96.
207. Valaes T, Karaklis A, Stravrakakis D, Bavela-Stravrakakis K, Perakis A, Doxiadis SA. Incidence and mechanism of neonatal jaundice related to glucose-6-phosphate dehydrogenase deficiency. Pediatr Res 1969;3:448-458.
208. Battistuzzi G, Morellini, M, Meloni T, Gandini E, Luzzatto L. Genetic factors in favism. In : Weatherall DJ, Fiorelli G, Gorini S (Eds.). Advances in Red Blood

Cell Biology. Raven Press, New York, 1982;339-346.
209. Belsey MA. The epidemiology of favism. Bull Wld Hlth Org 1973;48:1-13.
210. Orzalesi N, Sorcinelli R, Guiso G. Increased incidence of cataracts in male subjects deficient in glucose-6-phosphate dehydrogenase. Arch Pathol 1981;99:69-70.
211. Orzalesi N, Fossarello M, Sorcinelli R, Schlich U. The relationship between glucose-6-phosphate dehydrogenase deficiency and cataracts in Sardinia. An epidemiological and biochemical study. Doc Ophthal 1984;57:187-201.
212. Panich V, Na-Nakorn S. G6PD deficiency in senile cataracts. Hum Genet 1980;55:123-124.
213. Testa U, Meloni T, Lania A, Battistuzzi G, Cutillo S, Luzzatto L. Genetic heterogeneity of glucose-6-phosphate dehydrogenase deficiency in Sardinia. Hum Genet 1980; 56:99-105.
214. Fenu MP, Finazzi G, Manoussakis C, Valomba V, Fiorelli G. Glucose-6-phosphate dehydrogenase deficiency : genetic heterogeneity in Sardinia. Ann Hum Genet 1982;46:105-114.
215. Westring DW, Pisciotta AV. Anemia, cataracts, and seizures in patient with glucose-6-phosphate dehydrogenase deficiency. Arch Intern Med 1966; 118:385-390.
216. Helge H, Borner K. Kongenitale nichtsphaerozytaere haemolytische anaemie, Katarakt und Glucose-6-phosphat-dehydrogenase-mangel. Dtsch Med Wochenschr 1966; 91:1584-1589.
217. Harley JD, Agar NS, Yoshida A. Glucose-6-phosphate dehydrogenase variants : Gd(+) Alexandra associated with neonatal jaundice and Gd(-) Camperdown in a young man with lamellar cataracts. J Lab Clin Med 1978;91:295-300.
218. Panich V, Sungnate T, Pootrakul P. Gilbert's syndrome associated with glucose-6-phosphate dehydrogenase deficiency. J Med Assoc Thailand 1972;55:483-491.
219. Cooper MR, DeChatelet LR, McCall CE, LaVia MF, Spurr CL, Baehner RL. Leucocyte G6PD deficiency. Lancet 1970;ii:110.
220. Cooper MR, DeChatelet LR, McCall CE, LaVia MF, Spurr CL, Baehner RL. Complete deficiency of leukocyte glucose-6-phosphate dehydrogenase with defective bactericidal activity. J Clin Invest 1972;51:769-778.
221. Baehner RL, Johnston RB Jr, Nathan DG. Comparative study of the metabolic and bactericidal characteristics of severely glucose-6-phosphate dehydrogenase-deficient polymorphonuclear leukocytes and leukocytes from children with chronic granulomatous disease. J Reticuloendoth Soc 1972;12:150-169.
222. Gray GR, Klebanoff SJ, Stamatoyannopoulos G, Austin T, Naiman SC, Yoshida A, Kliman MR, Robinson GCF. Neutrophil dysfunction, chronic granulomatous disease,

and nonspherocytic haemolytic anaemia caused by complete deficiency of glucose-6-phosphate dehydrogenase. Lancet 1973;ii:530-534.

223. Vives Corrons JL, Feliu E, Pujades MA, Cardellach F, Rozman C, Carreras A, Jou JM, Vallespi MT, Zuazu FJ. Severe glucose-6-phosphate dehydrogenase (G6PD) deficiency associated with chronic hemolytic anemia, granulocyte dysfunction, and increased susceptibility to infections : Description of a new molecular variant (G6PD Barcelona). Blood 1982;59:428-434.
224. Clark M, Root RK. Glucose-6-phosphate dehydrogenase deficiency and infection : A study of hospitalized patients in Iran. Yale J Biol Med 1979;52:169-179.
225. Lampe RM, Kirdpon S, Mansuwan P, Benenson MW. Glucose-6-phosphate dehydrogenase deficiency in Thai children with typhoid fever. J Pediatr 1975;87:576-578.

GLUCOSE-6-PHOSPHATE DEHYDROGENASE DEFICIENCY AND MALARIA IN CENTRAL THAILAND

K.G. Blume[1], G. Flatz[2], H.W. Goedde[3] and W. Schloot[4]

[1]City of Hope National Medical Center, Duarte, CA
[2]Institute for Genetics, University of Hannover, FRG
[3]Institute for Human Genetics, University of Hamburg, FRG
[4]Institute for Human Genetics, University of Bremen, FRG

In 1966, we studied red cell glucose-6-phosphate dehydrogenase (G6PD) activity in 1260 young Thai males during their draft examination. This investigation was carried out in the province of Pitsanulok in Central Thailand. The province of Pitsanulok is separated into three western districts which are only slightly above sea level and three eastern districts which have altitudes approaching 800 meters (figure 1). The malaria parasite rate as determined by the World Health Organization in 1957-1959 was 10 to 20 times higher in the eastern (9.8 - 20.2%) compared to the western districts (0.9 - 3.1%). G6PD deficiency was significantly more frequent in the high altitude districts in the east (10-15%) compared to the lower altitude districts in the west (6-7%). Gene frequencies for hemoglobin E and β-thalassemia paralleled those of G6PD. Other genetic markers tested showed similar patterns for the populations living in the foothills and at higher altitudes indicating that the groups examined represented comparable cohorts.

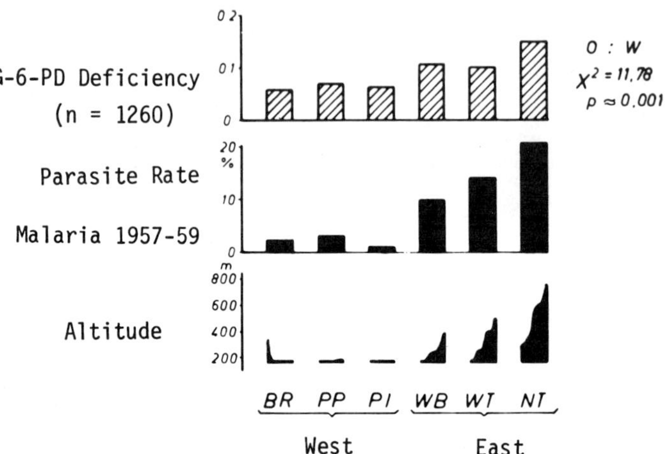

FIGURE 1. G6PD deficiency, malaria parasite rate and altitude in the province of Pitsanulok in Central Thailand.

G6PD VARIATION IN INDIA[1]

P. Meera Khan and J.Th. Wijnen

Department of Human Genetics, University of Leiden
The Netherlands

I. INTRODUCTION

The occurrence of G6PD deficiency in India was observed more than twenty years ago (Baxi et al., 1963; Meera Khan, 1964). Various screening procedures recommended by the WHO Scientific Group (Betke et al., 1967) were employed by different investigators to screen diverse caste, ethnic and linguistic groups belonging to different regions of the country. The deficiency in the males ranged from 0%-15.7%, the highest being

[1]The study was performed within the frame work of the project 5.A.12 ("Genetics of Human Enzymes") of the Medical Faculty at the State University of Leiden and supported by grants from WHO, IBP (the Dutch Branch) and FUNGO which was subsidized by the Netherlands Organization for Advancement of Pure Research (ZWO).

TABLE I

VARIANTS PREVIOUSLY REPORTED IN ASIATIC INDIANS

Number screened	Variants identified		Authors
	Name	No.	
97 males	G6PD B	94	Azevedo et al., 1968.
	G6PD Kerala	1	
	G6PD West Bengal	1	
	G6PD Mediterranean	1	
?	G6PD Jammu	1	Beutler, 1975.
?	G6PD Porbandar	1	Cayanis et al., 1977.
30 subjects	G6PD Kalyan	1	Ishwad and Naik, 1984.
40 males	G6PD B	38	Verma et al., 1984.
	G6PD Mediterranean	1	
	G6PD Punjab	1	
61 females	G6PD B	61	

among the Parsi's (Baxi, 1974; Kate et al., 1974). The North Indian populations were reported to exhibit wide differences, but different groups of investigators have employed diverse screening procedures (Agarwal et al., 1974). Further quantitative, electrophoretic or more elaborate physicochemical characterization studies were limited to a few relatively small sets of samples (Table I). The study of Azevedo et al. (1968) on 101 unrelated males originating from various areas in India comprises the largest of these series. Considering the size of the country and complex diversity of its population Azevedo's sample represents an extremely heterogeneous group. Nevertheless, her study clearly indicated for the first time, that considerable variation occurs at the G6PD locus among the Asiatic Indians. A systematic search or a routine screening in only about 150 males has lead to the identificaion of no less than 7 different G6PD variants in India (Table I).

Here we wish to present a preliminary report on our observations during a systematic search for G6PD variation and/or deficiency in a number of ethnic groups in Andhra Pradesh (The Andhras of South India), and heterogeneous populations of Punjabi immigrants to England and Surinami Hindustanis, the offspring of the people migrated from the Bihar-Uttar Pradesh region of North India to Surinam. A few samples from Hindus of Maharashtra (Wardha region)(kindly provided by Dr. R.de Vries, University Department of Immunohematology, Leiden) and a few Muslims from Pakistan were also included in the present study.

II. MATERIAL AND METHODS

A. The Population

Andhra Pradesh is a state in the north-eastern region of Peninsular India and has about 70 million people speaking Telugu. About 10% of its population comprises of tribals and Harijans. The tribals are largely forest dwellers while the Harijans live in the plains and traditionally engage in the lowest ranking professions. More than 80% of the Andhra population is made up of a multitude of strictly endogamous non-Harijan Hindu castes. The present study included two tribal populations called Koya Dora and Konda Reddi in the districts of East Godavari, West Godavari and Khammam. Konda Reddis in general, live in the interior forest villages in isolated single abodes or groups of huts dispersed on the hills and their slopes on either side of the river Godavari. The Koya Doras live in the fringes of these forests. The Koya and the Reddi may sometimes live in the same village but in widely separated neighbourhoods. The study covered about a thousand square kilometers of tribal area known to be endemic for malaria (V. Führer-Haimendorf, 1945). Van Führer-Haimendorf (1945) mentions that Konda Reddis used the digging stick for tilling the earth and shift cultivated after slashing and burning the forest. The Koya Doras engaged in more advanced type of agriculture by employing the cattle for ploughing the land. The Reddis and the Koyas do not intermarry. The Harijans by tradition were "untouchables" to the other Hindus since the inception of the concept of caste pollution in India. The majority of Harijans in Andhra Pradesh are represented by the two endogamous castes called Mala and Madiga. The Malas in social hierarchy are superior to the Madigas. The non-Harijan Hindus of the present Andhra Pradesh series comprise of at least 36 endogamous castes.

The castes among the Punjabis and the Maharashtrians were not recorded. The Surinami Hindustanis do not practise casteism.

B. Quantitative and Qualitative Studies

Blood samples were collected by venepuncture in EDTA or ACD vacutainers and maintained at about $4^{o}C$ till processed. Eighty percent of the samples were screened for G6PD deficiency, within 10 hours of collection and the rest within 1 week, by employing the known controls and using the respective methods indicated in Table II. Hemolysates were prepared using the lysis buffer (omitting DFP) (Meera Khan, 1971; Meera Khan et al., 1982). Samples from 45 Punjabi males, 100 male Surinami Hindustanis, 52 Maharastrians and 188 Andhras were screened for electrophoretic variants in Cellogel (Rattazzi et al., 1967).

C. Further Electrophoretic Characterization of the Variants

With the starch gel, 2 buffer systems, TEB (or EBT) (Porter et al., 1964) and phosphate (PO_4)(Mathai et al., 1966), were employed. For Cellogel respectively the TECl (Meera Khan, 1971) and TEMM (Wijnen et al., 1977) buffers were used.

D. Other Characterization Studies

These were performed as recommended in the WHO Report (Betke et al., 1967), on the partially purified preparations of G6PD obtained by the first two steps of the procedure described by Yoshida (1966). A known G6PD B was used as a control throughout. A direct cross-comparison of different variants was possible.

III. RESULTS

Observations on G6PD deficiency in various populations of the present study are listed in Table II. A number of electrophoretic variants including G6PD Andhra Pradesh (Rattazzi, 1966) detected and further characterized in the Koya Dora tribe will be described in detail, in a separate publication. All the deficients found in the rest of the populations resembled the Mediterranean variants in the levels of their activity and

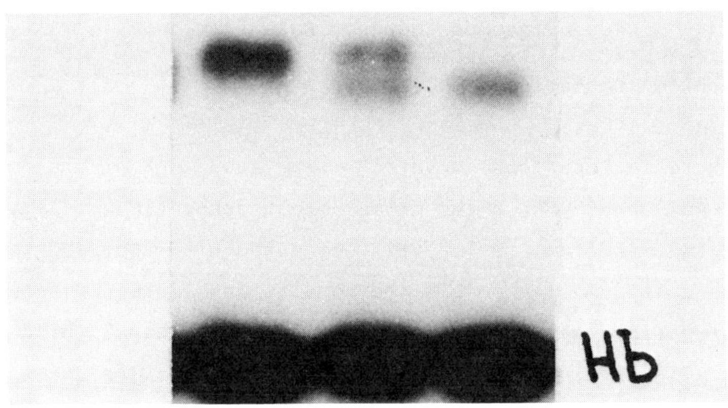

Fig.1. Zymograms of G6PD in Cellogel (TECl buffer, pH 7.5). Hb = hemoglobin A. The origin of the samples was below. The most anodal band (left) is of G6PD B. The sample on the right is G6PD Punjab and that in the middle is a heterozygote. The migration of Hb was 54% and that of G6PD Punjab was 95.25% respectively of G6PD B in this gel.

TABLE II

G6PD DEFICIENCY IN CERTAIN INDIAN POPULATIONS[a]

Region	No.males tested	Deficients[b] No.	Deficients[b] (%)	Test
South India				
Andhra Pradesh				
Tribes: Koya Dora	516	60	11.6	BCB dye
Konda Reddy	309	1	0.3	BCB dye
Harijans: Mala	27	0	0	BCB dye
Madiga	18	2	11.1	BCB dye
Hindu Castes (Misc.)	174[c]	0	0	BCB dye & spectro.
North India				
Punjab (Misc.)[d]	40	1	2.5	Fl. spot
Uttar Pradesh--Bihar (Misc.)	100	1	1.0	Spectro.

[a]Details of the data of this summary will be discussed elsewhere (P. Meera Khan, J.Th. Wijnen, N.V.S. Nayudu, R.M. Sukul and C. Verma, in preparation). [b]In BCB dye test 100 minutes was chosen as a cut off point. In fact, the normals took less than 60 min. to decolourize, with the exception of 4 that took 61-75 minutes. More than one half of the deficients took longer than 3 hours. [c]From at least 36 non-Harijan Hindu endogamous castes. [d]Data from Verma et al. (1984). Key: Misc. = miscellaneous; spectro. = spectrophotometric assay; Fl. = fluorescent.

electrophoretic behaviour. A variant with about 50% of the normal activity and about 96% of the normal electrophoretic mobility in Cellogel (TECl buffer, pH 7.5)(Fig. 1) appears to occur in polymorphic frequencies in all the populations listed in Table III. This variant was originally described in a Punjabi male by Verma et al. (1984) and designated as G6PD Punjab. Its characteristics compared to those of similar variants are assembled in Table IV. It is comparable, if not identical, to G6PD Kerala. Their electrophoretic mobilities in the starch gel were clearly different. It may be stated that in the absence of a direct comparison on the same gels it is hard to exclude their

TABLE III

G6PD PUNJAB IN DIVERSE POPULATIONS OF INDIA

Population	No.genot.	No. tested	Freq.
North			
Punjabi (Sikh)	1 hemi.	40 (Misc.) Males	0.044
Punjabi (Muslim)	1 hemi.	5 (Misc.) Males	
North-East			
U.P. and Bihar	3 hemi	100 (Misc.) Males	0.030
West			
Maharashtra	1 hemi	52 (Misc.) Males & Females	0.010
South			
Andhra Pradesh[a]		68 (Misc.) Females &120 (Misc.) Males	0.016
Harijan: Mala	1 hetero.		
Madiga	1 hetero.		
Castes: Eediga	1 hemi.		
Chakali	1 hemi.		

[a]The 188 persons were from 32 different Hindu endogamous castes including the Mala, Madiga, Eediga and Chakali communities. Key: No. = number; genot. = genotype; hemi. = hemizygote; hetero. = heterozygote; Misc. = miscellaneous; U.P. = Uttar Pradesh.

identity. However, till proven contrary, we consider that G6PD Punjab is a distinct variant. One striking feature of G6PD Punjab is the characteristic change in its relative electrophoretic behaviour following dialysis (Table V)(Fig. 2).

IV. DISCUSSION

A. G6PD Deficiency in India

The red cell G6PD deficiency was reported in a number of populations in different regions of India (See Baxi et al., 1974; Kate et al., 1974; Agarwal et al., 1974; Mutalik et al; 1974;

TABLE IV

COMPARISON OF G6PD PUNJAB WITH SIMILAR VARIANTS

Characteristics	Punjab[a] (B)[b]	Kerala (B$_w$)[b]	Porbandar	Kalyan
Activity (% of B)	52	50	78	25–35
Electr.mobil.(% of B)				
Cellogel:				
1. TECl (7.5)	96			
2. TEMM (7.4)	inhibited			
3. PO_4 (7.0)	92			
Starch gel:				
1. Tris (7.5)	75	90	74	85
2. TEB (8.6)	84	75	68	75
3. PO_4 (7.0)	48		83	63
Kinetics				
Km for NADP (μM)	4.0 ± 1.0 (3.7)	1.5 (3.9)	1.2	
Km for G6PD (μM)	22.6 ± 6.1 (73.8 ± 18.2)	23.0 (50.78)	4.80	20.0
Substr.anal.utln.				
2d-G6P (%G6P)	10.5 ± 3.6 (3.9 ± 1.3)	7.4 (2.3–3.7)	12–19	16–17

Table IV (Continued)

Characteristics	Punjab[a] (B)[b]	Kerala (B$_w$)[b]	Porbandar	Kalyan
Substr.anal.utln.(Continued)				
Gal-6-P (% of G6P)	12.3 + 5.1 (5.4 ± 1.3)	5.5 (1.4-2.5)		
deamino-NADP (% of NADP)	83.2 ± 27.3 (51.8 ± 4.6)		110 (79)	113
Thermostability	Normal	Normal	Normal	Labile
pH optimun	Biphasic (Trunc.)	Biphasic (Trunc.)	Trunc.	Trunc.

[a]Based upon Verma et al. (1984) for G6PD Punjab and our unpublished data from partially purified samples of "Punjab" variants in 2 Hindus from Andhra Pradesh, 3 Hindustanis of Uttar Pradesh-Bihar origin and 1 Pakistani Muslim. Since the sample was too little to be purified the variant in the Maharashtrian was characterized in crude hemolysate for its activity, electrophoretic behaviour and substrate analogue utilizations. [b]The results in the brackets are those in the control G6PD B observed in the present (B) and in the other (B$_w$) studies (Betke et al., 1967; Azevedo et al., 1968) respectively.

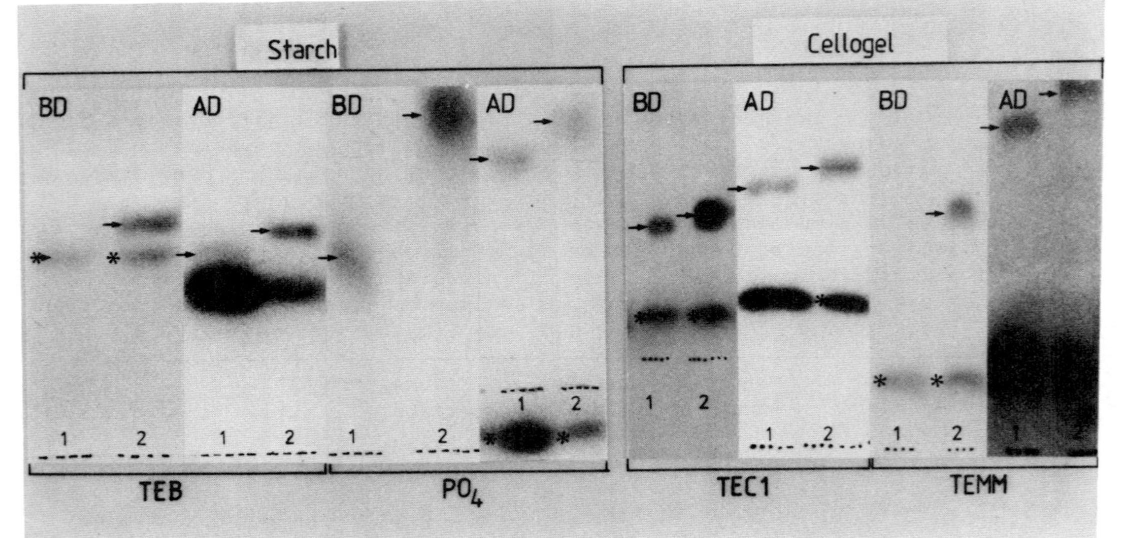

Fig.2. Patterns of electrophoresis of G6PD Punjab (channel 1) relative to G6PD B (channel 2) in the starch gel and Cellogel before and after dialysis (BD; AD). (See Table V for details about dialysis). The corresponding electrophoresis buffers were indicated below the respective gels. The arrows indicate the levels of individual G6PD bands and the stars those of hemoglobin (Hb) bands. Note that in TEB buffer (starch) the G6PD Punjab of the undialysed sample comigrates with Hb A, but that in the dialysed sample moves between Hb and G6PD B. The relative behaviour of G6PD Punjab in TEMM buffer in Cellogel is particularly characteristic (See Table V).

TABLE V

EFFECT OF DIALYSIS ON THE ELECTROPHORETIC MOBILITY OF G6PD PUNJAB[a]

System	Relative mobility (% of G6PD B)	
	Before dial.	After dial.
Starch gel		
1. TEB (8.6)	84.0	90.48
2. PO_4 (7.0)	48.0	91.18
Cellogel		
1. TECl (7.5)	96.0	94.61
2. TEMM (7.4)	inhibited	89.55

[a]Overnight dialysis with 2 changes in 0.025M tris-citric acid buffer, pH 7.5, containing 54mM EDTA, 1mM 2-mercaptoethanol and $2x10^{-5}$M NADP.

Bala and Seth, 1978; Agarwal and Bhalla, 1981; Chatterjea, 1966 for extensive tabulations and more exhaustive lists of references). In general, Parsis showed the highest incidence while the Muslims in various regions the lowest (Baxi et al., 1963; Hakim et al., 1973). The previous workers appear to have considered all the erstwhile "untouchables" in different regions of Maharashtra as a single caste, the Mahars, and reported the range of incidence being as wide as 0.5% to 10% (Parikh et al., 1969; Kher et al., 1967; Deshmukh and Sharma, 1968). This discrepancy may be due to the occurrence of different strictly endogamous groups among the "Mahars". Our study includes 2 such castes among the Harijans, the Mala and the Madiga. The Madigas are thought to be the Mahars of Andhra Pradesh and were branched into several strictly endogamous subcastes (Singh, 1969). During the present study one urban and one rural male Madiga were found to be deficient while none of the Malas was. Though the caste samples were small, this observation becomes more significant if we consider that the single individual with a total red cell G6PD deficiency found in our previous study on a miscellaneous non-tribal rural Andhra population was a Madiga male (Meera Khan, 1964; Meera Khan, unpublished observation).

The incidence of G6PD deficiency appears to be extremely low, if not absent, in the Andhra non-Harijan, non-tribal miscellaneous Hindu groups. So it appears also in the Konda Reddis among the tribals and the Malas among the Harijans. In the present study, the respective sizes of the Koya Dora and Konda Reddi samples collected from a relatively large geographical area, were large enough to suggest that the observed difference in the incidence of G6PD deficiency among them was most probably exact (Table II).

About 12% of the Koya males were found to be deficient. Von Führer-Haimendorf (1945) reasons that Koya Doras were probably more recent immigrants to the Reddi land. He observes that the Koyas represent the southernmost branch of the Gondi speaking people and show a good many of the typical features of Gond culture. The majority of the Koya retained their tribal tongue and use ploughs and bullocks for cultivation. It is interesting to note that Kher et al. (1967) observed that 10% of the Gond males were G6PD deficient. The Gonds of their study were from Nagpur area of Maharashtra, immediately north of Andhra Pradesh. In fact, Kher et al. (1967) found that the Gonds, Mahars and miscellaneous populations in Nagpur region had an incidence of 10%-12%. For heuristic purposes, we wish to propose that the present day Madigas are probably the descendants of Mahar immigrants of the past into Andhra region from the north.

Most of the North Indian populations exhibited the deficiency in appreciable frequencies (Baxi et al., 1974; Agarwal et al., 1974). It may be relevant to note that Azevedo et al. (1968) encountered 4 deficients in the north and none in the south though about one half of their 101 volunteers were from the south. The Punjabis and Surinami Hindustanis of the present study conform to this trend.

Most of the deficiency associated G6PD variants found in India have not been characterized. It is possible that there exists a considerable genetic heterogeneity among these deficients in the Indian Subcontinent (Azevedo et al., 1968; Ronald et al., 1968; McCurdy et al., 1970; Prchal et al., 1980).

The data on G6PD deficiency in tribal populations of India are very limited. It is interesting to note that Angami Nagas of Kohima show the highest incidence of G6PD deficiency (27.6%) so far reported in India (Seth and Seth, 1971).

B. G6PD Variants in India

In a relatively small number of Indian samples, the search by several groups of investigators has lead to the discovery of six variants (Table I). Recently, Verma et al. (1984) described in a Punjabi Sikh male from Amritsar, a slow moving variant of G6PD with an intermediate activity distinguishable from the other known variants, and designated it as G6PD Punjab (Table III). During the present study the G6PD Punjab was found also in a Muslim from Gujarat district in the Punjab state of Pakistan, 3 unrelated Surinami Hindustanis born to parents belonging to the families originated from the Uttar Pradesh-Bihar region in North India, and in a Maharashtrian Hindu male from Wardha area. Of particular interest is that in Andhra Pradesh individuals of four different endogamous castes possessed G6PD*Punjab - an Eediga and a Chakali were hemizygotes and a Madiga and a Mala were heterozygotes (Table III).

It is striking to note that G6PD Punjab occurs, probably in polymorphic frequencies, in most of the populations of the Indian subcontinent, irrespective of their regional, religious, linguistic and caste differences. It is possible that G6PD Punjab is limited to the lower strata of the caste hierarchy, at least in the south. It is therefore, conceivable that G6PD Punjab can be useful as one of the possible biological tracers in studies on the origin and dissemination of castes in ancient India.

ACKNOWLEDGMENTS

These studies were originally suggested and heavily supported by late Prof. J.B.S. Haldane. It is a delight for P.M.K. to recollect the friendly associations with Drs. M. Siniscalco and M.C. Rattazzi in Italy (Naples and Sardinia), India and The Netherlands during the initial stages.

We wish to thank Mrs. Prabha Khan and Mrs. Constance van Gogh-Davijt for preparing this manuscript.

REFERENCES

Agarwal,S., and Bhalla,V. (1981). Ind.J.Phys.Anthrop. and Hum.Genet. 7, 135.

Agarwal,S.S., Sharma,J.K., and Farooqui,J.Z. (1974). In "Human Population Genetics in India" (L.D.Sanghvi, V.Balakrishnan, H.M.Bhatia, P.K.Sukumaran, and J.V.Undevia, eds.) p.50. Orient-Longman Ltd., New Delhi.

Azevedo,E., Kirkman,H.N., Morrow,A.C., and Motulsky,A.G. (1968). Ann.Hum.Genet. 31,373.

Bala,S., and Seth,S. (1978). Ind. J. Phys. Anthrop. and Hum. Genet. 4, 117.

Baxi,A.J., Balakrishnan,V., Undevia,J.V., and Sanghvi,L.D. (1963). Ind. J. Med. Sci. 17, 493.

Baxi,A.J. (1974). In "Human Population Genetics in India" (L.D.Sanghvi, V.Balakrishnan, H.M. Bhatia, P.K.Sukumaran, and J.V.Undevia, eds.) p60. Orient-Longman Ltd., New Delhi.

Betke,K., Beutler,E., Brewer,G.J., Kirkman,H.N., Luzzatto,L., Ramot,B., and Siniscalco,M. (1967). Tech. Rep. Ser. Wld. Hlth. Org. no. 366.

Beutler,E. (1975). In "Trends in Hematology" (N.N. Basu and A.K. Sen, eds.),p. 279. Calcutta, India.

Cayanis,E., Lane,A.B., Jenkins,T., Nurse,G.T., and Balinsky,D. (1977). Biochem. Genet. 15, 765.
Chatterjea,J.B. (1966). Bull. Wld. Hlth. Org. 35, 837.
Deshmukh,V.V., and Sharma,K.D. (1968). Ind. J. Med. Res. 56, 821.
Hakim,S.M.A., Baxi,A.J., Balakrishnan,V., Kulkarni,K.V., Rao,S.S., and Jhala,H.I. (1973). Humangenetik 15, 90.
Ishwad,C.S., and Naik,S.N.(1984). Hum. Genet. 66, 171.
Kate,S.L., Mutalik,G.S., Phadke,M.A., Khedkar,V.A., and Shende,M.N. (1974). In "Human Population Genetics in India" (L.D. Sanghvi, V. Balakrishnan, H.M. Bhatia, P.K. Sukumaran, and J.V. Undevia, eds.) p.56. Orient-Longman Ltd., New Delhi.
Kher,M.M., Solanki,B.R., Parande, C.M., and Junnarkar,R.V. (1967). Indian Med. Gaz. 7, 34.
Mathai,C.K., Ohno,S., and Beutler,E. (1966). Nature 210, 115.
McCurdy,P.R., and Mahmood,L. (1970). J. Lab. Clin. Med. 76, 943.
Meera Khan,P. (1964). J. Genet. 59, 14.
Meera Khan,P., Rijken,H., Wijnen,J.Th., Wijnen,L.M.M., and de Boer,L.E.M. (1982). In. "The Orang Utan: Its Biology and Conservation". (L.E.M.de Boer, ed.)p.69. Junk, The Hague.
Meera Khan,P. (1971). Arch. Biochem. Biophys. 145, 470.
Mutalik,G.S., Kate,S.L., Malhotra,K.C., and Phadke,M.A. (1974). In "Human Population Genetics in India" (L.D.Sanghvi, V. Balakrishnan, H.M.Bhatia, P.K.Sukumaran, and J.V.Undevia, eds.)p.11. Orient-Longman Ltd., New Delhi.
Parikh,N.P., Baxi,A.J., Jhala,H.I., and Kulkarni,K.V. (1969). Ind. J. Med. Res. 57, 1467.
Prchal,J., Carroll,A.J., Prchal,J.F., Crist,W.M., Scalka,H.W., Gealy,W.J., Harley,J., and Mullah,A. (1980). Blood 56, 1048.
Porter,I.H., Boyer,S.H., Watson-Williams,E.J., Adam,A., Szeinberg,A., and Siniscalco,M. (1964). Lancet i, 895.
Rattazzi,M.C. (1966). Abstract 3rd Intl. Congr. Hum. Genet., Chicago, p.82 (quoted by Motulsky,A.G., and Yoshida,A., In "Biochemical Methods in Red Cell Genetics" (Yunis, ed.),p.51. Academic Press, New York.
Ronald,A.R., Underwood,B.A., and Woodward,T.E. (1968). Trans. Soc. Trop. Med. Hyg. 62, 531.
Seth,P.K., and Seth,S.(1971). Hum.Biol. 43, 557.
Singh.T.R. (1969). "The Madiga". Ethnographic & Folk Culture Society, UP, Lucknow.
Verma,C., Jairaj,M.B., Wijnen,J.Th., and Meera Khan,P. (1984). Anthropol. Contemporanea 7, 128.
Von Führer-Haimendorf. (1945). "The Reddis of the Bison Hills". Macmillan and Co.Ltd., London.
Wijnen,L.M.M., Grzeschik,K.-H., Pearson,P.L., and Meera Khan,P. (1977). Hum. Genet. 37, 271.
Yoshida,A. (1966). J.Biol. Chem. 241, 4966.

GLUCOSE-6-PHOSPHATE DEHYDROGENASE VARIANTS IN JAPAN

Shiro Miwa and Hisaichi Fujii

Department of Internal Medicine, Institute of Medical Science, University of Tokyo, Tokyo, Japan.

ABSTRACT

Thirty nine out of a total of 40,082 Japanese males (0.1%) screened using Beutler's spot test or a new screening method had glucose 6-phosphate dehydrogenase (G6PD) deficiency. They had mild enzyme deficiency and no symptom. Seven variants, namely G6PD Hofu, B(-)Chinese, Ube, Konan, Kamiube, Kiwa and Washington-like, were identified. All these variants belonged to Class 3. G6PD Konan and Ube appear to be relatively common in Japan. From the blood samples which were sent to our laboratory with the diagnosis of either chronic nonspherocytic hemolytic anemia or drug-induced hemolysis, 18 variants were characterized. G6PD Tokyo, Tokushima, Ogikubo, Kurume, Fukushima, Yokohama, Yamaguchi, Wakayama, Akita, Gifu, Kobe, Sapporo, Nagano, Sendagi, and Asahikawa belonged to Class 1, because all these cases showed chronic hemolytic anemia. G6PD Mediterranean-like, Ogori and Fukuoka belonged to Class 2.

This study was supported in part by research grants from the Ministry of Education, Science and Culture and Ministry of Health and Welfare, Japan.

The thermostability test seems to be reliable in indicating whether G6PD deficiency is associated with acute or chronic hemolysis.

I. INTRODUCTION

In Japan located in the temperate zone and where malaria is not endemic, glucose 6-phosphate dehydrogenase (G6PD) deficiency appears to be rare. Until recently, little was known about the frequency among Japanese. Using the spot test of Beutler and Mitchell (1) or a new screening method recently established by us (2), males living in various places of Japan were screened for G6PD deficiency. We also found 18 G6PD variants from the blood samples which were sent to our laboratory with the request of red cell enzyme study. Herein, we summarize the results of both of these studies, and also report on the new simple screening method (formazan ring test) briefly.

II. MATERIALS AND METHODS

A total of 40,082 Japanese males were screened for the deficiency using Beutler's spot test (1) or a formazan ring test (2). Cases suspected to have G6PD deficiency were confirmed by enzyme assays and starch-gel electrophoresis. Three buffer systems, namely, a phosphate buffer system at pH 7.0 (3), a Tris-EDTA-borate buffer system at pH 8.6 (4), and a Tris-HCl buffer system at pH 8.8 (4) were used for the electrophoresis. The new screening method was performed by detecting the enzyme activity of dried blood absorbed on the cation-exchange cellulose paper (Whatman, P 81) on a gel containing the reagents for the detection of G6PD activity. As pretreatment, P 81 paper was saturated with 100 mM Tris-HCl buffer, pH 6.5 containing 10 mM

$MgCl_2$ and dried at room temperature. Whole blood (10 μl) was applied to the treated P 81 paper and dried at room temperature. The gel was prepared by mixing 25 mg of glucose 6-phosphate (G6P), 5 mg of nicotinamide adenine dinucleotide phosphate (NADP), 5 mg of 3(4,5 dimethylthiazolyl 1-2)2,5 diphenyltetrazolium bromide (MTT), 5 mg of phenazine methosulfate, and 150 mg of Noble agar in 20 ml of 100 mM Tris-HCl buffer, pH 6.5 containing 10 mM $MgCl_2$ and 1 mM sodium azide. The mixture was kept at 55 °C and poured into the plastic box (13 X 6 X 2 cm) to give a gel thickness of 2.6 mm. After the gel hardened, circular pieces, 6 mm in diameter, cut from the paper were pressed onto the gel surface. The box was covered with aluminum foil. When NADP is reduced to NADPH through the reaction of G6PD, MTT is reduced to a blue insoluble formazan in the presence of phenazine methosulfate. A blue ring of formazan is produced around the circular paper. After 8 hr of incubation at 37 °C, the colored area was calculated by measuring the spot diameter.

Partially purified G6PD, free of hemoglobin and 6-phosphogluconate dehydrogenase activity, was characterized by methods recommended by the WHO Scientific Group (4). The inhibition constant (Ki) for NADPH was determined by the methods of Yoshida (5).

III. RESULTS

Figure 1 shows the blue formazan rings around the P 81 paper. G6PD Ube is the common variant in Japan showing half-normal enzyme activity without clinical symptoms. G6PD Fukushima is seen in chronic nonspherocytic hemolytic

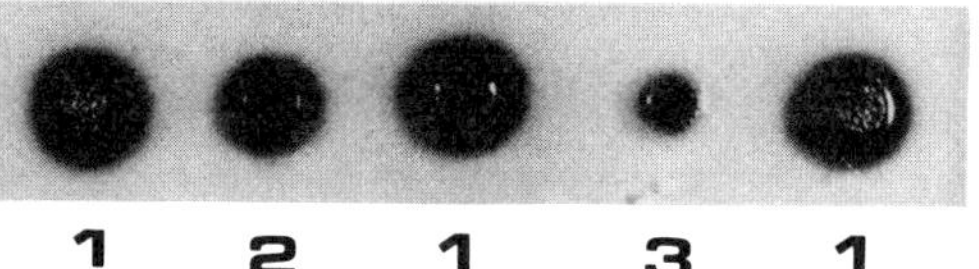

1 2 1 3 1

Fig.1 The blue formazan rings around P 81 paper: 1, normal control; 2, G6PD Ube; 3, G6PD Fukushima.

anemia. The enzyme activity of this variant was 2.8% of that of the normal enzyme. We could differentiate the size of the colored area of #2 (G6PD Ube) and #3 (G6PD Fukushima) from that of #1 (normal G6PD activity). To quantitatively compare the blue ring of formazan prodeced by this method with enzyme activity, a hemolysate with a normal range of G6PD activity (n=20, 11.27±2.98 U/gHb) was diluted variously with a solution consisting of 2.7 mM disodium ethylenediamine-tetraacetate, pH 7.0 and 0.7 mM 2-mercaptoethanol. There was a significant positive relation between quantitatively determined enzyme levels and colored area using 38 different samples (Y=6.27X-4.80, r=0.94, $p < 0.005$) (Fig. 2). The colored area produced by normal blood samples was more than 50 mm^2.

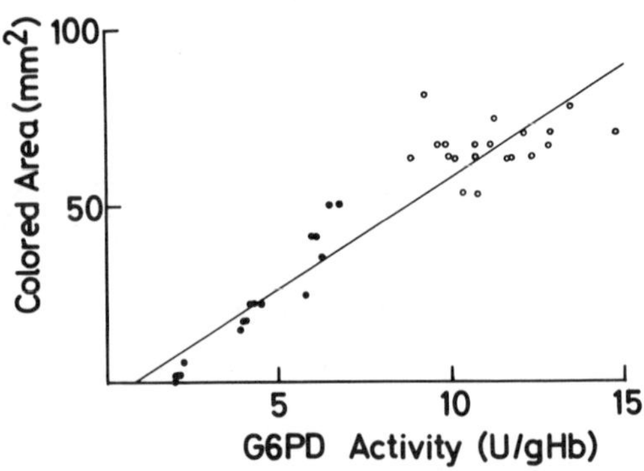

Fig.2 The relationship between quantitatively determined enzyme levels and colored area of 20 normal samples (o) and 18 samples prepared with various dilution of normal hemolysate(●).

G6PD deficiency was detected in 39 of the 40,082 Japanese males using the spot test or formazan ring test which gives an average frequency of 0.097%. Figure 3 shows the incidence of G6PD deficiency in various places of Japan. The

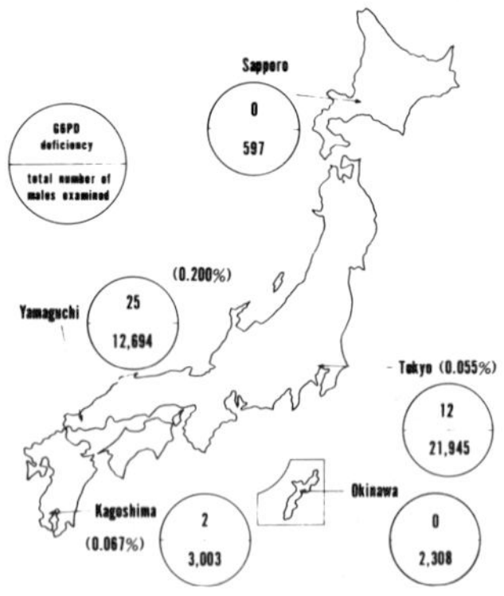

Fig.3 Incidence of G6PD deficiency in various places of Japan.

incidence in Yamaguchi prefecture, south-west of main island, seemed to be slightly higher than that in Tokyo or Kagoshima. Further screening in Sapporo and Okinawa is in progress. The red cell G6PD of all the subjects with G6PD deficiency were characterized and the 7 variants, G6PD Hofu (6), B(-)Chinese (7), Ube (8), Konan, Kamiube, Kiwa (9), and Washington-like, were differentiated. All these variants had moderately decreased G6PD activity, ranging from 12 to 64%, and did not cause clinical signs of hemolysis, though very slightly increased reticulocyte counts were seen in G6PD Hofu and B(-) Chinese for unknown reasons. Thus, all these variants were classified as Class 3. Geographic distribution of these variants is shown in figure 4. Fifteen subjects had the same variant, G6PD Konan, 11 others had G6PD Ube and 3 had G6PD Kamiube. One subject with G6PD Ube was a Korean. G6PD Hofu, B(-)Chinese, Kiwa, and Washington-like were sporadic.

Fig.4 Geographic distribution of G6PD variants in Japan: 1,Class 1 variant; 2,Class 2 variant; 3,Class 3 variant.

From the blood samples which were sent to our laboratory with the diagnosis of either chronic nonspherocytic hemolytic anemia or drug-induced acute hemolysis, 18 variants were characterized (Tables I and II).

G6PD Tokyo, Tokushima (10), Ogikubo, Yokohama, Akita (11), Kurume, Fukushima, Yamaguchi, Wakayama (12), Gifu (13), Kobe, Sapporo (14), Nagano (15), Sendagi (16), and Asahikawa (17) belonged to Class 1, because all showed chronic hemolytic anemia with severe enzyme deficiency. As shown in Table I, neonatal hyperbilirubinemia was noted in 3 variants. Six of the 15 Class 1 variants showed hemolytic crises solely after upper-respiratory

Table I. Main Features of G6PD Variants Found in Japan

	Hb (g/100 ml)	Retics (%)	Indirect Bilirubin (mg/dl)	Neonatal Jaundice	Hemolytic Crisis	
					Infection	Drug
<u>Class 1</u>						
Tokyo	12.6	4.5	1.4	-	+	sulfisomedine, sulfa-methoxypyridazine, chloramphenicol
Tokushima	12.1	22.0	2.6	-	-	
Ogikubo	9.2	8.4	2.3	+	-	acetylsalicylic acid
Kurume	10.6	4.0	3.5	-	-	
Fukushima	6.5	18.5	1.6	-	-	sulpyrine, aminopyrine
Yokohama	7.5	35.2	1.2	-	+	
Yamaguchi	6.9	32.4	1.2	-	+	aminopyrine, sulpyrine
Wakayama	10.6	3.6	1.9	-	+	sulpyrine, sulfisomezole
Akita	11.1	15.0	1.6	-	-	sulpyrine
Gifu	11.1	10.7	0.8	-	+	
Kobe	8.8	39.9	6.1	-	+	
Sapporo	8.6	20.2	1.1	+	-	phenacetin, aminopyrine erythromycin
Nagano	10.0	33.8	1.6	+	+	
Sendagi	5.8	49.6	1.4	-	+	
Asahikawa	4.9	13.1	1.7	-	+	
<u>Class 2</u>						
Mediterr-anean-like	9.2	2.0	0.3*	-	-	chloramphenicol
Ogori	15.9	3.3?	0.8*	-	-	
Fukuoka	11.1	1.7	0.2	-	-	sulpyrine, erythromycin
<u>Class 3</u>						
Hofu	15.1	3.2	-	-	-	
B(-) Chinese	13.1	4.3	0.6	-	-	
Ube	14.0	1.0	0.6*	-	-	
Konan	13.7	0.8	-	-	-	
Kamiube	12.5	1.1	-	-	-	
Kiwa	13.0	1.0	-	-	-	
Washington -like	14.7	1.6	-	-	-	

* total bilirubin

Table II. Enzymatic Properties of G6PD Variants Found in Japan

Class	Variant	RBC G6PD activity (%)	Electro-phoretic mobility (%)	Km G6P (μM)	Km NADP (μM)	Ki NADPH (μM)	Utilization 2dG6P % of G6P	Utilization dNADP % of NADP	Heat stability (20 min)	pH optimum	Reference
	Normal B	100	100	31-71	2.6-6.6	4.9-12.9	1.6-6.4	51-69	66-99	normal	
1	Tokyo	4.4	90(Tris[a]) 90(TEB[b]) 70(ph[c])	65	5.5	normal	2.5	55	0	normal	10
	Tokushima	3.0	100	50	27.0	normal	2.0	52	4.8	normal	10
	Ogikubo	3.0	100	47	3.0	11.5	7.5	52	2.6	normal	11
	Kurume	0.8	96(Tris) 92(TEB) 87(ph)	43	5.7	1.9	5.0	66	0	biphasic	12
	Fukushima	2.8	86(Tris) 88(TEB) 74(ph)	31	5.0	4.4	3.0	37	22.0	normal	12
	Yokohama	1.9	100	70	6.1	2.9	4.0	63	30.0	normal	11
	Yamaguchi	3.5	57(Tris) 73(TEB) 48(ph)	37	15.2	7.6	4.1	94	0	narrow peak at pH 8.76	12
	Wakayama	4.5	89(Tris) 81(TEB) 78(ph)	46	6.5	3.2	1.7	66	3.7	normal	12
	Akita	0	100	33	3.3	7.9	3.7	59	12.0	normal	11
	Gifu	2.9	100	48	3.1	7.1	4.8	121	52.1	normal	13
	Kobe	21.7	82(Tris) 89(TEB) 61(ph)	143	4.7	8.2	4.6	80	4.0	acidic	14
	Sapporo	3.6	100	51	3.3	40.3	6.6	75	63.0	normal	14
	Nagano	5.5	123(Tris) 121(TEB) 132(ph)	28	9.1	3.1	3.0	65	33.0	normal	15
	Sendagi	8.4	97(Tris) 98(TEB) 84(ph)	11	4.4	15.3	8.4	142	0	alkaline	16
	Asahikawa	3.8	98(Tris) 94(TEB) 94(ph)	30	18.3	2.1	7.1	109	9.1	biphasic	17

(CONTINUED)

Class	Variant	RBC G6PD activity (%)	Electro-phoretic mobility (%)	Km G6P (μM)	Km NADP (μM)	Ki NADPH (μM)	Utilization 2dG6P dNADP % of G6P	NADP	Heat stability (20 min)	pH optimum	Reference
	Normal B	100	100	31-71	2.6-6.6	4.9-12.9	1.6-6.4	51-69	66-99	normal	
2	Mediterra-nean-like	3.5	100	20	2.1	-	12.6	180	60.3	slightly biphasic	6
	Ogori	3.5	100	40	5.0	-	3.4	68	90.9	normal	6
	Fukuoka	6.4	109(Tris) 105(TEB) 113(ph)	92	6.0	7.9	2.1	79	27.8	biphasic	13
3	Hofu	12.1	100	25	5.0	-	5.2	105	89.5	normal	6
	B(-)Chinese	25.1	100	51	5.6	21.0	3.1	67	92.0	normal	7
	Ube	46.4	108(Tris) 107(TEB) 113(ph)	48	6.1	17.0	2.4	53	90.0	normal	8
	Konan	42.1	108(Tris) 107(TEB) 112(ph)	41	5.9	8.7	4.5	57	90.0	normal	9
	Kamiube	45	100	37	3.9	7.9	3.5	53	88.0	normal	9
	Kiwa	64	102(Tris) 103(TEB) 105(ph)	42	6.1	8.6	5.7	68	88.0	normal	9
	Washington-like	17.5	90(Tris) 92(TEB) 88(ph)	32	2.3	8.3	5.1	80	89.0	normal	

a; Tris-hydrochloride buffer, pH8.8, b; Tris-EDTA borate buffer, pH8.6, c; phosphate buffer, pH7.0

infection, and in 3 cases, hemolysis was noted after infection and/or drug administration. Drug-induced hemolysis, mainly due to sulfonamides and antipyretics, was noted in 7 cases. G6PD Mediterranean-like, Ogori (6), and Fukuoka (13) belonged to Class 2, because they all showed severe enzyme deficiency. G6PD Mediterranean-like and Fukuoka were accompanied with drug-induced hemolysis.

IV. DISCUSSION

A new screening method based upon detection of the enzyme activity of dried blood absorbed on the cation-exchange cellulose paper on a gel containing the reagents for the detection of G6PD activity was used. Net charge of hemoglobin and G6PD are positive and negative at pH 6.5, respectively (18). Under this condition, cation-exchange cellulose paper strongly binds hemoglobin while G6PD is spread out. Therefore, the diffusion area of G6PD is visible without interference of hemoglobin. We could easily detect the enzyme deficient-subjects of less than 50% normal enzyme activity by this method. Although a slight decrease in enzyme activity was noted during 7 days of storage at room temperature, it was still diagnostic. When anemia is severe, enough plasma should be removed after centrifugation to bring the packed cell volume to 40 $\pm$ 5%. Although the fluorescent spot test reported by Beutler and Mitchell (1) is useful and reliable, our method has several advantages: a) The procedure is very simple without the need for special technical experience. b) The gel is stable for several months if kept in the dark. c) Hundreds of samples can be examined at the same time. d) The results can be judged clearly by measuring the spot diameter, and no false results have been obtained, at least by comparison with the known positive and negative samples. e) This method requires no special equipment, such as a spectrophotometer of UV light. For these reasons, our method is recommended for clinical purpose and for genetic surveys.

Our screening tests showed that the frequency of G6PD deficiency in Japanese is about 0.1%. All these cases thus discovered had asymptomatic Class 3 variants. Thus it is certain that the frequency of G6PD deficiency which may cause drug-induced acute hemolysis in the Japanese population is much less than 0.1%. Interestingly, one boy with G6PD Ube was a Korean, suggesting a gene flow between the two races. The frequency of 0.1% is quite different from that found in southern Chinese (5.5%) (19), in the Taiwan-Hakka population (5.5%) (20), and in the Philippines (6.0%) (21). The variants found in Chinese and the Philippines are different from those found in Japanese except for G6PD B(-)Chinese (7). G6PD Mahidol and Canton which are common and widespread in Asia had not been discovered in Japan. Furthermore, the incidence in Kagoshima and Okinawa, southern parts of Japan, is not much higher than that in other places in Japan. Therefore, there may have been little gene-flow of G6PD deficiency from southeastern Asia into Japan. The mutation might have occurred after the Japanese were isolated in the present islands. In addition, the frequency in Japan could not increase probably because of lack of a positive selective pressure of endemic malaria.

The determination of the primary structure of human G6PD is now in progress (Yoshida, pers.comm.). Therefore, the correlation between the molecular abnormality and clinical expression in G6PD deficiency is not known. Yoshida (22) emphasized particularly the role of the affinities of each variant for NADP and NADPH in determining whether it is associated with congenital hemolytic anemia. The hemolytic variants associated with relatively mild enzyme deficiency were very strongly inhibited by the physiological concentration of NADPH, while nonhemolytic variants associated with severe enzyme deficiency were resistant to the inhibition by NADPH. In our cases, 6 out of 15 Class 1 variants showed low Ki NADPH.

The thermostability of 25 different G6PD variants have been studied (figure 5). All of the Class 1 variants associated with chronic hemolytic anemia without drug administration showed marked thermostability.

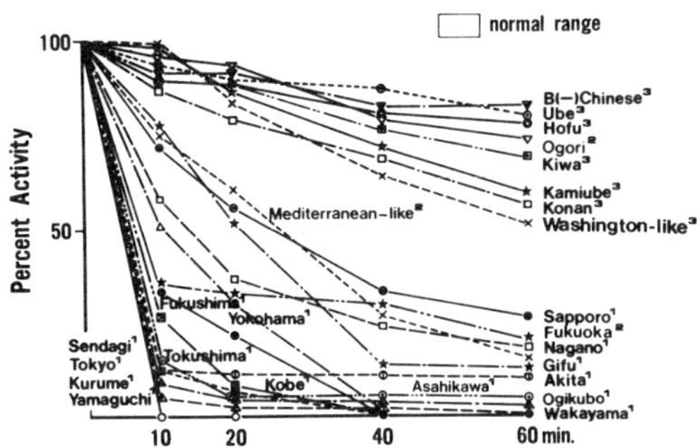

Fig.5 Thermostability test of G6PD variants characterized by us: 1, Class 1 variant; 2, Class 2 variant; 3, Class 3 variant.

Among the Class 2 variants, G6PD Mediterranean-like and Fukuoka, a thermolabile variant, were associated with drug-induced hemolysis, whereas G6PD Ogori, a thermostable variant, was not associated with any clinical symptoms. Thermostability was normal in all the Class 3 variants which showed mild to moderate enzyme deficiency without clinical symptoms. Thus, the thermostability test seems to be reliable in indicating whether G6PD deficiency is associated with acute or chronic hemolysis.

REFERENCES

1. Beutler, E., and Mitchell, M. (1968). Blood 32, 816.
2. Fujii, H., Takahashi, K., and Miwa, S. (1984). Acta Haematol. Jpn. 47, 185.
3. Mathai, C.K., Ohno, S., and Beutler, E. (1966). Nature 210, 115.
4. Report of WHO Scientific Group (1967). WHO Tech. Rep. Ser. 366, 1.
5. Yoshida, A. (1973). Science 179, 532.
6. Miwa, S., Nakashima, K., Ono, J., Fujii, H., and Suzuki, E. (1977). Hum. Genet. 36, 327.
7. Beutler, E., (1978). In "Hemolytic Anemia in Disorders of Red Cell Metabolism", Plenum Medical Book Co., New York.
8. Nakashima, K., Ono, J., Abe, S., Miwa, S., and Yoshida, A.

(1977). Am. J. Hum. Genet. 29, 24.
9. Nakatsuji, T., and Miwa, S. (1979). Hum. Genet. 51, 297.
10. Miwa, S., Ono, J., Nakashima, K., Abe, S., Kageoka, T., Shinohara, K., Isobe, J., and Yamaguchi, H. (1976). Amer. J. Hemat. 1, 433.
11. Miwa, S., Fujii, H., Nakashima, K., Miura, Y., Yamada, K., Hagiwara, T., and Fukuda, M. (1978). Hum. Genet. 45, 11.
12. Miwa, S., Fujii, H., Nakatsuji, T., Ishida, Y., Oda, E., Kaneto, A., Motokawa, M., Ariga, Y., Fukuchi, S., Sasai, S., Hiraoka, K., Kashii, H., Kodama, T., and Miura, Y. (1978). Amer. J. Hemat. 5, 131.
13. Fujii, H., Miwa, S., Takegawa, S., Takahashi, K., Hirono, A., Takizawa, T., Morisaki, T., Kanno, H., Taguchi, T., and Okamura, J., (1984). Hum. Genet. 66, 276.
14. Fujii, H., Miwa, S., Tani, K., Takegawa, S., Fujinami, N., Takahashi, K., Nakayama, S., Konno, M., and Sato, T. (1981). Hum. Genet. 58, 405.
15. Takahashi, K., Fujii, H., Takegawa, S., Tani, K., Hirono, A., Takizawa, T., Kawakatsu, T., and Miwa, S. (1982). Hum. Genet. 62, 368,
16. Morisaki, T., Fujii, H., Takegawa, S., Tani, K., Hirono, A., Takizawa, T., Takahashi, K., Shinogi, M., Teshirogi, T., and Miwa, S. (1983). Hum. Genet. 65, 214.
17. Takizawa, T., Fujii, H., Takegawa, S., Takahashi, K., Hirono, A., Morisaki, T., Kanno, H., Oka, R., Yoshioka, H., and Miwa, S. (1984). Hum. Genet. 68, 70.
18. Yoshida, A., (1966). J. Biol. Chem. 241, 4966.
19. Chan, T. K., Todd, D., and Wong, C. C. (1964), Br. Med. J. 2, 102.
20. Lee, T. C., Shin, L. Y., Huang, P. C., Lin, C. C., Blackwell, B. N., Blackwell, R. Q., and Hsia, D. Y. Y. (1963). Am. J. Hum. Genet. 15,126.
21. Motulsky, A. G., Stransky, E., and Fraser, G. R. (1964). J. Med. Genet. 1, 102.
22. Yoshida, A., (1977). Acta. Biol. Med. Germ. 36, 689.

DISCUSSION

Paul R. McCurdy, M.D.

American Red Cross
Washington Regional Blood Services
Washington, D.C., USA

My interest in red cell G-6-PD deficiency stemmed from an early opportunity to collaborate with Dr. Neil Kirkman in the partial purification and characterization of a variant of this enzyme from the red cells of a Chinese male found to have G-6-PD deficiency (1). The geographic distribution of this genetic polymorphism was just being mapped and I set out to fill in a few of the areas with biochemical descriptions of the variant or variants found. The simple description of a new variant seemed to me to be incomplete unless it was put in the perspectives of biochemical and population genetics. Being a bit of a "jury-rigged" biochemist, I was nervous about ascribing too much to what appeared to me to be small differences in biochemical characteristics. I still feel constrained to ask, "How much of a difference need there be to separate one similar variant from another?"

Additionally, I was intrigued by this clear example of the interaction between heredity and environment. Although I had been introduced to genetic polymorphisms through my exposure to sickle cell disease and other abnormal hemoglobins, I found G-6-PD deficiency to be a model for how structural changes in proteins might alter molecular function, even to the point of changing substrate specificity and forming a "new" enzyme. This is illustrated in Figure 1 for G-6-PD Benevento, a variant from Italy. The reaction is started with the normal substrate, glucose-6-phosphate; at this point, 2-deoxy-G-6-P is added and the reaction speeds up. The enzyme prefers the abnormal substrate over the normal one. Having a predominantly clinical background, I wanted to correlate the biochemical characteristics of a variant enzyme with the clinical syndrome found in the patient. I believe we saw this morning that this correlation still eludes us to a significant degree.

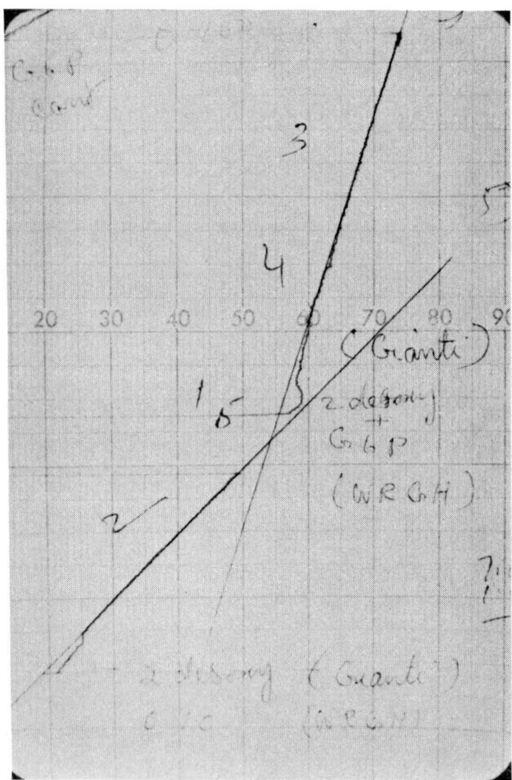

FIGURE 1: Tracing of OD changes (340μ) as NADPH is generated. Initially the only hexose substrate is glucose-6-phosphate. When 2 deoxy G-6-P is added, the slope abruptly steepens as the reaction rate increases.

From my involvement with the abnormal hemoglobins, it came as no surprise to me when Yoshida demonstrated single amino acid sustitutions in two G-6-PD variants, G-6-PD A^{+} and a high activity variant, Hektoen (2,3). We shall learn more about structural changes later in this conference. One might expect amino acid substitutions and perhaps even thalassemia-like mutations, since there are now good examples of each in a number of proteins, including some blood group antigens (4) and clotting proteins such as Factor VIII (5).

Our opportunities to characterize G-6-PD variants came both from chance appearance of subjects from various ethnic groups and from systematic studies. The former were often house staff or subjects for other studies. Some of our opportunities of both types involved the Near East. As one might expect from history, there is considerable heterogeneity there. G-6-PD Mediterranean is widespread; we found examples in an Ashkenazi Jewish man whose family came from Kiev (6), in Iranian subjects, in Saudi Arabia and in Egypt (7). G-6-PD Athens (or "Athens-like") has also been found widely dispersed in the Near East. We found it in Pakistan, Saudi Arabia, Abu Dahbi and Egypt. Our study of Egyptians resulted in the discovery of a veritable hodge-podge (7). Fourteen previously recognized variants were found and four new ones described among 38 families. We were able to get specimens from several oases, as well as from Cairo and the Nile Delta region (Figure 2). The lack of a pattern disturbed me but may result from Egypt's role as a crossroads in war and commerce throughout history. Our Saudi* studies are incomplete; samples were small and repeats not available. There appeared to have been examples of G-6-PD's Mediterranean, "Athens-like", A- and possibly Clichy. Studies of 11 individuals from Abu Dahbi*

* *These studies were done in conjunction with Dr. Karim Kamel, Cairo, Egypt.*

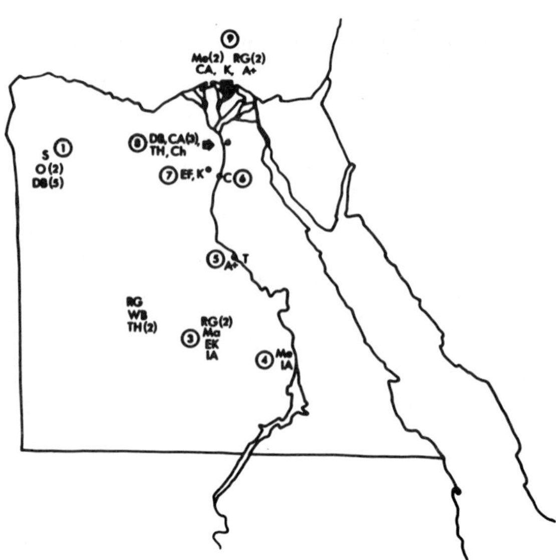

FIGURE 2: Distribution of G-6-PD variants within Egypt. Areas are numbered: *(1) Siwa Oasis; (2) El Dakhla Oasis; (3) El-Kharga Oasis; (4) Upper Egypt (Aswan Governorate); (5) Tahta; (6) Bani Suef; (7) El-Fayoum Governorate; (8) Cairo; and (9) Delta region.* Variants are signified by initials: *S = G-6-PD Siwa; O = G-6-PD Ohio; DB = G-6-PD Debrousse/A- (one of those from the Siwa Oasis is probably A-; the others seem more homogeneously Debrousse - see text); R-G = G-6-PD Ramat-Gan; WB = G-6-PD West Bengal; TH = G-6-PD Tel-Hashomer; Ma = G-6-PD Madrona; EK = G-6-PD El-Kharga; IA = G-6-PD Ibadan-Austin; Me = G-6-PD Mediterranean; A+ = G-6-PD A+; T = G-6-PD Tahta; C = G-6-PD Columbus; EF = G-6-PD El-Fayoum; K= G-6-PD Kabyle; CA = G-6-PD Corinth-Athens ("Athens-like"); and Ch = G-6-PD Chicago.*

revealed examples of Siwa, Athens, Kabyle and probably El-Kargha (7). In the far East, we have been able to find additional examples of G-6-PD Canton, as well as describe some previously unknown variants among the aborigines of Taiwan (8).

From studies presented this afternoon, it seems that genetic variablilty in the red cell enzyme G-6-PD is very large. But it is rather peculiar, I think, that there is one variant, A-, that is overwhelmingly predominant in subjects of African descent. Why, I wonder, is the variability elsewhere so great? Or in

Equatorial Africa so small? Is it more apparent than real? Is there a polymorphism peculiar to G-6-PD A- (and even A+)? I look forward to hearing about the molecular genetic studies to be presented later in this conference. Perhaps from those studies will come better understanding.

Before I close, I would like to add one additional bit of information, perhaps even more pertinent to this morning's presentations than to those after lunch. It has often been assumed that subjects with G-6-PD A- have a normal red cell life span when not subjected to oxidative stress. Some data has been published to the contrary (9). In previously unpublished studies, shown in Figure 3, a red cell survival curve that is not different from normal subjects was obtained in each of five black male subjects with G-6-PD A-. For other deficient variables, the data are limited, but there is reason to suggest that unstressed red cell survival may be shortened. This is probably true for subjects with G-6-PD Mediterranean.

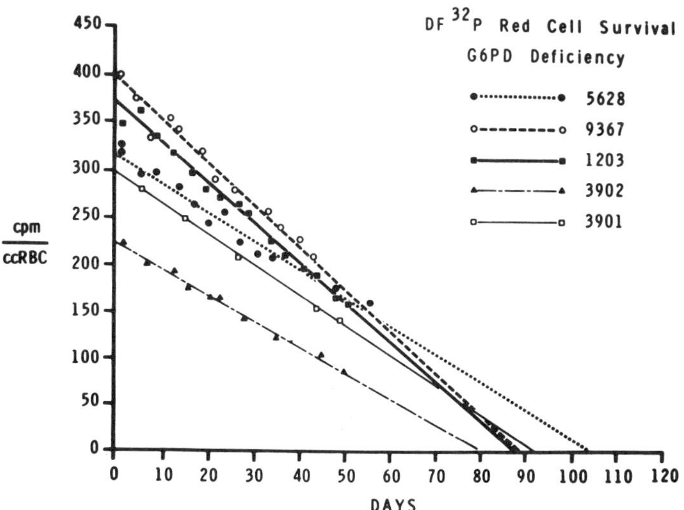

FIGURE 3: Red cell survival studies on males with G-6-PD A-. There was no apparent source of oxidative stress. Tag was DF^{32}P. Extrapolation to the base-line is not greatly different for a small number of studies done in normal individuals and in those with sickle cell trait (10).

REFERENCES

1. McCurdy, P.R., Kirkman, H.N., Naiman, J.L., Jim, R.T.S., and Pickard, B.M. (1966). J. Lab. and Clin. Med. 67: 374.
2. Yoshida, A. (1967) Proc. Nat. Acad. Sci. 57: 835.
3. Yoshida, A. (1970) J. Mol. Biol. 52: 483.
4. McCurdy, P.R. and Luban, N.L.C. (1984) *In* "Current Hematology and Oncology" (V.F. Fairbanks, ed.), Vol. 3, p. 103. John Wiley & Sons, New York.
5. Antonarakis, S.E., Waber, P.G., Kittur, S.D., *et al.* (1985). New Eng. J. Med. 313: 842.
6. George, J.N., Sears, D.A., McCurdy, P.R., and Conrad, M.E. (1967). J. Lab. and Clin. Med. 70: 80.
7. McCurdy, P.R., Kamel, K., and Selim, O. (1974). J. Lab and Clin. Med. 75: 788.
8. McCurdy, P.R., Blackwell, R.Q., Todd, D., Tso, S.C., and Tuchinda, S. (1970) J. Lab. and Clin. Med. 75: 788.
9. Brewer, G.J., Tarlov, A.R., and Kellermeyer, R.W. (1961) J. Lab. and Clin. Med. 58: 217.
10. Barbedo, M.M.R., and McCurdy, P.R. (1974) Acta Haematologica 51: 339.

GLUCOSE-6-PHOSPHATE DEHYDROGENASE FROM *LEUCONOSTOC MESENTEROIDES*[1]

H. Richard Levy

Department of Biology, Syracuse University
Syracuse, New York, USA

I. INTRODUCTION

In the context of this symposium, it is of interest that both glucose 6-phosphate dehydrogenase (G6PD) and $NADP^+$ were first discovered in erythrocytes (1,2). During the subsequent two decades, G6PD was isolated from a variety of animals, plants and microorganisms (3). These enzymes all utilized $NADP^+$ as the coenzyme for the oxidation of glucose 6-phosphate. As other NADP-linked enzymes were discovered and as the metabolic roles of the oxidized and reduced nicotinamide coenzymes were recognized, it became clear that despite their great structural similarity, NAD and NADP functioned quite differently in metabolism. In particular, the oxidation of NADH is generally associated with catabolic, ATP- generating processes, such as mitochondrial electron transport, whereas NADPH oxidation accompanies

---

[1]Unpublished studies cited from the author's laboratory were supported by grant PCM 8309379 from the National Science Foundation and by BSRG grant SO7 RR0770G8-19 awarded by the Biomedical Research Support Grant Program, Division of Research Resources, National Institutes of Health.

ATP-utilizing, anabolic reactions such as fatty acid synthesis (4). With very few exceptions (e.g. glutamate dehydrogenase in animal tissues (5) and homoserine dehydrogenase in plants (6)) enzymes that utilize nicotinamide coenzymes are specific for either NAD or for NADP.

In 1953, DeMoss *et al* described a G6PD from the bacterium *Leuconostoc mesenteroides* that could utilize both NAD^+ and $NADP^+$ at physiological concentrations (7). Subsequently, other microbial G6PDs with dual nucleotide specificity were isolated (3). Our initial interest in the *L. mesenteroides* enzyme arose as a result of our observations that G6PD from various mammalian tissues could react with high concentrations of NAD^+ (8) and that the NAD- and NADP-linked activities of mammary G6PD appeared to be associated with different conformational states of the enzyme (9,10). This weak, NAD-linked activity, which was also observed in human erythrocyte G6PD by Kirkman (11), is not known to have a physiological function, but we were interested in the structural basis of the dual nucleotide specificity and so we turned to the *L. mesenteroides* G6PD as a model for mammalian G6PDs. What was initially a diversion from our main research interest subsequently became a primary concern. This chapter summarizes the current status of our understanding of *L. mesenteroides* G6PD.

L. mesenteroides is a facultative anaerobe that metabolizes glucose *via* heterolactic fermentation to yield ethanol, lactate and CO_2 (12). Kemp and Rose showed that this G6PD utilizes NAD^+ and $NADP^+$ *in vivo* (13). Using glucose specifically labeled with ^3H, these authors presented data indicating that the NADH generated via the G6PD-catalyzed reaction is utilized primarily for the synthesis of ethanol and lactate, whereas NADPH is utilized for fatty acid synthesis

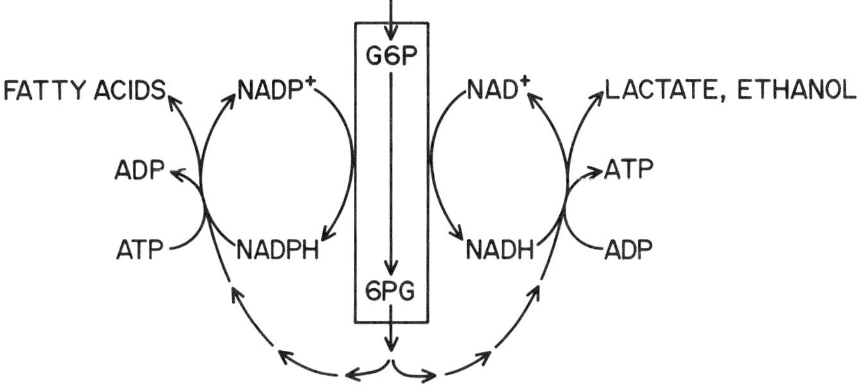

Fig. 1 The amphibolic role of G6PD in the metabolism of *L. mesenteroides.* The oxidation of glucose 6-phosphate (G6P) to 6-phosphogluconolactone (6PG) is linked to the catabolic production of ATP, lactate and ethanol when G6PD utilizes NAD^+, and to the anabolic, ATP-dependent biosynthesis of fatty acids when G6PD utilizes $NADP^+$.

(13). No NADPH-generating enzymes, other than G6PD, have been described in *L. mesenteroides*; 6-phosphogluconate dehydrogenase is NAD-specific in this organism (12). Thus, *L. mesenteroides* G6PD is an amphibolic enzyme whose NAD- and NADP-linked activities are linked to catabolic and anabolic reactions, respectively (Fig. 1). This fact underscores the importance of regulation of the enzyme in *L. mesenteroides*, and this is one aspect of the dual nucleotide-linked activity that has interested us. A second aspect that has concerned us is the delineation of protein structural features that are involved in binding NAD^+ and $NADP^+$ and in the enzyme's ability to select between the two coenzymes.

II. MOLECULAR PROPERTIES

L. mesenteroides G6PD was extensively purified in 1967 (14) and subsequently shown to be homogeneous (15). A rapid purification procedure involving dye-ligand chromatography has been published (16). The molecular weight of the native enzymes is 103,700 (15); it consists of two identical subunits (17) of molecular weight 54,800 (18) and, unlike GGPDs from animal tissues or brewer's yeast, it shows no propensity to form tetramers (15). The isoelectric point of the enzyme is approximately 4.6 (15).

The amino acid composition of *L. mesenteroides* G6PD is unusual as this protein contains no cysteine (18). The absence of cysteine has not been reported in other G6PDs and is extremely rare among dehydrogenases (3). The enzyme has valine and glycine residues at its N- and C-terminus, respectively (17). The complete amino acid sequence is currently being determined in a collaborative project by Dr. T. Geoffrey Flynn, at Queen's University in Kingston, Ontario, and myself. Using cyanogen bromide cleavage, tryptic hydrolysis following maleylation, and thermolysin digestion, we have prepared a large number of peptides. These have been purified to homogeneity by high-performance liquid chromatography and many have now been sequenced, enabling us to account for over half the enzyme's amino acids, including the 50-amino acid N-terminal region.

Of the enzyme's 37 lysines per subunit, only one reacts with pyridoxal 5'-phosphate (19). The function of this lysine appears to be to bind glucose 6-phosphate (19). A peptide containing this lysine was isolated and sequenced (17); it conforms to those structural features, described by Minchiotti *et al* (20), that are typical for pyridoxal-5'-phosphate binding sites in a number of enzymes.

Two of the enzyme's 18 arginine residues per subunit are modified with 2,3-butanedione, resulting in complete inactivation. Neither of these arginines functions in coenzyme binding; the inactivated enzyme still binds both NAD^+ and NADPH (21). We do not yet know the function of these arginines; one of them may interact with the phosphate group of the substrate (21).

The fact that $NADP^+$ binds to *L. mesenteroides* G6PD a thousand-fold tighter than NAD^+ (see below) suggests that there is a specific interaction between its 2'-phosphate group and an amino acid. Brenda White (unpublished) has shown that interaction of the enzyme with diethylpyrocarbonate leads to modification of one histidine and one or more lysine residues. The histidine modification does not inactivate the enzyme or affect its binding properties but modification of lysine(s) leads to inactivation and inability to bind $NADP^+$ or NADPH. Photooxidation of the enzyme in the presence of Rose Bengal causes inactivation, loss of histidine and tyrosine residues, and inability to bind $NADP^+$ or NADPH. These experiments did not provide unequivocal evidence for the involvement of specific amino acid residues in NADP binding. Ms. White also prepared the dialdehyde analog of $NADP^+$ ($oNADP^+$) and examined its interaction with *L. mesenteroides* G6PD (White, B., and Levy, H. R., manuscript in preparation). $oNADP^+$ satisfies all the criteria for an affinity analog of *L. mesenteroides* G6PD. It is an effective coenzyme in kinetic assays. Upon incubation with the enzyme it causes time-dependent, irreversible inactivation that involves the initial formation of a reversible complex followed by covalent modification, with a pseudo first-order kinetic dependence on $oNADP^+$ concentration. $NADP^+$ protects against $oNADP^+$ modification, but NAD^+ does not. The pH dependence of inactivation exhibits a pK of 7.2. The binding of $oNADP^+$ involves the formation, not of a Schiff base but of a dimorpholino derivative. The $oNADP^+$- modified enzyme does not bind NADPH. Using radioactive $oNADP^+$, Ms. White has modified _L. mesenteroides_ G6PD and subjected it to thermolysin digestion. The isolation and sequencing of the modified peptide is currently in progress.

In 1983 we reported preliminary X-ray data on _L. mesenteroides_ G6PD (22). Since that time, X-ray structural studies on the enzyme have been pursued at Oxford in Dr. Margaret Adams' laboratory. A complete native data set has been collected to 3.5Å resolution. The enzyme crystallizes with a dimer in the asymmetric unit and there is a peak corresponding to a non-crystallographic 2-fold axis relating the two monomers in the asymmetric unit. Data have also been collected to 3.5Å on two heavy atom derivatives. From these, a good solution has been obtained recently for a major heavy atom site (Bhadbhade, M. and Adams, M. J., personal communication).

The active enzyme dimer is dissociated to inactive subunits in 8M urea; removal of the urea by dilution leads to reassociation of the enzyme to the catalytically active dimer which is indistinguishable from the native enzyme by a variety of criteria (23). The kinetics of reassociation were monitored by measuring regain of catalytic activity, fluorescence, and susceptibility of histidine residues to diethylpyrocarbonate modification. These studies enabled us to propose a model in which the inactive subunits first undergo one or more rapid conformational changes, followed by a slower, second-order process that results in restoration of the enzyme's native structure and catalytic activity (23).

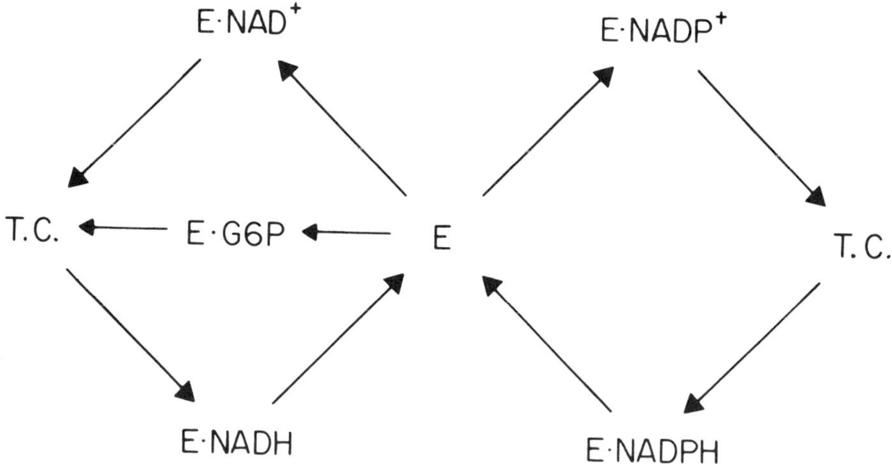

Fig. 2 The kinetic mechanisms of the NAD- and NADP-linked reactions catalyzed by *L. mesenteroides* G6PD. E, enzyme; G6P, glucose 6-phosphate; T. C., ternary complex.

III. KINETIC STUDIES

The steady state kinetic mechanisms of the NADP- and the NAD-linked reactions catalyzed by *L. mesenteroides* G6PD differ (Fig. 2). The NADP-linked reaction proceeds *via* an ordered sequential mechanism in which the coenzyme binds first and is released last (24). The NAD-linked reaction was, at first, postulated to proceed via an ordered sequential mechanism that includes an obligatory enzyme isomerization step (24), but more extensive investigation has shown that the mechanism is random (25). Kinetic constants for the substrate, coenzymes and thionicotinamide analogs for the coenzymes are summarized in Table I (24, 26). The V_{max} for the NAD-linked reaction is greater than that of the NADP-linked reaction, but the K_m for $NADP^+$ is nearly 20 times lower than that for NAD^+. The thionicotinamide coenzyme analogs, which proved to be useful in regulatory studies described below, displayed K_m values that are substantially lower than those of the natural coenzymes (26).

TABLE I   Kinetic Constants for *L. mesenteroides* G6PD[a]

	NAD^+	$NADP^+$	S-NAD^+	S-$NADP^+$
k_m (mM)	106	5.69	11.0	1.23
k_{cat} (min^{-1})	31,100	17,300	6,480	3,500
k_{cat}/K_m (min^{-1} mM^{-1})	293	3,040	589	2,850
K_m for G6P (mM)[b]	52.7	81.0	7.43	9.24

[a]Data from references 24 and 26

[b]K_m for glucose 6-phosphate with each of coenzymes or coenzyme analogs

Kinetic studies, at various pH values, on the NAD-linked reaction revealed that V_{max} is invariant between pH 6.5 and pH 9.9 (24). Viola's studies on the NADP-linked reaction revealed two pK values in the V_{max} profile, at pH 6.3 and pH 8.7 (27). The latter was postulated to reflect a group on the enzyme which, in its unionized form, limited the rate of product release (27). The V/K_{G6P} profile for the NAD-linked reaction provided evidence for a group with a pK of 8.9 that is involved in substrate binding (24); this group was subsequently shown to be a lysine residue (see above). The V/K_{G6P} profile of the NADP-linked reaction showed that ionization of two groups, with pK values of 9.3 and 10.3, and protonation of two groups, with pK values of 4.5 and 5.5, caused decreased activity (27). The pK of 5.5 was identified as that of glucose 6-phosphate, demonstrating that the substrate must be in the dianionic form to bind to the enzyme. The group with a pK of 9.3 is, presumably, the lysine residue identified in the experiments with NAD^+ (14). Viola suggested that the group with a pK of 4.5 is involved in catalysis, functioning as the general base which accepts the proton, from the hydroxyl group on the anomeric carbon atom, that is released when the substrate is oxidized to the lactone (27). It would be useful to attempt to identify such a group by chemical modification.

IV. BINDING STUDIES

Dissociation constants for various ligands are summarized in Table II. The K_D for $NADP^+$, determined from fluorescence quenching measurements (28) and equilibrium dialysis measurements (29) in different buffers, is 3μM or 6.5μM, respectively. From equilibrium dialysis studies it appears that there are two $NADP^+$ binding sites per molecule of

TABLE II Dissociation Constants for Various Ligands

Ligand	K_D, mM	Method of Determination
NAD^+	2,500	Fluorescence quenching, Tris acetate pH 7.8 (28)
	7,200	Fluorescence quenching, phosphate pH 7.6 (29)
$NADP^+$	3.0	Fluorescence quenching, Tris acetate pH 7.8 (28)
	6.5	Equilibrium dialysis, phosphate pH 7.6 (29)
NADPH	37.6	Kinetics, Tris chloride pH 7.2 (24)
	24.0	Fluorescence enhancement, Tris acetate pH 7.8 (28)
S-NADPH	16.4	Fluorescence quenching, phosphate pH 7.6 (29)
Glucose 6-phosphate	630	Kinetics, phosphate pH 7.8 (19)

dimeric enzyme (29). The K_D for NAD^+ is 2.5mM (23) or 7.2mM (29), depending on the buffer used. The difference between the ratio of K_m and K_D values for the two coenzymes reflects two aspects of the interaction between the enzyme and coenzymes. First, the much larger K_D value for NAD^+, compared to that for $NADP^+$, suggests that a significant portion of the free energy of NAD^+ binding is utilized in altering the enzyme's conformation (30). Evidence to support this idea is presented below. Second, the binding of NAD^+, but not $NADP^+$, is promoted by glucose 6-phosphate binding. This is illustrated by the fact that the K_m values for $NADP^+$ and glucose 6-phosphate are invariant with the concentration of glucose 6-phosphate and $NADP^+$, respectively, whereas the apparent K_m values for NAD^+ and glucose 6-phosphate decrease with increasing concentration of the other ligand (24). Furthermore, the K_D of NAD^+, but not of $NADP^+$, is decreased in the presence of glucose 1-phosphate, a substrate-competitive inhibitor (28) and is also decreased in the enzyme that has been pyridoxylated at the lysine residue which appears to bind glucose 6-phosphate (29). The importance of glucose 6-phosphate for NAD^+ binding is also exemplified by its role in the regulation of coenzyme utilization, discussed below. The binding constant for glucose 6-phosphate to _L. mesenteroides_ G6PD has been measured only indirectly, using kinetic studies, and found to be 0.63 mM (19).

An aspect of ligand binding to the enzyme that has occupied us for some time is the conformational changes associated with the binding of NAD^+, $NADP^+$ and glucose 6-phosphate. In particular, we are concerned with establishing whether NAD^+ and $NADP^+$ induce different conformational changes in the enzyme. This is of interest in understanding how the same enzyme can utilize both coenzymes efficiently for catalysis.

From the demonstration that NADPH inhibits competitively with respect to $NADP^+$ (24) and that the binding of NADPH is competitive with NAD^+ binding (28), it is evident that NAD^+ and $NADP^+$ bind to the same site on the enzyme, although it appears likely that their precise mode of interaction differs. The fact that $NADP^+$ binds so much more tightly than NAD^+ suggests that one or more amino acid side chains interact specifically with the 2'-phosphate of $NADP^+$, but, so far, we have been unable to identify such a group on the enzyme. Alternatively, the binding of NAD^+ necessitates a rearrangement of amino acid side chains in the coenzyme binding region to accommodate the smaller coenzyme, perhaps in an orientation that is not optimal for catalysis unless glucose 6-phosphate is also present. This would not only be in keeping with the argument noted above, concerning the utilization of a portion of the free energy of NAD^+ binding to induce a conformational change in the enzyme, it would also be consistent with recent transferred nuclear Overhauser enhancement (TRNOE) studies on the conformations of NAD^+ and $NADP^+$ bound to _L. mesenteroides_ G6PD in solution (31). These proton nuclear magnetic resonance studies showed that, whereas for $NADP^+$ bound to the enzyme the conformation of the nicotinamide-ribose bond is syn, in a significant proportion of the bound NAD^+ molecules, the conformation of this bond is anti. With the pyridoxylated enzyme, however, the conformation of the nicotinamide-ribose bond of the bound NAD^+ is syn, suggesting that pyridoxal phosphate, and, by implication, glucose 6-phosphate, helps to place NAD^+ into a conformation that assures strict β-stereospecificity for H transfer (32).

Several analogs of the adenine and nicotinamide portions of the coenzyme inhibit the NAD-linked reaction competitively with respect to NAD^+ (28). All the adenine analogs have lower K_i values than the nicotinamide analogs, indicating that the adenine moiety of the coenzyme binds tighter to its subsite on the enzyme than the nicotinamide portion binds to its subsite. Multiple inhibitor studies showed that binding of the two types of analogs was mutually exclusive, and the simultaneous addition of nicotinamide mononucleotide and ADP-ribose, unlike NAD^+ addition, caused no quenching of the enzyme's fluorescence (28). These results suggest that the adenine and nicotinamide binding regions of the NAD^+ binding site must be moved apart by NAD^+ as it binds and that this necessitates that the two halves of the molecule be covalently linked.

There is now extensive, albeit indirect evidence, that the binding of NAD^+ to _L. mesenteroides_ G6PD induces a conformational change that is different from that produced by $NADP^+$ binding. First, the 2,3-butanedione-induced inactivation of the enzyme can be completely prevented by NAD^+, whereas even high concentrations of $NADP^+$ provide no more than 50% protection against this reagent (21). Second, the enzyme's intrinsic fluorescence can be extensively quenched by NAD^+ binding, whereas $NADP^+$ binding produces only small fluorescence quenching (28,29). Third, the fluorescence of the pyridoxyl group of pyridoxylated enzyme is quenched to a greater extent by $NADP^+$ binding than by NAD^+ binding (29). Fourth, the fluorescence of S-NADPH bound to one subunit of _L. mesenteroides_ G6PD is quenched when NAD^+ binds to the other subunit, whereas it is enhanced when $NADP^+$ binds to the other subunit (29). Finally, the inactivation of _L. mesenteroides_ G6PD by heat, urea and four different proteolytic enzymes is protected to a greater extent by 10 K_D NAD^+ than by 10 K_D

Table III Effects of $NADP^+$ and NAD^+ on Enzyme Inactivation

Inactivation Procedure	$k_{inact.}$ (min^{-1})		
	No ligand	$10K_D NADP^+$	$10K_D NAD^+$
Trypsin	0.025	0.025	0.015
Chymotrypsin	0.020	0.013	0.006
Thermolysin	0.057	0.051	0.013
Pronase	0.012	0.010	0.006
Urea, 4M	0.019	0.014	0.011
Heat, 49°C	0.050	0.043	0.020

$NADP^+$ (Hilburger, A., White, B. and Levy, H. R., manuscript in preparation). These experiments were conducted under conditions which resulted in first-order kinetics of inactivation; the rate constants are summarized in Table III. The fact that for all six inactivation processes, as well as for 2,3-butanedione inactivation (21), NAD^+ protected better than $NADP^+$, suggests that NAD^+ can induce a greater conformational change of the enzyme than $NADP^+$ (33). This, in turn, may help to account for the fact that the K_D for NAD^+ is a thousand-fold greater than that for $NADP^+$: a substantial portion of the intrinsic free energy of NAD^+ binding may be used to bring about the conformational changes in the enzyme (30).

V. REGULATION

The fact that NADH and NADPH, generated by the G6PD-catalyzed reaction in growing *L. mesenteroides*, are utilized for catabolic and anabolic reactions, respectively (see Introduction), strongly suggests that the choice of coenzyme utilized by G6PD must be strictly controlled. An extensive search for likely metabolic intermediates and other factors that could differentially modulate the NAD- and NADP-linked reactions catalyzed by *L. mesenteroides* G6PD *in vitro* proved unsuccessful (34). In all these experiments, conventional assays of either the NAD- or the NADP-linked reaction were performed under various conditions. A novel method was then designed, that enabled us to search for possible regulatory mechanisms under conditions where both the NAD- and NADP-linked reactions could proceed simultaneously, i.e. a procedure that more closely approached physiological conditions than did conventional assays. The thionicotinamide analogs of NAD^+ and $NADP^+$, upon enzymatic reduction, absorb maximally at 398 nm (35). This wave length is sufficiently far from 340 nm, the absorbance maximum for NADH and NADPH, that it is possible to monitor the reduction of the natural coenzyme and its thionicotinamide analog simultaneously. We established that S-NAD^+ and S-$NADP^+$ were highly reactive with L. mesenteroides G6PD (see Table I) and that they behaved, kinetically, like NAD^+ and $NADP^+$, respectively (26). This enabled us to perform dual wave length assays at 340 nm and 400 nm which measured the simultaneous reduction of S-NAD^+ and $NADP^+$, or NAD^+ and S-$NADP^+$, by glucose 6-phosphate in the presence of L. mesenteroides G6PD (26). The replacement of NAD^+ or $NADP^+$ by its thionicotinamide analog affects not only the kinetic constant of the coenzyme, but also of glucose 6-phosphate (see Table I). These effects must be taken into account in interpreting the reaction rates. The dual wave length assays enabled us to uncover effects on reaction rates that are impossible to detect using conventional assays (26,36). The most significant of these effects are the simultaneous stimulation of the NAD-linked and inhibition of the NADP-linked reaction by two different conditions: increased glucose 6-phosphate concentration (26) and increased ratios of NADPH to $NADP^+$ concentrations (36) (Fig. 3). Both these effects are readily explained by the difference between the kinetic mechanisms for the NAD- and NADP-linked reactions (26) and are the direct result of the fact that the enzyme-glucose 6-phosphate complex can react with NAD^+, but not with $NADP^+$, to give a kinetically competent ternary complex. It is also clear, from examining the kinetic mechanisms (Fig. 2), that the demonstration of these effects is completely dependent upon the simultaneous catalysis of both the NAD- and NADP-linked reactions.

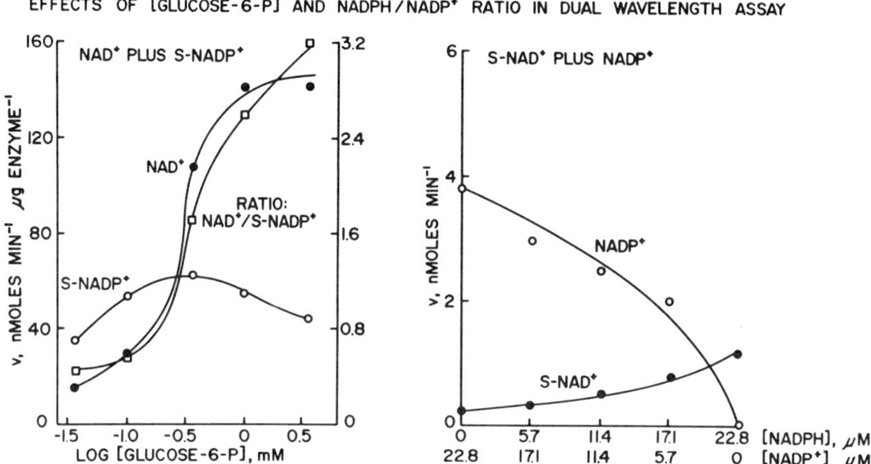

Fig. 3 The effects of glucose 6-phosphate concentration (left) and the NADPH/NADP+ concentration ratio (right) on the NAD- and NADP-linked reactions catalyzed by *L. mesenteroides* G6PD. The dual wavelength assay was used (26). In the experiment illustrated on the left, NAD+ and S-NADP+ were each present at 4.0 times their respective K_m concentrations (data from reference 26). In the experiment illustrated on the right, S-NAD+ was present at 4.0 times its K_m concentration and the concentration of glucose 6-phosphate was 0.06 mM (Archives of Biochemistry and Biophysics 198:411, 1979. Reprinted with permission).

Enhancement of the NAD-linked, at the expense of the NADP-linked reaction, by high concentration of glucose 6-phosphate and by a high NADPH to NADP+ concentration ratio appears to be logical in the context of the metabolism of *L. mesenteroides.* Whether these effects actually operate *in vivo* has not been determined.

VI. CONCLUSIONS

The G6PD from L. mesenteroides differs in a number of respects from other G6PDs, especially from human and other mammalian G6PDs. Most importantly, and related to its physiological role, L. mesenteroides G6PD reacts with both NAD^+ and $NADP^+$ under physiological conditions. Although it is not unique in this respect (3), L. mesenteroides G6PD has been investigated in sufficient detail to allow us to reach a broad understanding about how this enzyme can utilize two coenzymes that are structurally so similar, but whose physiological roles are so different. It appears that the enzyme can exist in two different conformations; one, induced by the binding of either NAD^+ or glucose 6-phosphate, in which it catalyzes the NAD-linked reaction, and one, induced by $NADP^+$, in which it catalyzes the NADP-linked reaction. In addition, the kinetic mechanisms of the two reactions differ in a manner that enables the enzyme to respond to changes in glucose 6-phosphate concentration and NADPH/$NADP^+$ concentration ratios by selecting either NAD^+ or $NADP^+$ as the coenzyme for oxidizing glucose 6-phosphate.

Although mammalian G6PDs utilize $NADP^+$ and not NAD^+ under physiological conditions, it is a curious fact that they display a weak NAD-linked activity which appears to be associated with a different enzyme conformation than the NADP-linked activity (8-11). The NAD-linked activity does not appear to be physiologically significant per se, but a conformational transition associated with NAD^+ binding might reflect an unrecognized regulatory role for these enzymes.

There are several other differences between the L. mesenteroides G6PD and mammalian G6PDs, the significance of which is not yet clear. L. mesenteroides G6PD lacks cysteine residues. It does not readily form tetramers. It is not inhibited by dehydroepiandrosterone (14), a potent inhibitor of mammalian G6PDs (37,38). And it is only weakly inhibited by palmitoyl-CoA, without affecting its quaternary structure (14,39), whereas palmitoyl-CoA is a potent inhibitor of G6PDs from *Saccharomyces cerevisiae* (39) and rat mammary gland (3), enzymes that it dissociates to dimers (39) and monomers (3), respectively. We hope that these and other questions will be clarified when the X-ray structures of *L. mesenteroides* G6PD and some of its complexes are solved, and that the ongoing studies with this enzyme will deepen our understanding of the interactions between dehydrogenases and the nicotinamide coenzymes.

REFERENCES

1. Warburg, O. and Christian, W. (1931). Biochem. Z. 238, 131.

2. Warburg, O. and Christian, W. (1936). Biochem. Z. 287, 291.

3. Levy,H. R. (1979). Adv. Enzymol. Relat. Areas Mol. Biol. 48, 97.

4. Kaplan, N. O., Swartz, M. N., Frech, M. N. and Ciotti, M. M. (1956). Proc. Nat. Acad. Sci. U.S. 42, 481.

5. Smith, E. L., Austen, B. M., Blumenthal, K. M. and Nyc, J. F. (1975). Enzymes, 3rd. Ed. 11, 293.

6. Bryan, J. K. (1980). *In*: "The Biochemistry of Plants; A Comprehensive Treatise" (P. K. Stumpf, and E. E. Conn, eds.) Vol. 5, pp. 403, Academic Press, New York.

7. DeMoss, R. D., Gunsalus, I. C. and Bard, R. C. (1953). J. Bacteriol. 66, 10.

8. Levy, H. R. (1961). Biochem. Biophys. Res. Commun. 6, 49.

9. Levy, H. R. (1963). J. Biol. Chem. 238, 775.

10. Levy, H. R., Raineri, R. R. and Nevaldine, B. H. (1966). J. Biol. Chem. 241, 2181.

11. Kirkman, H. N. (1962). J. Biol. Chem. 237, 2364.

12. DeMoss, R. D., Bard, R. C. and Gunsalus, I. C. (1951). J. Bacteriol. 62, 499.

13. Kemp, R. G. and Rose, I. A. (1964). J. Biol. Chem. 239, 2998.

14. Olive, C. and Levy, H. R. (1967). Biochemistry 6, 730.

15. Olive, C. and Levy, H. R. (1971). J. Biol. Chem. 246, 2043.

16. Hey, Y. and Dean, P. D. G. (1983). Biochem. J. 209, 363.

17. Haghighi, B., Flynn, T. G. and Levy, H. R. (1982). Biochemistry 21, 6415.

18. Ishaque, A., Milhausen, M. and Levy, H. R. (1974). Biochem. Biophys. Res. Commun. 59, 894.

19. Milhausen, M. and Levy, H. R. (1975). Eur. J. Biochem. 50, 453.

20. Minchiotti, L., Ronchi S. and Rippa, M. (1981). Biochim. Biophys. Acta 657, 232.

21. Levy, H. R., Ingulli, J. and Afolayan, A. (1977). J. Biol. Chem. 252, 3745.

22. Adams, M. J., Levy, H. R. and Moffat, K. (1983). J. Biol. Chem. 258, 5867.

23. Haghighi, B. and Levy, H. R. (1982). Biochemistry 21, 6429.

24. Olive, C., Geroch, M. E. and Levy, H. R. (1971). J. Biol. Chem. 246, 2047.

25. Levy, H. R., Christoff, M. E., Ingulli, J. and Ho, E. M. L. (1983). Arch. Biochem. Biophys. 222, 473.

26. Levy, H. R. and Daouk, G. H. (1979). J. Biol. Chem. 254, 4843.

27. Viola, R. E. (1984). Arch. Biochem. Biophys. 228, 415.

28. Grove, T. H., Ishaque, A. and Levy, H. R. (1976). Arch. Biochem. Biophys. 177, 307.

29. Haghighi, B. and Levy, H. R. (1982). Biochemistry 21, 6421.

30. Jencks, W. P. (1975). Adv. Enzymol. Relat. Areas Mol. Biol. 43, 219.

31. Levy, H. R., Ejchart, A. and Levy, G. L. (1983). Biochemistry 22, 2792.

32. Arnold, J. J., Jr., You, K., Allison, W. S. and Kaplan, N. O. (1976). Biochemistry 15, 4844.

33. Citri, N. (1973). Adv. Enzymol. Relat. Areas Mol. Biol. 37, 397.

34. Olive, C. (1969). Ph.D. Dissertation, Syracuse University.

35. Stein, A. M., Lee, J. K., Anderson, C. D. and Anderson, B. M. (1963). Biochemistry 2, 1015.

36. Levy, H. R., Daouk, G. H. and Katopes, M. A. (1979). Arch. Biochem. Biophys. 198, 406.

37. Marks, P. A., and Banks, J. (1960). Proc. Nat. Acad. Sci. U. S. 46, 447.

38. McKerns, K. W., and Kaleita, E. (1960). Biochem. Biophys. Res. Commun. 2, 344.

39. Kawaguchi, A., and Bloch, K. (1974). J. Biol.Chem. 249, 5793.

G6PD OF *DROSOPHILIA MELANOGASTER*

John H. Williamson[1]

Department of Biology
Davidson College
Davidson, North Carolina

I. INTRODUCTION

Glucose-6-phosphate dehydrogenase (G6PD, EC 1.1.1.49) is routinely recognized as a ubiquitous enzyme, widely distributed in plants, animals, fungi and microorganisms (Levy, 1979). The enzyme occurs in several forms generally considered to be multimers of a basic subunit and catalyzes the initial reaction in the oxidative component of the pentose phosphate pathway.

Drosophila melanogaster has been recognized since 1910 (Morgan, 1910) as an excellent model organism for genetic analysis. G6PD was first described in mammalian erythrocytes in 1931 and in brewer's yeast in 1932 (Warburg and Christian, 1931a, b, 1932). However, it was not until 1964 that genetic variants of G6PD were detected in *Drosophila melanogaster* (Young *et al*. 1964). Since that initial description of G6PD in *Drosophila* this gene-enzyme system has received detailed attention in studies of gene dosage effects (Bowman and

----

[1]Supported in part by NSF Grant #PCM-8407680.

Simmons, 1973; Seecof *et al*., 1969; Stewart and Merriam, 1974), dosage compensation in this non-mammalian system (Lucchesi, 1978), in nutritional studies in insect systems (Geer *et al*. 1974, 1976, 1978, 1979, 1981) and in studies on the genetic structure of natural populations (Laurie-Ahlberg *et al*. 1980, 1981; Voelker *et al*. 1980). *Drosophila* G6PD has been purified and described in several laboratories. Recently the structural gene for G6PD has been isolated from genomic clones of *Drosophila* and is currently receiving detailed molecular attention (Ganguly *et al*., 1985). Therefore, it seems appropriate that in a symposium on G6PD we should summarize what is known about the *Drosophila* gene enzyme system.

II. *DROSOPHILA* G6PD

A. Genetics

Glucose-6-phosphate dehydrogenase was first detected in *Drosophila* in 1964 (Young *et al*., 1964). Arguing that the qualitative polymorphism of this enzyme that had been found in humans should also be found in a more genetically tractable organism such as *Drosophila*, Young and his coworkers proceeded to use starch gel electrophoresis to screen wild-type populations for variants of glucose-6-phosphate dehydrogenase. They found two electrophoretically variant forms of this enzyme system in *Drosophila* populations. Two alleles were determined to code for a fast form of G6PD (G6PD-A) and for a slow form of G6PD (G6PD-B). Females homozygous for one of these alleles had only one of the two forms of the enzyme. Females heterozygous for these two alleles had both forms of the enzyme but, more or less surprisingly, exhibited no intermediate form of G6PD. Males had one or the other of the two electrophoretic forms.

After careful consideration of these results, Young concluded that, as in man, the *Drosophila* gene for G6PD is sex-linked --- males have one dose and females have two doses of this gene. Carrying this logic further, Young (1966) reported two years later that 6-phosphogluconate dehydrogenase (6PGD; EC 1.1.1.44) in *Drosophila* was also sex-linked. On the argument that G6PD and 6PGD are catalytically closely linked in the pentose phosphate pathway, it may not be surprising that the genes for these two enzymes are both located on the same chromosome. One might even have predicted that these two genes would be closely linked on the X chromosome. While the genes for G6PD and 6PGD are sex-linked in *Drosophila*, it turned out that they are not at all closely linked but occur some 60 map units apart (*Pgd*: 0.64; *Zw*: 63; Lindsley and Grell, 1968). This turns out, in fact, to be the general case in *Drosophila*, whereby enzymes which are closely linked in a functional manner are not at all closely linked in the genome of this organism. The one known exception involves the gene-enzyme system regulating pyrimidine biosynthesis in *Drosophila* (Seagraves *et al*., 1984).

B. Cellular Localization

In animal tissues, G6PD is generally considered to be a cytosolic enzyme (Levy, 1979). *Drosophila* is no exception to this rule (Table I). Under the conditions of our cell fractionation procedure, 100% of the total G6PD activity was found to be cytosolic while 70% of the total succinnic dehydrogenase activity was found in particulate fractions (Williamson and Bentley, 1983).

Table I. Cellular distribution of G6PD and succinate dehydrogenase in *Drosophila*.

Cell Fraction	G6PD Sp.Act.	G6PD % of Total	SDH Sp.Act.	SDH % of Total
Crude Homogenate	23.8	100	67.5	100
600 g Pellet	0	0	14.2	2.2
Mitochondrial	0	0	166.5	28.5
Microsomal	0	0	138.5	37.8
100,000 Supernatant	81	100	122.1	31.8

Williamson and Bentley, 1983.

C. Molecular Weight

Subsequent to Young's demonstration of G6PD polymorphism in wild-type populations, additional screens detected only three naturally occurring major variants of G6PD in *Drosophila* - G6PD-A, G6PD-B and the low activity variant, G6PD-lo. G6PD-lo was initially described as a null-activity mutant. These allelic variations not only affect the electrophoretic mobility of this enzyme but indeed seem to effect its tertiary structure. It is now clear that the electrophoretic variations in G6PD in this organism are consequences of variation in molecular weights of the two most active forms of this enzyme.

In the original work on *Drosophila* G6PD, Steele *et al*. (1968, 1969) estimated the molecular weight of G6PD-A to be 147,000 and that of G6PD-B to be 317,000 (Table II). Subsequently, Lee *et al*. (1978) reported the molecular weight of G6PD-B to be 240,000 with subunits of 55,000. This form of the enzyme then seems to be a tetramer. Using techniques of affinity chromatography to purify G6PD we (Williamson and Bentley, 1983) estimated the molecular weight of G6PD-B to be 320,000, very close to that of Steel *et al*., but different from that of Lee *et al*. We estimated the molecular weight of G6PD-A to be 146,000 with subunit weights of 55,000. The third allele coded for a form of G6PD, G6PD-lo, that had a molecular weight of 54,000. Thus, it seems that G6PD-lo is a monomeric form of the enzyme, G6PD-A is a dimeric form of the enzyme, while G6PD-B is a tetrameric form.

Table II. Molecular Weights of Drosophila G6PD.

G6PD-A	G6PD-B	Subunit	Reference
147,000	317,000	-	Steele *et al.*, 1968
	240,000	55,000	Lee, *et al.*, 1978
146,000	320,000	55,000	Williamson and Bentley, 1983
115,000	283,000	69,000	Hori & Tanda, 1980

While interconversion of G6PD-A and G6PD-B has been reported (Steele *et al*. 1968, 1969; Hori and Tanda, 1981), we saw no evidence that this can be obtained under standard routine laboratory conditions. When we extracted G6PD from *Drosophila* with high levels of NADP maintained in the extraction buffer, the estimated molecular weight of G6PD-A was 335,000 similar to that of G6PD-B (362,000). Under these conditions the molecular weight of G6PD-lo was estimated to be 165,000 (Williamson and Bentley, 1983). These observations are consistent with the hypothesis that G6PD-B is a tetrameric molecule with high stability in low concentrations of NADP, that G6PD-A is generally considered to be a dimeric molecule in low concentrations of NADP and that G6PD-lo is a monomeric form of the enzyme in low concentrations of NADP. It seems then that at high concentrations of NADP, G6PD-lo assumes a dimeric configuration and G6PD-A and B both assume a tetrameric configuration. The alleles that code for these three forms of G6PD must alter amino acid sequences that are critical to subunit interactions.

D. Kinetic Parameters

Kinetic properties of the three forms of G6PD from *Drosophila melanogaster* have been determined (Table III). The apparent K_M's for NADP of purified preparations of the three forms of the enzyme are all very closely related, while the apparent K_M's for G6P demonstrate a four-fold difference (Williamson and Bentley, 1983). G6PD-B has the highest affinity for substrate while G6PD-lo has the lowest affinity for G6P. Thus, it seems that not only are the genetic variants affecting the tertiary structure of the enzyme, but substrate binding abilities are also different for the three forms of the enzyme.

Since *Zw-lo* zygotes produce a monomeric form of G6PD, characterized by a low affinity for the substrate, we next used rocket immunoelectrophoresis to monitor G6PD-CRM in females homozygous for each of the three *Zw* alleles and in the three possible heterozygotes. Using *Zw-A* homozygotes as a control, we determined *Zw-B* homozygotes to have approximately 65% as much antigenically cross-reacting material. *Zw-lo* homozygotes had approximately 76% as much CRM as did *Zw-A* controls. Each of the three heterozygous classes had CRM levels intermediate between the two parental strains. Thus, it appears that *Zw-lo* individuals produce essentially normal levels of G6PD subunits, but that these subunits do not associate into one of the most active forms of the enzyme. Whatever the true situation may be, it seems apparent that *Zw-lo* flies synthesize much more G6PD-CRM than their G6PD activity would indicate.

E. Dosage Compensation

Dosage compensation is a compensatory mechanism, operative in diploid sexually reproducing organisms with heteromorphic sex chromosomes. In such organisms, the sex carrying

Table III. Kinetic parameters of Drosophila G6PD

Enzyme	G6P(M)	NADP(M)	Reference
G6PD-A	3×10^{-4}	2.5×10^{-5}	Steele *et al.*, 1969
G6PD-B	1.7×10^{-4}	5.8×10^{-5}	Steele *et al.*, 1969
G6PD-A, Males	5.6×10^{-4}	-	Komma, 1968
G6PD-A, Females	3.5×10^{-4}	-	Komma, 1968
G6PD-B, Males	5.4×10^{-4}	-	Komma, 1968
G6PD-B, Females	3.6×10^{-4}	-	Komma, 1968
G6PD-A	$5.1\text{-}6 \times 10^{-5}$	$1.1\text{-}1.4 \times 10^{-5}$	Hori & Tanda, 1981
G6PD-B	$5\text{-}6 \times 10^{-5}$	$1.0\text{-}1.3 \times 10^{-5}$	Hori & Tanda, 1981
G6PD-A	11.9×10^{-5}	1.7×10^{-5}	Williamson and Bentley, 1983
G6PD-B	5.8×10^{-5}	1.4×10^{-5}	Williamson and Bentley, 1983
G6PD-lo	20.1×10^{-5}	1.3×10^{-5}	Williamson and Bentley, 1983

the heteromorphic pair of chromosomes has one copy of each sex-linked gene while the sex with homomorphic chromosomes has two copies of each sex-linked genes. While many genes on the sex chromosomes may be involved with sex determination, another large subset of sex-linked genes mediate basic functions not related to sex differentiation (Lucchesi, 1978). The phenotypic product of such genes, however, is usually found to be equivalent in the two sexes. This phenomenon was first described by Bridges (1922) and termed dosage compensation by Muller (1950). Glucose-6-phosphate dehydrogenase has been very instrumental in demonstrating some of the processess of dosage compensation in *Drosophila*.

When Young first described the polymorphic nature of G6PD in *Drosophila* he noted that heterozygous females expressed both forms of G6PD, but that these heterozygous females did not carry the intermediate form. This observation is consistent with regulation of sex-linked genes in mammals where dosage compensation is a consequence of inactivation of one X chromosome in each somatic cell of females. However, it turns out that in *Drosophila* dosage compensation does not involve chromosome inactivation. Instead, both X's in females are active. This was first noted by Young (1966) when he picked up polymorphisms for 6PGD in wild-type populations of *Drosophila*. Females heterozygous for electrophoretically different forms of 6PGD contained both forms of this enzyme as well as the intermediate form. That the intermediate form, the hybrid form, of the enzyme was not a artifact was demonstrated by *in vitro* mixing of the two different forms of the enzyme and the failure of these mixing experiments to demonstrate a hybrid enzyme. These observations were consistent with those made on sex-linked body color differences in *Drosophila* that were either mutant or non-mutant but not spotted as one would expect to find if dosage compensation in *Drosophila* were the consequence of X-chromosome inactivation.

F. Genetic Rescue

The pentose phosphate pathway in *Drosophila melanogaster* has been used as a model system with which to study the rescue of lethal mutations by genetic means (Hughes and Lucchesi, 1977, 1978). Although fairly common among prokaryotes and lower eukaryotes, this type of genetic interaction is relatively undocumented in higher eukaryotes. Null activity mutants of the structural gene (*Pgd*) for 6PGD are lethal in *Drosophila* (Bewley and Lucchesi, 1975). This suggests that the absence of the pentose pathway is a lethal situation in higher organisms. While human geneticists would know that this is not true, it may well be that the pentose pathway is an essential function in *Drosophila*. Null activity levels of glucose-6-phosphate dehydrogenase, however, are perfectly viable genetic conditions in *Drosophila* as in man. Thus, it is obvious that the pentose pathway can be dispensed and viability maintained, but it seems that the pathway must be dismissed at its beginning and not somewhere in the middle. Apparently an accumulation of some product or side event of the pentose pathway accumulates lethal components, whereas if the pathway is missing *in toto* there is no accumulation of these lethal substances. When one has a double mutant, *i.e.*, neither glucose-6-phosphate dehydrogenase nor 6-phosphogluconate dehydrogenase are present, viability ensues. Such a mechanism may well exist in humans because many null or low activity mutants have been discovered for G6PD although very, very few and probably no independent mutants at 6PGD have been found in human populations.

G. Nutritional Studies

Genetic variants modifying the oxidative component of the pentose phosphate pathway have been used by Geer in his studies of nutritional physiology in insects (Geer *et al*. 1976, 1978, 1979, 1981). Activities of G6PD and 6PGD increase according to the concentration of dietary sucrose (Geer *et al*., 1976). Geer (1978) fed larvae a variety of dietary supplements that can be converted into glycolytic intermediates. D-glycerate was found to be an effective coordinate inducer of G6PD and 6PGD activities. Strains genetically defective for the pentose phosphate pathway activity were non-inducible.

Gvozdev (1976, 1977) has suggested that 6PGD null activity mutations behave as lethals as a consequence of 6-

phosphogluconate accumulation and of deficient pentose-phosphate biosynthesis. The addition of a null-activity allele of the gene for G6PD to the 6PGD-deficient genotype would prevent the accumulation of 6-phosphogluconate as well as enhance the flux of glucose-6-phosphate to glyceraldehyde 3-phosphate via glycolysis and to ribulose-5-phosphate via reversal of the nonoxidative portion of the shunt (Cavener and Clegg, 1981; Lucchesi et al. 1979). Support for this hypothesis comes from the observation that PGD-null larvae can be rescued by the addition of fructose or fatty acid to the diet (Hughes and Lucchesi, 1978; Geer et al. 1981).

Studies in the area of nutritional genetics has just begun and will not be simple. For example, we already know that dietary modulation of G6PD and 6PGD activities in Drosophila are regulated by autosomal genes in addition to the respective sex-linked structural genes for these enzymes (Cochrane et al. 1983). Chromosome III carries at least one and probably several genes with major effects on this system.

REFERENCES

Bewley, G.C. and J.C. Lucchesi, 1975. Genetics 79:451-466.

Bowman, J.L. and J.R. Simmons, 1973. Biochem. Genetics 10:319-331.

Bridges, C.B., 1922. Amer. Nat. 56:51-63.

Cavener, D.R. and M.T. Glegg, 1981. Proc. Natl. Acad. Sci. U.S. 78:4444-4447.

Cochrane, B.J., J.C. Lucchesi and C.C. Laurie-Ahlberg, 1983. Genetics 102:601-613.

Ganguly, R., N. Ganguly and J.E. Manning, 1985. Gene 35:91-101.

Geer, B.W., J.T. Bowman and J.R. Simmons, 1974. J. Exp. Zool. 187:77-86.

Geer, B.W., S.N. Kamiak, K.R. Kidd, R.A. Nishimura and S.J. Yemm, 1976. J. Exp. Zool. 195:15-32.

Geer, B.W., D.L. Lindel and D.M. Lindel, 1979. Biochem. Genetics 17:881-895.

Geer, B.W., J.H. Williamson, D.R. Cavener and B.J. Cochrane, 1981. In: Current Topics in Insect Endocrinology and Nutrition, pp. 253-281. Bhaskaren, G., S. Friedman and

J.G. Rodriguez, eds. Plenum Publishing Co.
Geer, B.W., C.G. Woodward and S.D. Marshall, 1978. J. Exp. Zool. 203:391-402.
Gvozdev, V.A., T.I. Gerassimova, G.L. Kogan and O.Y. Braslavskaya, 1976. FEBS Letters 64:85-88.
Gvozdev, V.A., T.I. Gerasimova, G.L. Kogan and J.M. Rosovsky, 1977. Molec. Gen. Genetics 153:191-198.
Hori, S.H. and S. Tanda, 1980. Japan. J. Genetics 55:211-223.
Hori, S.H. and S. Tanda, 1981. Japan. J. Genetics 56: 257-277.
Hughes, M.B. and J.C. Lucchesi, 1977. Science 196:1114-1115.
Hughes, M.B. and J.C. Lucchesi, 1978. Biochem. Genetics 16:469-475.
Komma, D.J., 1968. Biochem. Genetics 1:337-346.
Laurie-Ahlberg, C.C., G. Maroni, G.C. Bewley, J.C. Lucchesi and B.S. Weir, 1980. Proc. Natl. Acad. Sci. U.S. 77:1073-1077.
Laurie-Ahlberg, C.C., J.H. Williamson, B.J. Cochrance, A.N. Wilton and F.I. Chasalow, 1981. Genetics 99:127-150.
Lee, C.Y., C.H. Langley and J. Burkhart, 1978. Anal. Biochem. 86:697-706.
Levy, H.R., 1979. In: Advances in Enzymology 48:97-192.
Lindsley, D.L. and E.H. Grell, 1968. Genetic Variations of Drosophila melanogaster. Carnegie Inst. Wash. Pub. 627.
Lucchesi, J.C., 1978. Science 202:711-716.
Luchesi, J.C., M.B. Hughes and B.W. Geer, 1979. In: Current Topics in Cellular Regulation 15:143-154. Academic Press.
Morgan, T.H., 1910. Science 32:120-122.
Muller, H.J., 1950. Harvey Lect. 43:188-229.
Seagraves, W.A., C. Louis, S. Tsubota, P. Schedl, J.M. Rawls and B.P Jerry, 1984. J. Mol. Biol. 175:1-17.
Seecof, R.L., W.D. Kaplan and D.G. Futch, 1969. Proc. Natl. Acad. Sci. U.S. 62:528-535.
Steele, M.W., W.J. Young and B. Childs, 1968. Biochem. Genetics 2:159-175.
Steele, M.W., W.J. Young and B. Childs, 1969. Biochem. Genetics 3:359-370.
Stewart, B.R. and J.R. Merriam, 1974. Genetics 76:301-309.
Voelker, R.A., C.H. Langley, A.J.L. Brown, S. Ohnishi, B. Dickson, E. Montgomery and S.C. Smith, 1980. Proc. Natl. Acad. Science U.S. 77:1091-1095.
Warburg, O. and W. Christian, 1931a. Biochem. Z. 238:131-134.
Warburg, O. and W. Christian, 1931b. Biochem. Z. 242:206-227.
Warburg, O. and W. Christian, 1932. Biochem. Z. 254:438-458.
Williamson, J.H. and M.M. Bentley, 1983. Biochem. Genetics 21:1153-1166.

Williamson, J.H. and M.M. Bentley, 1983. Genetics 103:649-658.
Young, W.J., 1966. J. Heredity 57:58-60.
Young, W.J., J.E. Porter and B. Childs, 1964. Science 143:140-141.

GLUCOSE-6-PHOSPHATE DEHYDROGENASE AND HEXOSE-6-PHOSPHATE DEHYDROGENASE: AN EVOLUTIONARY ASPECT

S.H.Hori

Department of Zoology, Faculty of Science,
Hokkaido University, Sapporo, JAPAN

ABSTRACT. G6PD occurs in a wide variety of animals, while H6PD is found in some echinoderms and vertebrates. Both enzymes act on G6P and NADP, but H6PD has a broader substrate specificity. Their structural similarity was demonstrated by immunological cross-reaction and amino acid analyses; rabbit antibody to starfish H6PD binds to starfish and sea urchin G6PDs, while antibody to sea urchin G6PD binds to starfish H6PD. Although such cross-reactions were not observed between G6PDs and H6PDs from crucian carp and rat, their amino acid compositions were found to be very similar to each other. This suggests that the two enzymes may have diverged from a common prototype G6PD possibly at the time of or before echinoderm evolution.

H6PD is almost exclusively located on microsomal membranes and its activity is high in steroid- and drug-metabolizing cells, such as hepatic parenchymal cells, testicular interstitial cells, ovarian luteal cells, adrenal cortex and P_3 segments of renal proximal convolutions. H6PD activity is also high in plasma cells and in striated duct and serous tubular cells of submandibular glands, which have recently been found to possess polycyclic hydrocarbon metabolizing activities. In view of these and other findings, it is conjectured that functional differentiation of G6PD and H6PD might be related to steroid and drug metabolism.

I. INTRODUCTION

Glucose-6-phosphate dehydrogenase (G6PD) is widely distributed throughout the animal and plant kingdoms, and its main function is thought to be to provide reduced nicotinamide adenine dinucleotide phosphate (NADPH) for various reductive biosynthetic reactions along with its partner in the pentose phosphate pathway, 6-phosphogluconate dehydrogenase (6PGD). The activities of these enzymes are very high in adipose tissue, liver and steroidogenic tissues, whereas they are low in skeletal muscle and brain, thus supporting the idea that a major role of these oxidative enzymes is to generate NADPH for reductive biosyntheses.

Gene duplication is thought to be one of the basic processes of evolution. An extra copy resulting from a chromosomal aberration can accept mutations, because the other still plays the original roles. Diversification of genes and functions thus results; globins and isozymes are well known, but only a few of such examples. Footprints of gene duplication can be found on the primary structure of protein; delta- and beta-chains of hemoglobins are much alike, but they are also like gamma-chains and similar to alpha-chains in amino acid sequence. Such diversification of globins enables mammals to perform sophisticated functions. G6PD is not an exception. There are good reasons to believe that the gene coding for G6PD duplicated during invertebrate evolution, and that one of the copies diverged from the other in structure and function.

G6PD has been known to exist almost exclusively in cytosol (Glock and McLean, 1953; Newburgh and Cheldelin, 1956). However, some investigators claimed that the enzyme did exist also in a large particulate fraction, possibly in mitochondria, and that the enzyme of this fraction differed from that of the cytosol fraction in catalytic, immunological, and other properties(Bagdasarian and Hulanicka, 1956; Baquer and McLean, 1972; Baquer et al., 1972; Razumovskaya et al., 1970; Watanabe and Taketa, 1973; Yamada and Shimazono, 1961; Zaheer et al., 1967). Particularly interesting to us were the findings of Baquer and McLean (1972) that the non-oxidative enzymes as well as the oxidative enzymes of the pentose phosphate pathway exist within mitochondria, particularly in steroidogenic organs, and that their activities are latent and change independently of the activities of the cytosol counterparts. If this holds, there must be some reason why mitochondria have their own pentose phosphate pathway. In order to examine this reason, we purified G6PD 1,300-fold from rat liver mitochondrial fraction, and analyzed its properties. The results showed that the enzymes solubilized with 1M NaCl and with 1% Triton X-100 were both indistinguishable from the cytosol enzyme with respect to Km values, pH optimum, sensitivity to several chemicals and immunological response to anti-cytosol G6PD antibody. Effects of washing, some nucleotides, sugar phosphates and inorganic ions on the solubilization of mitochondrial G6PD were also investigated, but no positive proof was obtained that the enzyme is intrinsic to mitochondria. We therefore concluded that mitochondrial location of G6PD is artefactual, due to a contamination of cytosol enzyme which occurred during homogenization and is not ascribed to a distinct molecular form as predicted by some investigators(Kodama et al., 1981). Similar data were also presented by Grunwald and Hill (1976).

In the meantime, data have been presented that a distinct form of G6PD is present in the microsomal fraction of mammalian livers (Ohno et al., 1966; Shaw, 1966; Ruddle et al.,1968; Hori and Matsui, 1968). Ohno et al. designated the enzyme as hexose-6-phosphate dehydrogenase (H6PD) inasmuch as it was as active on galactose-6-phosphate (Gal6P) as on G6P. Subsequent works on H6PD have revealed that this enzyme is resistant to inhibitors of G6PD, has a broad substrate and coenzyme specificity, is not precipitated by antibody against G6PD, and is present also in fish (Beutler and Morrison, 1967; Srivastava and Beutler, 1969; Mandula et al., 1970; Shatton et al., 1971; Stegeman and Goldberg, 1971, 1972; Srivastava et al., 1972). Beutler and Morrison (1967) stated that H6PD might be identical with the enzyme formerly known as glucose dehydrogenase, since the H6PD activity could not be separated from the glucose dehydrogenase activity during purification of H6PD.

Whatever the designation of this enzyme having H6PD and glucose dehydrogenase activities may be, there is no doubt concerning the presence of two kinds of enzyme molecules having G6PD activity in vertebrates; one has a narrow and the other has a broad substrate specificity. Are they homologous or analogous molecules? This question could be answered by comparing the properties of as many G6PDs from both invertebrates and vertebrates as possible. Such comparative studies might provide some valuable information on the evolution of G6PD.

The present report is a summary of our studies on many G6PDs which were initiated with the characterization of electrophoretically distinguishable forms of rat G6PD, followed by comparison of G6PDs and H6PDs from a variety of species and inquiries into their possible physiological functions.

II. DIVERGENCE OF H6PD FROM G6PD

A. Multiple Molecular Forms of Rat Liver G6PD

Homogenates of rat liver exhibit seven bands of G6PD activity on polyacrylamide gels when stained with G6P and NADP as substrates. These are designated as A, B, C, D, E, E_1 and F in the order of increasing mobility. Among these, the B enzyme is magnesium-dependent and reduces NADP in the absence of G6P. Therefore, it appears to be a so-called nothing dehydrogenase. The rest may be classified into two categories; one including A, C, E and E_1, and the other, D and F. The former is located in microsomes and is active on Gal6P and 2-deoxyglucose-6-phosphate (dG6P) as well as on G6P, while the latter is present in cytosol and is specific for G6P. They also differ in sensitivity to inhibitors and heat (Hori and Matsui, 1968; Kamada and Hori,1970; Mochizuki and Hori, 1973). The enzymes D and F are more active in females than in males, while such a sex difference is not observed in newborn rats. The cause of this difference was demonstrated to be estrogens; injection of estradiol benzoate into orchidectomized young rats increased the total G6PD activity to the level of adult females and the change was attributed to the D and F enzymes (Hori and Matsui, 1967). Induction of these enzymes is also observed when rats are fed on a high carbohydrate diet. Furthermore, the two enzymes are interconvertible; the D enzyme can be formed *in vitro* from the F enzyme after treatment with cystamine HCl, while conversion of the D to the F enzyme occurs in the presence of SH reagents (Hori and Yonezawa, 1972). The molecular weights of the D and F enzymes are approximately 22×10^4 and 11×10^4 daltons, respectively. Since the F enzyme is a dimer, the D enzyme would be a tetramer.

The A enzyme is manifested only in liver, and is convertible to the C enzyme; the A enzyme extracted from electrophoretic gels showed both the A and C enzymes upon re-electrophoresis. Since the molecular weight of the A enzyme is about twice as large as that of the C enzyme, the A may be a tetrameric form.

In an attempt to characterize the A and C enzymes we happened to find that the C enzyme was degraded by the action of trypsin into two enzymatically active molecular forms which showed the same electrophoretic mobility as the E and E_1 enzymes, and that lysosomal fractions could mimic the trypsin effect. The possibility is thus suggested that the E and E_1 enzymes might be degradation products of the C enzyme produced by lysosomal proteinases (Hori and Noda, 1971). All these findings suggest that there are only two distinct genes encoding G6PD isozymes in rat, one for the isozyme located in cytosol and the other for the isozyme in

TABLE I

PROPERTIES OF RAT G6PD AND H6PD

	Glucose-6-P dehydrogenase	Hexose-6-P dehydrogenase
Km (μM)		
Glucose-6-P	32	1.3
Galactose-6-P	5.9×10^3	29
2-deoxy-G6P	910	1.1×10^3
Glucose (NAD)	---	5×10^6
NADP	6	1.2
NAD	---	1
Effects of*:		
$MgCl_2$, 10 mM	109	69
PCMB, 1 mM	42	82
DEA, 70 μM	22	100
Heat, 50°C, 5 min	37	76
Molecular weight		
intact enzyme (dimer)	110×10^3	190×10^3
trypsin-treated enzyme	---	144×10^3 115×10^3

* Activity expressed in percentage of untreated control. Data from Hori and Sado (1974), Takahashi and Hori (1978) and Oka et al. (1981).

microsomes. In the following description, the former is designated as G6PD and the latter, H6PD.

B. Properties of Animal G6PDs and H6PDs

Comparison of the kinetic and other properties of G6PD and H6PD was first made with the enzymes purified from rat liver. The results shown in Table I may be summarized as follows: 1, H6PD has low Km values for G6P and Gal6P in comparison with G6PD; 2, H6PD is active on glucose and NAD, but G6PD is not; 3, H6PD is inhibited by magnesium ions, but G6PD is stimulated, though slightly; 4, H6PD is relatively resistant to *p*-chloromercuribenzoate (PCMB), dehydroepiandrosterone (DEA) and heat; 5, both enzymes exist in dimeric forms, but the subunit of H6PD is about twice as large as that of G6PD; 6, upon trypsin digestion, H6PD is degraded into

TABLE II

PROPERTIES OF ANIMAL G6PDs AND H6PDs

	Glucose-6-phosphate dehydrogenase		Hexose-6-phosphate dehydrogenase	
	Average ± S.E. of six vertebrate species	Average ± S.E. of fourteen invertebrate species	Average ± S.E. of four vertebrate species	Starfish
Km (μM)				
Glucose-6-P	29 ± 3.4	32.5 ± 6.44	2 ± 0.4	8
Galactose-6-P	$(6.1 \pm 0.86) \times 10^3$	$(5.6 \pm 1.17) \times 10^3$	16 ± 4.5	230
NADP	8.1 ± 1.89	10.8 ± 3.05	2 ± 0.2	1
NAD (G6P)	---	---	1 ± 0	1
Glucose (NAD)	---	---	$(2.2 \pm 1.1) \times 10^6$	1.6×10^6
Effects of :*				
$MgCl_2$, 10 mM	109 ± 3	112 ± 2.0	65.3 ± 2.8	79
PCMB, 1 mM	29 ± 10	29 ± 6.0	76.0 ± 3.2**	100
DEA, 70 μM	38 ± 7	78 ± 5.8	100 ± 0	100
Heat, 50°C, 5 min	37***	16 ± 7.6	76***	4

* Activity expressed in percentage of untreated control.

** Excluding one exceptional case, toad H6PD which is stimulated.

*** Data of rat.

Data from Sado and Hori (1976), Mochizuki and Hori (1976), Ohnishi and Hori (1977), Matsuoka et al. (1977), Matsuoka and Hori (1980), Hori and Tanda (1980), Sado (1980), and Oka et al. (1981).

enzymatically active, smaller forms, but G6PD is not.

Are these characteristics of G6PD and H6PD common to all animal G6PDs and H6PDs? Is H6PD ubiquitous in vertebrate species but not in invertebrate species? To answer these questions, we analyzed the substrate specificity of G6PD isozymes which were separated by polyacrylamide gel electrophoresis (Kamada and Hori, 1970; Mochizuki and Hori,1973). A survey of G6PDs from more than a hundred animal species revealed an interesting fact that a G6PD isozyme occurs in starfish and sea cucumber which strikingly resembles rat H6PD in various respects (Mochizuki and Hori, 1973, 1976; Matsuoka et al.,1977). Characterization of G6PDs and H6PDs from a wide variety of animals had also been performed with partially purified enzymes (Ohnishi and Hori, 1977; Sado and Hori, 1976). Before going into the description of interesting echinoderm enzymes, the data obtained with vertebrate and invertebrate G6PDs and vertebrate H6PDs will be summarized: the G6PDs from six vertebrate and fourteen invertebrate species were found to have very similar properties with each other as shown in Table II; i.e., Km values for G6P, Gal6P and NADP fall within narrow ranges, and all enzymes are slightly stimulated with magnesium ions, but are considerably inhibited with PCMB as is the rat G6PD. It is noteworthy that the invertebrate G6PDs are much more heat-labile, and more resistant to DEA than the vertebrate G6PDs.

Studies on molecular weight showed that reptilian, avian and mammalian G6PDs are mostly present in small forms (10-12 x 10^4 daltons), while fish and amphibian G6PDs tend to exist in large forms (20-26 x 10^4 daltons). In invertebrate species, the enzyme exists in either small or large forms or both. In vitro conversion of large forms into small forms was observed with earthworm and hydra G6PDs in the presence of excess 2-mercaptoethanol, and also with crayfish G6PD at a high ionic strength (Ohnishi and Hori, 1977). In the case of *Drosophila* G6PD, the wild-type enzyme exists predominantly as a small form, whereas a mutant enzyme exists mostly as a large form (Hori and Tanda, 1980).

Taking into account that all the enzymes purified to a homogeneous state from rat, crucian carp, sea urchin and *Drosophila* consist of identical subunits having molecular weights of about 5-6 x 10^4 daltons, it may be concluded that the small and large forms observed in various species could be dimers and tetramers of identical subunits, respectively. It has been reported that the human G6PD exists as a homodimer in the presence of both G6P and NADP, but in the absence of the substrate resumes tetrameric or dimeric form depending upon the pH and ionic strength of the solution (Cohen and Rosemeyer, 1969; Yoshida and Hoagland, 1970). It is thus likely that the *in vivo* equilibrium between the dimer and tetramer of this enzyme may shift toward dimer in some species, but toward tetramer in other species, depending upon the amino acid substitutions they received.

As in the case of H6PD, the properties of H6PDs from four vertebrate species were very much alike as shown in Table II. The only exception was the reactivity of toad H6PD to PCMB; this enzyme was twice as active in the presence of PCMB as in its absence. H6PDs from some other vertebrates also resemble rat H6PD in that the enzymes are degraded into smaller, enzymatically active molecular forms upon trypsin digestion (Hori et al.,1975). It is worthy of noting here that this can not be regarded as a criterion of H6PD; i.e., G6PDs from *Radix*, *Musca* and *Drosophila* are also modified with proteinases without loss of activity (Hori et al.,1975; Gasperi et al., 1978; Hori and Tanda, 1980), thus suggesting a structural similarity of G6PD and H6PD.

C. Starfish H6PD

Discovery of an H6PD-like enzyme in some echinoderm species prompted us to purify this enzyme for detailed characterization. We chose pyloric caeca of the starfish, *Asterias amurensis*, as enzyme source, since this was rich in the enzyme and also available in quantity. However, purification of this enzyme was extremely difficult for two reasons: (1) the organ contains a large amount of proteinases which can not be completely separated from the enzyme by any means tested, and (2) the enzyme itself is unstable in the absence of excess ammonium ions. The easiest and most effective way of removing proteinases was treatment with 20% ethanol. We therefore homogenized the pyloric caeca in the presence of soybean trypsin inhibitor, and treated the homogenate with 0.25 volume of ethanol. Approximately 80% of the enzyme activity was precipitated by this treatment, while more than 95% of total casein-hydrolyzing activity remained in the supernatant. The precipitate was then subjected successively to ammonium sulfate fractionation, affinity chromatography on hemoglobin-Sepharose 4B, ion-exchange chromatography on DEAE-Sephadex and gel filtration on Ultrogel AcA 34. The enzyme, purified 100-fold, lost about 50% of its activity when freed from ammonium sulfate by overnight dialysis. Upon standing for 1 day at 4°C, it further lost 70% of the residual activity. However, loss of activity was prevented by the addition of ammonium sulfate or ammonium acetate at a concentration of 0.5 M or more; the enzyme dissolved in 0.5M ammonium sulfate retained its activity at least for a month. Electrophoretic analyses showed however that the enzyme thus prepared was gradually degraded into a smaller form with time, though without loss of activity. This suggests that a trace of proteinases was still contaminating such samples (Matsuoka et al.,1977).

The properties of partially purified starfish H6PD are given in Table II; they clearly indicate functional homology of starfish H6PD and vertebrate H6PDs. The native and degraded starfish H6PDs have molecular weights of 198,000 and 108,000, respectively.

There is plenty of evidence which suggests that functionally similar enzymes were developed from a common ancestral protein in the process of evolution. Examples of such enzymes are transaminases requiring pyridoxal phosphate, dehydrogenases requiring NAD, NADP or flavin, esterases and proteases with different substrate specificities (Smith, 1970). However, similarity in substrate specificity does not necessarily mean homology, as has been reported with a second gene for beta-galactosidase in *Escherichia coli* (Campbell et al., 1973) and with fructose diphosphate aldolases (Rutter, 1965). Therefore, comparisons of the structure of enzymes are badly needed for the establishment of their homology. Structural homology of starfish and vertebrate H6PDs was tested with rabbit antibody elicited to crucian carp H6PD (Hori et al.,1977) as follows: 1. Mixture of starfish H6PD and antibody against crucian carp H6PD was electrophoresed and stained for H6PD activity. If they formed immune complexes, then the H6PD would move at a much slower rate than the intact H6PD. The result was what we expected; the mobility of starfish H6PD relative to that of bromphenol blue was 0.19 in the presence or absence of normal IgG, but was zero in the presence of immune IgG. 2. The enzyme and antibody were mixed, allowed to stand for 3 hours and subjected to gel filtration on Ultrogel. The elution diagram of the mixture showed an activity peak in the position distinctly different from that with the enzyme alone or with the mixture of the enzyme and normal IgG, thus enabling us to separate the immune complex. The molecular weight of this complex was estimated to be approximately 5×10^5. 3. The immune complex thus obtained was heated at 56°C for 30 min in order to denature the enzyme and the resulting solution was tested for its potency against crucian carp H6PD. The result showed that the solution had a potency for modifying the activity of fish H6PD, thus demonstrating that antibody bound to starfish H6PD was the one specific for crucian carp H6PD. 4. Rabbit antibody against crucian carp H6PD, and that against rat H6PD as well have multiple effects on H6PD activity; the activities of H6PD on Gal6P and dG6P are inhibited progressively by increasing amounts of antibody, but the activity on G6P is stimulated in an antigen excess area (Takahashi and Hori, 1978; Hori et al., 1977; Matsuoka and Hori, 1980). It was therefore tested if antibody to crucian carp H6PD had similar effects on starfish H6PD. As a result, a slight, but significant increase in the G6P activity, and decreases in the Gal6P and dG6P activities of starfish H6PD was observed when treated with antibody. All these findings clearly indicate that starfish and crucian carp H6PDs are immunologically related, and hence share at least one antigenic determinant with each other.

D. Starfish G6PD

When homogenates of starfish pyloric caeca were electrophoresed and stained with G6P and NADP as substrates, there appeared on gels only a single band of H6PD without a G6PD band (Mochizuki and Hori, 1976). However, this would probably mean that starfish G6PD is extremely unstable and is not capable of withstanding electrophoretic conditions, since it is hard to think of a possibility where H6PD exists in place of G6PD in this particular group of organisms. Attempts were therefore made to purify G6PD from pyloric caeca, digestive tracts and ovaries. As a result, G6PD was purified 300-fold from ovaries by affinity chromatography on anti-crucian carp H6PD antibody-coupled Sepharose 4B (this removes H6PD) and on 2',5'-ADP-Sepharose 4B. The enzyme thus purified was quite similar to sea urchin G6PD with respect to all the parameters listed in Table II.

E. Homology of Echinoderm G6PD and H6PD

It is reasonable to assume that H6PD may be functioning as a G6PD *in vivo*, because Gal6P and dG6P are not available *in vivo* and its affinity to glucose is too low to be of physiological significance. It is thus highly probable that these functionally similar enzymes originated from a common ancestral protein. Ubiquity of G6PD in all organisms from bacteria to man and a restricted occurrence of H6PD in some echinoderm and vertebrate species further suggest that H6PD diverged from an ancestral G6PD at the time of or before echinoderm evolution. If so, it may be possible to demonstrate the structural similarity of echinoderm G6PD and H6PD by immunological means. Attempts were therefore made to examine if echinoderm G6PD would cross-react with anti-echinoderm H6PD antibody, and if echinoderm H6PD would cross-react with anti-echinoderm G6PD antibody (Matsuoka and Hori, 1980).

As mentioned above, neither G6PD nor H6PD of starfish could be purified to a homogeneous state in our hands, and hence no specific antibodies to starfish enzymes were available. To overcome this problem, we devised the following methods: (1) sea urchin G6PD was purified to a homogeneous state, and rabbit antibody to this enzyme was prepared and used in place of antibody to starfish G6PD. (2) The partially purified starfish H6PD was mixed with antibody to crucian carp H6PD, and the immune complex was separated by gel filtration to prepare mono-specific antibody to starfish H6PD. The immune complex was then heated at 56°C in order to denature starfish H6PD and centrifuged to remove denatured proteins. Antibody present in the resulting solution was coupled to CNBr-activated Sepharose 4B. To this immunoabsorbent the partially purified starfish H6PD was applied. The amount of enzyme was so limited that all the enzyme molecules were absorbed to the gel. The unabsorbed fraction was then mixed with antibody elicited to the partially purified starfish H6PD and allowed to stand overnight. The precipitate formed was removed by centrifugation. The supernatant prepared in this way was found to contain only the antibody specific to starfish H6PD.

The monospecific antibody against starfish H6PD thus prepared precipitated the specific antigen effectively, but not starfish G6PD at all. The activity of G6PD was also unaffected. However, electrophoretic analysis of the mixture of antibody and G6PD unequivocally demonstrated that they reacted to form a soluble complex (Fig.1). A similar phenomenon was also observed with sea urchin G6PD and antibody to starfish H6PD. In a reciprocal test, anti-sea urchin G6PD antibody was mixed with starfish H6PD. This antibody neither precipitated starfish H6PD nor affected its enzyme activity. However, it was demonstrated electrophoretically that the antibody does bind to starfish H6PD (Fig.1). These results indicate that starfish G6PD and H6PD, and sea urchin G6PD possess at least one antigenic determinant in common. Taking into

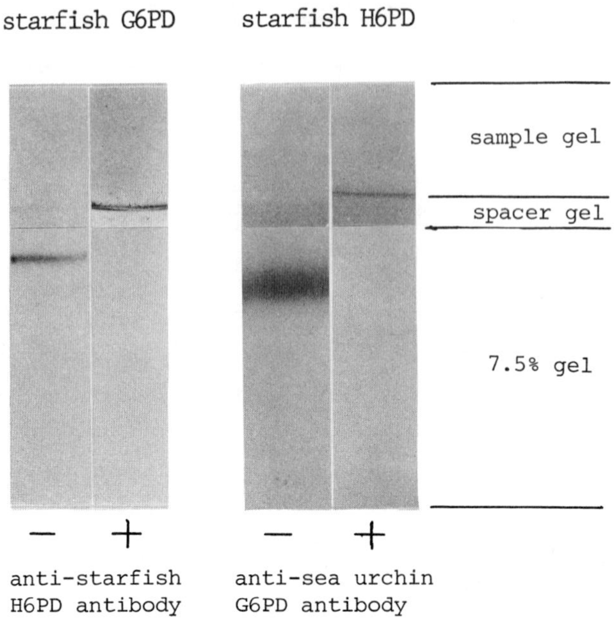

Fig.1. Effects of antibodies on starfish G6PD and H6PD. Left, starfish G6PD was electrophoresed before and after incubation with anti-starfish H6PD antibody, and the gels were stained for G6PD activity. Note that the G6PD does not move beyond the sample gel-spacer gel interface after incubation with the antibody. Right, starfish H6PD was electrophoresed before and after incubation with anti-sea urchin G6PD antibody, and the gels were stained for H6PD activity. As in the case of the G6PD, the H6PD does not move into the spacer gel after incubation with the antibody.

account that more than 60% homology in sequence is required for the two molecules to immunologically cross-react (Prager and Wilson, 1971), the present findings strongly suggest the structural homology of echinoderm G6PD and H6PD.

F. Homology of Vertebrate G6PD and H6PD

In contrast to starfish enzymes, G6PDs and H6PDs from crucian carp and rat would not cross-react. However, this does not necessarily mean that they are not homologous, since immunological cross-reactivity is frequently one of the earliest common features to be lost in evolution as pointed out by Metzger et al. (1968). Therefore, we analyzed amino acid compositions of vertebrate G6PDs and H6PDs to ascertain their structural homology. It has been known that the overall amino acid compositions of evolutionarily related proteins remain quite similar to each other regardless of changes in their primary structures (Metzger et al., 1968; Walsh and Neurath, 1964; Fondy and Holohan, 1971; Marchalonis and Weltman, 1971; Harris and Teller, 1973).

Figure 2 presents the distributional profiles of each amino acid of crucian carp and rat G6PDs and H6PDs, where the distance from the center circle (not shown) to each edge point corresponds to the content (mol%) of individual amino acids. It can be seen that the amino acid compositions of the two enzymes are strikingly similar to each other in both species, particularly in crucian carp. In order to make a more objective comparison of the amino acid compositions, we calculated the index of the "composition divergence (D)" proposed by Harris and Teller, 1973). It is defined as

$$D = [\Sigma (X_{i,A} - X_{i,B})^2]^{\frac{1}{2}}$$

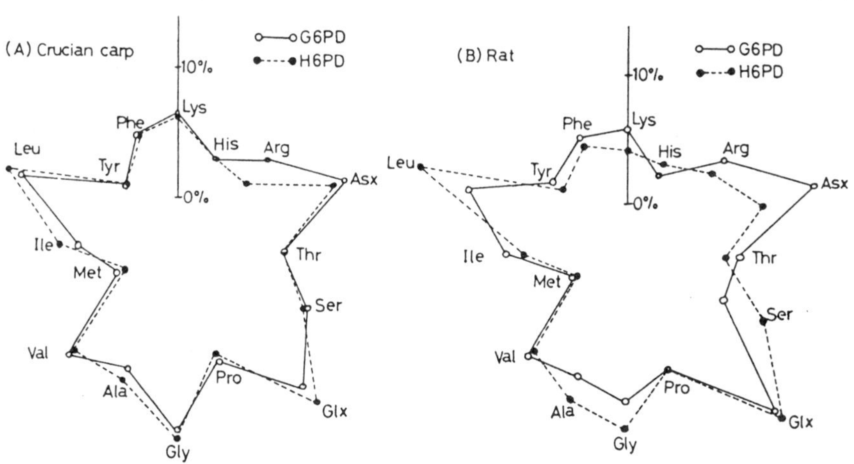

Fig.2. Comparisons of amino acid compositions between G6PDs and H6PDs from crucian carp and rat.

TABLE III

COMPOSITION DIVERGENCE (D) OF G6PDs AND H6PDs

	Sea urchin G6PD	Crucian carp G6PD	Crucian carp H6PD	Rat G6PD	Rat H6PD
Sea urchin G6PD	---	0.062	0.062	0.085	0.099
Crucian carp G6PD		---	0.038	0.059	0.069
H6PD			---	0.068	0.054
Rat G6PD				---	0.085

where $X_{i,A}$ is the mole fraction of amino acid i in protein A and $X_{i,B}$, the same amino acid in protein B. This value is a measure of the divergence of amino acid composition. Table III shows the results of such calculations. The lowest composition divergence was seen between G6PD and H6PD of crucian carp, and the highest value between sea-urchin G6PD and rat H6PD. The values obtained from the orthologous comparisons in crucian carp and rat were approximately equal (D = 0.059 for G6PD and 0.054 for H6PD). In all these calculations, the values for cysteine and tryptophan were neglected. As described by Marchalonis and Weltman (1971) these amino acids constitute only a small fraction of most proteins, and thus the effects of these amino acids on the parameter (D) would be insignificant.

Harris and Teller (1973) compared the amino acid compositions of many sequenced proteins, and found that a correlation exists between sequence dissimilarity and composition dissimilarity. They demonstrated that the mean and standard deviation of D values were found to be 0.052 ± 0.019 for homologous proteins and 0.130 ± 0.042 for non-homologous proteins, and values in the range of 0.07-0.09 were subject to much uncertainty. In comparing these values with the present data, the D value obtained in the cross comparison of G6PD/H6PD in crucian carp (0.038), and the values obtained in the orthologous comparisons of G6PD/G6PD and H6PD/H6PD from crucian carp and rat (0.059, 0.054) are in the range of those for homologous proteins, whereas the value from the cross comparison in rat is in the range of much uncertainty (0.085). However, this value is comparable to those obtained in the cross comparisons of bovine chymotrypsin/trypsin (0.091), of human amylase α/β (0.07-0.13), of human hemoglobin α/β (0.097) and of pig malate dehydrogenase mitochondrial/cytoplasmic forms (0.074) which are all assumed to be evolutionarily related proteins (Harris and Teller,1973; Marchalonis and Weltman, 1971; Fondy and Holohan, 1971). Accordingly, it may be safe to conclude that G6PD and H6PD are homologous proteins derived from a common ancestral protein and that their divergence might have occurred at the time of or before echinoderm evolution.

As to the mechanism of H6PD divergence, two independent phenomena will be considered; one is the broadening of substrate specificity and the other is the increment of molecular size. Studies on genetic variants of human G6PD have demonstrated that a significant change in the substrate specificity of G6PD can be induced by a simple mutation; i.e., among more than a hundred variants so far reported, some have a substrate specificity quite similar to H6PD; e.g., G6PD Markham and G6PD Union can oxidize dG6P and Gal6P as well as G6P, and the former can even utilize NAD as a coenzyme just like H6PD (Kirkman et al.,1968; Yoshida et al. 1970). It is thus possible to speculate that accumulation of many point mutations might not have been required for the divergence of H6PD and G6PD as far as the substrate specificity is concerned. However, point mutations can not explain the increase of the subunit size. Accordingly, a probable sequence of events occurred during the divergence of H6PD from G6PD might be as follows: the G6PD gene was first duplicated, one of the copies was again duplicated and the copies fused to form a single gene. The new gene thus formed accumulated mutations as time went by. The mutations accumulated in the catalytic sites might not necessarily be numerous as already mentioned, but some other portions would have accepted mutations to a considerable extent. These portions include antigenic determinants and those required for the enzyme to bind microsomal membranes.

This poses a question as to why H6PD and G6PD cross-react in echinoderm, but not in fish, rat and human (Srivastava et al., 1972). It has been assumed that the rate of amino acid substitution in the primary structure of a protein during evolution is approximately constant per site per year as long as the function and the tertiary structure of the protein remain essentially unaltered (Kimura, 1968; King and Jukes,1969; Kimura and Ohta,1974). If this holds, the absence of immunological cross-reactivity between vertebrate G6PDs and H6PDs implies that the functional constraint on G6PD or H6PD or both may be different between the echinoderm and vertebrate lineages, thus resulting in alterations of evolutionary rate. In addition, the D value analyses suggest the difference of evolutionary rate between fish and mammalian lineages. When the results presented in Table II are viewed from this standpoint, it appears to be of physiological importance that invertebrate G6PDs are relatively insensitive, but vertebrate G6PDs are sensitive to DEA, whereas starfish and vertebrate H6PDs are invariably resistant to this hormone. Hence, it may be conjectured that there was only one species of G6PD before the divergence of H6PD when G6PD was not required to be under the control of DEA. However, once DEA-resistant H6PD diverged, the prototype G6PD was so altered as to come under the control of steroid hormones in order to meet a requirement for diversification of biosynthetic pathways in which G6PD had been involved. The Ki for DEA in human G6PD has been reported to be 6.2×10^{-6}M (Benes et al., 1970) and that for DEA sulfatide as 6×10^{-7}M (Oertel and Rebelein, 1970), thus they are low enough to be of physiological significance. Other steroids having an oxo group at C-17 or C-20 position and a hydroxyl group at C-3 position are also potent inhibitors of G6PD. Accordingly, it is plausible that the acquisition of susceptibility to DEA in vertebrate G6PDs might have something to do with diversification of steroid metabolizing systems.

III. FUNCTIONAL DIFFERENTIATION OF G6PD AND H6PD

How did nature elaborate two G6PD isozymes? How are they functionally differentiated? As mentioned above, it is likely that the two enzymes have been evolved at different rates in echinoderm and vertebrate lineages, and that their differentiation may be related to steroid metabolism in one way or another. Almost exclusive localization of H6PD in microsomal membranes also suggests its involvement in microsomal electron transport systems, and hence, in steroid and drug metabolism.

Oxygenation of steroids, fatty acids, carcinogens and many other lipid-soluble xenobiotics is one of the important roles of microsomes, and this is accomplished by the mixed-function oxidase systems consisting of NADPH-cytochrome c reductase and cytochrome P-450. The systems appear to be rate-limited by the amount of NADPH-cytochrome c reductase (Davies et al.,1969; Gigon et al., 1969; Miwa et al.,1978), and hence by the supply of NADPH (Moldeus et al., 1974; Thurman et al., 1977; Weigl and Sies, 1977; Estabrook et al., 1971), but not necessarily by the amount of P-450. It has been assumed that NADPH necessary for the microsomal electron transport systems comes from cytosol, probably by way of the pentose phosphate pathway (Sies and Brauser, 1970; Moldeus et al., 1974; Junge and Brand, 1975; Kauffman et al., 1977). However, the presence of H6PD in microsomal membranes suggests another possibility that this enzyme may be the actual NADPH supplier to the electron transport system (Mandula et al., 1970). The results of our studies which were undertaken in order to clarify the physiological role of H6PD are summarized as follows.

A. Changes of G6PD and H6PD Activities under Physiological and Experimental Conditions

The specific activity of hepatic G6PD is at the same level in male and female suckling rats, but is about twice as high in adult females as in adult males. The cause of such sex difference may be estrogens, since ovariectomy hinders the activity increase during growth, while injection of estradiol into normal and ovariectomized rats induces the activity increase (Hori and Matsui,1967; Kodama and Hori, 1982). In contrast to G6PD, hepatic H6PD is more active in males than in females, decreases after castration and is restored to the normal level by androgen administration. Curiously, estradiol is a more powerful inducer of H6PD than methyltestosterone (Takahashi and Hori, 1981; Kodama and Hori, 1982). Feeding rats on a high carbohydrate diet causes about ten-fold increase in hepatic G6PD activity without effect on H6PD activity (Oka et al., 1981), while phenobarbital induces a slight, but significant increase of hepatic and renal H6PD (Kodama and Hori, 1982; Hori and Takahashi, 1974). Methylcholanthrene stimulates activities of hepatic and renal G6PDs and of renal H6PD but not hepatic H6PD (Kodama and Hori, 1982).

The specific activities of microsomal H6PD are high in steroid metabolizing organs as well as those of G6PD; the H6PD activities expressed as milliunits per mg protein are: liver 11.9; testicular interstitial cells 8.3; ovary 6.0; adrenal 4.8; kidney 3.2, while the values are 0.6-1.7 for intestine, brain, lung and spleen. In testis, both G6PD and H6PD activities are concentrated in the interstitial cells and increase during maturation. When stimulated with human chorionic gonadotrophin, the specific activities of microsomal H6PD and cytosol G6PD increase about 50% (Takahashi et al., 1979).

B. Distribution of H6PD in Various Organs of Rat

It is important to know a precise localization of H6PD in various tissues and cells, since information of this sort may be useful to infer the physiological role of H6PD. For this purpose, we examined histochemically the cytological localization of this enzyme utilizing a peroxidase-labeled antibody method (Tanahashi and Hori, 1980). As predicted, hepatic cells gave the most intense reaction. Steroidogenic cells were all positive, although the adrenal cortex reacted less intensely than testicular interstitial cells, ovarian theca interna and lutein cells. Besides these, the tubular secretory cells and striated ducts of the submandibular gland, plasma cells and the P_3 segments of proximal convolutions and collecting tubules in the cortex and inner medulla of the kidney exhibited a moderate staining. Cerebrum, cerebellum, seminal vesicle, urinary bladder and heart were negative. Electron microscopic observations were made on the liver, intestine, kidney, ovary and submandibular gland. In all positive cells, H6PD was located on the membranes of smooth- and rough-surfaced endoplasmic reticulum, but not on the plasma and Golgi membranes. The nuclear membrane was positive in liver cells, but not in other cells. Fibroblasts of various organs and mucous epithelial cells of the small intestine occasionally showed fragmented staining of the endoplasmic reticulum.

Kidney has been known as a steroid metabolizing organ as well as a drug metabolizing organ. It has also been reported that renal tubules are segmentally differentiated with respect to both morphology and function, and that the P_3 segment is characterized by an abundance of smooth-surfaced endoplasmic reticulum, high activities of G6PD and malate dehydrogenase and a predominance of 3α-hydroxysteroid dehydrogenase and of NADPH-dependent polycyclic hydrocarbon metabolizing systems, thus suggesting the concentration of steroid- and drug-metabolizing activities in this particular segment (see Tanahashi and Hori, 1980 for references). Accordingly, the above findings that H6PD is rich in this segment of renal tubules as well as in hepatic parenchymal cells, testicular interstitial cells and ovarian theca and lutein cells appear to be favorable to a hypothesis that H6PD might be involved in drug and steroid metabolism. Implication of the presence of H6PD in plasma cells and in striated ducts and serous tubular portions of the submandibular gland remains to be explored.

C. Perylene Metabolizing Activity in Renal P_3 Segments, Plasma Cells and Submandibular Gland Tubular Cells

As far as the author is aware, literature is not available that plasma cells and salivary gland cells are capable of metabolizing steroids and xenobiotics. However, if our assumption is correct that H6PD might be involved in microsomal steroid and drug metabolism, then it is probable that plasma cells and the salivary gland cells might also possess steroid or drug metabolizing activities. To test this, histochemical and quantitative studies were performed with normal and immunologically stimulated lymph nodes, and normal and dihydrotestosterone-stimulated submandibular glands of the rat. For histochemical studies, frozen-sections were stained for drug-metabolizing activity by a modification of the method of Wattenberg and Leong (1962) utilizing benzo(a)pyrene, methylcholanthrene and perylene as substrates (Tanahashi and Hori, unpubl.).

Among the three hydrocarbons tested, perylene gave most intense and clear-cut reactions; the cells having a high enzyme activity, such as liver, adrenal, renal P_3 segments and ovarian luteal cells exhibited green to blue-green fluorescence against dark background when stained with perylene as a substrate. On the other hand, only hepatic cells fluoresced bright blue when stained with benzo(a)pyrene, and none gave intense reactions to methylcholanthrene. In the submandibular gland, striated ducts and serous tubular cells showed a blue-green fluorescence with perylene. Following dihydrotestosterone treatment, a striking hypertrophy of tubular cells occurred, but the fluorescent intensity of such cells appeared similar to that of unstimulated cells. In the submandibular lymph nodes, a varying intensity of fluorescence was observed in lymphocytes of various sizes located mainly in the medullary cords, whereas lymphocytes in the cortical nodules were negative. In the medulla of immunologically stimulated lymph nodes, there appeared a number of B lymphocytes at varying stages of differentiation into plasma cells, which were characterized by a proliferating rough-surfaced endoplasmic reticulum containing an immunohistochemically demonstrable amount of IgG in the lumen. These cells showed a positive fluorescent reaction. In the kidney, the P_3 segments of proximal convolutions showed a most intense green fluorescence, while the distal convolutions and collecting tubules fluoresced blue-green indicating a moderate level of enzyme activity. Thus, it became clear that the distribution of perylene metabolizing activity coincides well with that of H6PD in various organs of rat.

For quantitative studies, microsomal fractions from various organs were incubated with perylene and NADPH, and the reaction products were separated by silica gel chromatography and quantitated by fluorescence spectrophotometry. For comparison, benzo(a)-pyrene hydroxylase activity was also assayed by the method of Van Cantfort et al. (1977) using ^{3}H-benzo(a)pyrene as a substrate. The results are illustrated in Fig.3, which clearly demonstrate that there is an intimate correlation between H6PD activity and perylene metabolizing activity in various rat organs, and that the enzyme system involved in perylene metabolism may be different from that for benzo(a)pyrene metabolism. In any case, the findings that stimulated B lymphocytes and plasma cells, and striated ducts and serous tubular cells of the submandibular gland do possess polycyclic hydrocarbon metabolizing activities are consistent with our hypothesis that H6PD might be involved in microsomal drug metabolism. In this connection it is worthy of note that several workers have reported the induction of aryl hydrocarbon hydroxylase activity in mitogen-stimulated cultured human lymphocytes; resting lymphocytes have a very low level of the hydroxylase activity, but treatments with either mitogens or aryl hydrocarbons increase the enzyme activity, and the effects become several fold when treated with these substances in combination. Similar findings have also been reported with macrophages and monocytes (Busbee et al.,1972; Bast et al.,1974; Cantrell et al., 1973; Coomes et al., 1976; Fujino et al., 1982; Gurtoo et al.,

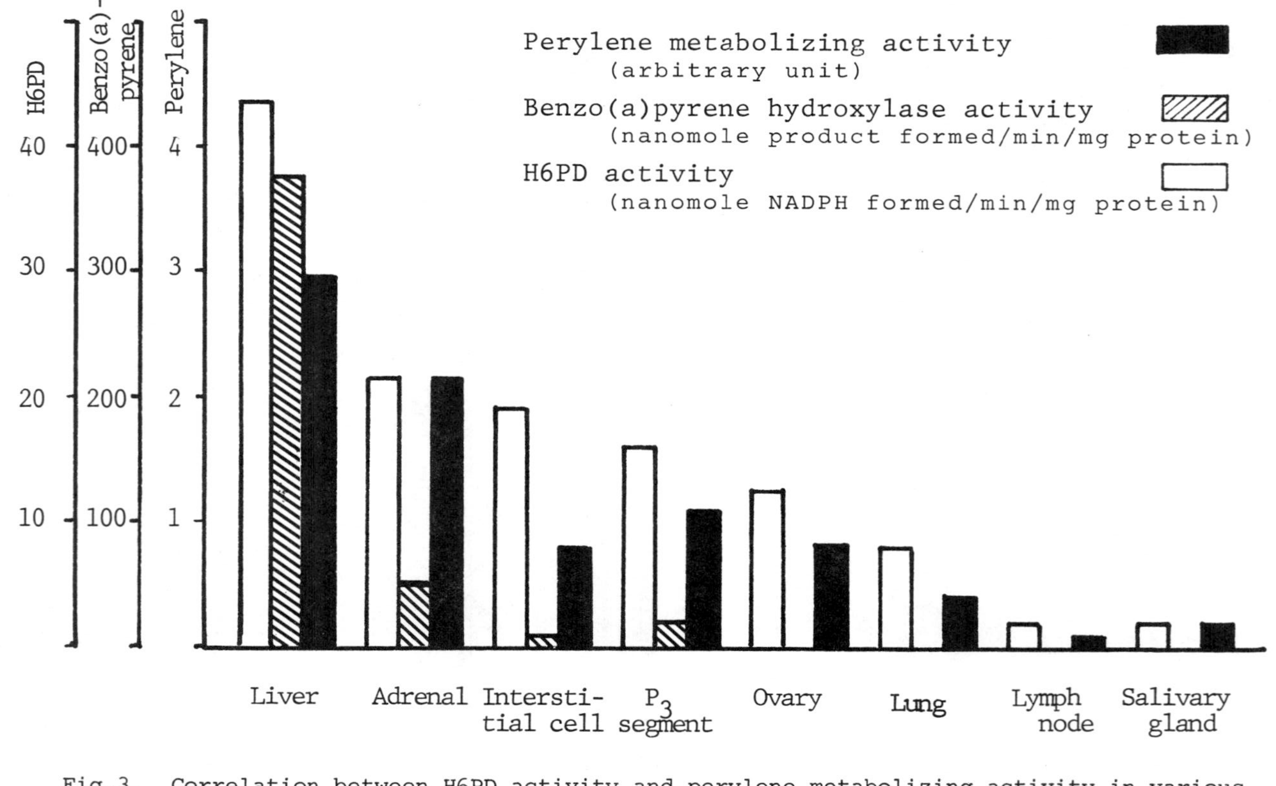

Fig.3. Correlation between H6PD activity and perylene metabolizing activity in various rat tissues.

1975; Kellermann et al., 1973; Whitlock et al., 1972). However, none of these studies referred to the *in vivo* hydroxylase activity of mature plasma cells.

D. Possible Functional Coupling of H6PD to Microsomal Electron Transport System

In order to look for still other evidence which suports our hypothesis, we attempted to examine if microsomes could perform benzo(a)pyrene hydroxylation or aminopyrene demethylation in the presence of NADP, dG6P and benzo(a)pyrene or aminopyrene. dG6P was used as a substrate, because H6PD is much more active on dG6P than on G6P and Gal6P at a physiological pH. As a result, we found that microsomes can metabolize benzo(a)pyrene or aminopyrene under the above conditions. At first glance this success appeared to be the most favorable evidence for our hypothesis, but we soon realized that the results of such experiments indicate nothing but the well-known fact that H6PD and drug metabolizing activities are both present in microsomes. In other words, it only demonstrates that H6PD may be utilized as a NADPH generator in place of a commonly used G6PD, but does not necessarily demonstrate a functional coupling of H6PD to drug metabolizing systems. As a matter of fact, the reaction rate of aminopyrene demethylation was found to be dependent on the amount of NADPH available at the start of reaction, and hence it was necessary to preincubate microsomes with NADP and dG6P for a while before adding aminopyrene in order to accelerate the reaction which was otherwise very slow.

Our next attempt was to examine whether H6PD and the cytochrome c reductase/cytochrome P-450 system behave in a similar fashion under various experimental conditions based on an assumption that similarity in their behaviors may reflect their functional coupling (Kodama and Hori, 1982). For this purpose, the P_3 segments of renal proximal convolutions were chosen as materials, since drug and steroid metabolizing activities in the kidney are concentrated in those segments, which therefore can be regarded as specifically differentiated tissues for these activities. This contrasts to multi-functional hepatic cells, and hence might facilitate the demonstration of functional coupling of H6PD and the electron transport system. For comparison, changes of G6PD activity were also examined. The results may be summarized as follows: (1) the specific activity of G6PD is highest in the inner medulla and lowest in the cortex, while H6PD, cytochrome c reductase and P-450 were much more concentrated in the cortex and the outer strip of medulla than in the rest of the kidney (the outer strip contains P_3 segments); (2) both H6PD and the reductase increased in the cortex and P_3 segments after treatments with phenobarbital, methylcholanthrene, methyltestosterone and estradiol, and after 3 days fasting, while increases of P-450 were observed only after fasting and methylcholanthrene treatment. In contrast, G6PD increased in the cortex after methylcholanthrene treatment and in P_3 segments after estradiol treatment, but it was not affected by other treatments. Thus, it became clear that there is an intimate correlation between H6PD and cytochrome c reductase activities under various experimental conditions, whereas no such correlation exists between G6PD and H6PD.

E. Differential Effects of the NADPH/NADP Ratio on H6PD and G6PD

It has been known that G6PD is inhibited by NADPH competitively with respect to NADP and hence its *in vivo* activity may be suppressed below 1% of its maximal *in vitro* activity under such physiological conditions that the coenzyme redox ratio is 110 (Eggleston and Krebs, 1974). It is not known however, if H6PD would be similarly inhibited by NADPH. We therefore studied the steady-state kinetics of rat liver H6PD using G6P and NADP as substrates (Oka et al., 1981). As a result, it was found that NADPH inhibits the enzyme noncompetitively with respect to NADP, and uncompetitively with respect to G6P. At a given concentration of G6P, the reaction follows the basic inhibition equation:

$$\frac{V}{v} = \frac{Ka}{A}\left(\frac{Q}{Kiq} + 1\right) + \frac{Q}{Kiq'} + 1$$

where A is NADP concentration; Q, NADPH concentration; Ka, Michaelis constant for NADP; Kiq, dissociation constant for NADPH; Kiq', dissociation constant for the enzyme-NADP-NADPH complex; V, maximal velocity.

This suggests the presence of the enzyme-NADP-NADPH complex in striking contrast to the competitive inhibition of G6PD by NADPH. An attempt was then made to estimate the *in vivo* activities of H6PD and G6PD in rat liver in the presence of NADPH at various coenzyme redox ratios. The results showed that the two enzymes appear to be at about the same level of activity in normal rat liver where the coenzyme redox ratio is 110, and the G6P concentration is 217 μM. Under the same conditions, but with 50 μM DEA (a potent inhibitor of G6PD, but not of H6PD), the latter enzyme is estimated to be 1.6 times as active as the former. Such differential effects of NADPH and steroids on the two enzymes support our hypothesis that H6PD may have advantages over G6PD in steroid- or drug-metabolizing tissues.

F. Problems to be Solved

Our data presented in this report are all consistent with the hypothesis that H6PD might be involved in microsomal steroid and drug metabolism. However, there are some important findings which can not be ignored when considering the physiological role of H6PD; one relates to the latency of H6PD activity, and the other to the NADPH-cytochrome c reductase. Rat liver microsomes isolated with isotonic sucrose exhibit no H6PD activity at all when assayed at 21°C. This is due to the impermeability of microsomal membranes to NADP and to the intramenbraneous location of H6PD (Takahashi and Hori, 1978). Parenthetically, the extent of latency differs significantly in different vertebrate species (Takahashi et al. 1979). However, rat microsomes show approximately 16% of the total H6PD activity when assayed at 37°C. Therefore, it is likely that H6PD is sufficiently active *in vivo* to support the microsomal electron transport system. On the other hand, it has been known that intact microsomes are capable of hydroxylating steroids and drugs *in vitro* in the presence of oxygen and exogeneous NADPH, and that the active site of NADPH-cytochrome c reductase is exposed on the cytoplasmic surface of microsomal membranes (Depierre and Ernster, 1977). It is supposed therefore that the microsomal electron transport system may be able to utilize NADPH present in cytosol. A question then arises why the microsomal membranes must have their own built-in NADPH generator.

H6PD has persisted for several hundreds of millions of years as a component of microsomal membranes with essential properties remaining constant throughout evolution of echinoderms and vertebrates. This would not be possible without strict functional constraints and selection pressure. H6PD may be functioning for a certain purpose which we can not specify with confidence at present. Mutant strains of laboratory animals or tissue culture cells lacking in H6PD should provide useful materials for this sort of study, but these are not available. Therefore, it appears that the only way to solve this problem is to accumulate more evidence.

REFERENCES

Bagdasarian,G.,and Hulanicka,D.(1965).Biochim.Biophys.Acta 99,369.

Baquer,N.Z.,and McLean,P.(1972).Biochem.Biophys.Res.Commun.46,167.

Baquer,N.Z., Sochor,M.,and McLean,P. (1972). Biochem.Biophys.Res. Commun. 47,218.

Bast,R.C., Whitlock,J.P., Miller,H., Rapp,H.J., and Gelboin,H.V. (1974). Nature 250,664.

Benes,P., Freund,R., Menzel,P., Starka,L. and Oertel,G.W. (1970). J. Steroid Biochem. 1,287.

Beutler,E., and Morrison,M. (1967). J.Biol.Chem. 242,5289.

Busbee,D.L., Shaw,C.R., and Cantrell,E.T. (1972). Science 178,315.

Campbell,J.H., Lengyel,J.A., and Langridge,J. (1973). Proc. Nat. Acad. Sci. U.S.A. 70, 1841.

Cantrell,E., Busbee,D., Warr,G. and Martin,R. (1973).Life Sci. 13, 1649.

Cohen,R., and Rosemeyer,M.A. (1969). Europ.J.Biochem. 8, 8.

Davies,D.S., Gigon,P.L., and Gillette,J.R. (1969). Life Sci. 8, Part II, 85.

Depierre,J.W., and Ernster,L. (1977). Ann.Rev.Biochem. 46,201.
Eggleston,L.V., and Krebs,H.A. (1974). Biochem.J. 138,425.
Estabrook,R.W., Franklin,M., Baron,J., Shigematsu,A. and Hildebrandt,A. (1971). In "Drugs and Cell Regulation" (E. Mihich, ed.), p.227. Academic Press, New York.
Fondy,T.P., and Holohan,P.D. (1971). J.Theor.Biol. 31,229.
Fujino,T., Park,S.S., West,D., and Gelboin,H.V. (1982). Proc.Nat. Acad.Sci.U.S.A. 79,3682.
Gasperi,G., Malacrida,A., Cima,L., Sacchi,L., and Grigolo,A.(1978) J.Histochem.Cytochem. 26,850.
Gigon,P.L., Gram,T.E., and Gillette,J.R. (1969). Mol.Pharmacol. 5,109.
Glock,G.E., and McLean,P. (1953). Biochem.J. 55,400.
Grunwald,M., and Hill,H.Z.(1976). Biochem.J. 159,683.
Gurtoo,H.L., Bejba,N., and Minowada,J. (1975). Cancer Res.35,1235.
Harris,C.E., and Teller,D.C. (1973). J.Theor.Biol.38,347.
Hori,S.H., and Matsui,S. (1967). J.Histochem.Cytochem. 15,530.
Hori,S.H., and Matsui,S. (1968). J.Histochem.Cytochem. 16,62.
Hori,S.H., and Noda,S. (1971). J.Histochem.Cytochem. 19,299.
Hori,S.H., and Yonezawa,S. (1972).J.Histochem.Cytochem. 20,804.
Hori,S.H., and Takahashi,T. (1974). Biochem.Biophys.Res.Commun. 61,1064.
Hori,S.H., Yonezawa,S., Mochizuki,Y., Sado,Y. and Kamada,T. (1975) In "Isozymes,IV" (C.Markert, ed.), p.839. Academic Press. New York.
Hori,S.H., and Takahashi,T. (1977). Biochim.Biophys.Acta 496,1.
Hori,S.H., Sado,Y., and Matsuoka,N. (1977). Japan.J.Genet.52,269.
Hori,S.H., and Tanda,S. (1980). Japan.J.Genet. 55,211.
Junge,O., and Brand,K. (1975). Arch.Biochem.Biophys. 171,398.
Kamada,T., and Hori,S.H. (1970). Japan.J.Genet. 45,319.
Kellermann,G., Cantrell,E., and Shaw,C.R. (1973). Cancer Res. 33, 1654.
Kauffman,F.C., Evans,R.K., and Thurman,R.G. (1977). Biochem.J. 166,583.
Kimura,M.(1968). Nature 217,624.
Kimura,M. and Ohta,A. (1974). Proc.Nat.Acad.Sci.U.A.S. 71,2848.
King,J.L., and Jukes,T.H. (1969). Science 164,788.
Kirkman,N.N., Kidson,C., and Kennedy,M. (1968). In"Studies of Samples from New Guinea in Hereditary Disorders of Erythrocyte Metabolism" (E.Beutler, ed.), P.126. Grune and Stratton, New York.
Kodama,T., Ohnishi,K., and Hori,S.H. (1981). J.Fac.Sci.Hokkaido Univ.Ser.6,Zool. 22,191.
Kodama,T., and Hori,S.H. (1982). Biochim.Biophys.Acta 715,151.
Mandula,B., Srivastava,S.K., and Beutler,E. (1970). Arch.Biochem. Biophys. 141,155.
Marchalonis,J.J., and Weltman,J.K. (1971). Comp.Biochem.Physiol. 38B,609.
Matsuoka,N., Mochizuki,Y., and Hori,S.H. (1977). J.Fac.Sci. Hokkaido Univ.Ser.6,Zool. 21,12.

Matsuoka,N. and Hori,S.H. (1980). Comp.Biochem.Physiol.65B,191.
Metzger,H., Shapiro,M.B., Mosimann,J.E., and Vinton,J.E. (1968). Nature, 219,1166.
Miwa,G.T., West,S.B., and Lu,A.Y.H. (1978). J.Biol.Chem.253,1921.
Mochizuki,Y., and Hori,S.H. (1973). J.Fac.Sci.Hokkaido Univ.Ser.6, Zool. 19,58.
Mochizuki,Y., and Hori,S.H. (1976) Comp.Biochem.Physiol.54B,489.
Moldeus,P., Grundin,P., Vadi,H., and Orrenius,S. (1974). Eur.J. Biochem. 46,351.
Newburgh,R.W., and Cheldelin,V.H. (1956). J.Biol.Chem. 218,89.
Oertel,G.W., and Rebelein,I. (1970). J.Steroid Biochem. 1,92.
Ohnishi,K., and Hori,S.H. (1977). Japan.J.Genet. 52,92.
Ohno,S., Payne,H.W., Morrison,M. and Beutler,E. (1966). Science 153,1015.
Oka,K., Takahashi,T., and Hori,S.H. (1981). Biochim.Biophys.Acta 662,318.
Prager,E.M., and Wilson,A.C. (1971). J.Biol.Chem.246,5978.
Razumovskaya,N.I., Pleskov,V.M., and Perova,T.L. (1970). Biochemistry 35,196.
Ruddle,F.H., Shows,T.B., and Roderick,T.H. (1968). Genetics 58, 599.
Rutter,W.J. (1965). In "Evolving Genes and Proteins" (V.Bryson, and H.J.Vogel, ed.), p.279. Academic Press, New York.
Sado,Y. (1980). J.Fac.Sci.Hokkaido Univ.Ser.6,Zool. 22,156.
Sado,Y., and Hori,S.H. (1976). J.Fac.Sci.Hokkaido Univ.Ser.6, Zool. 20,277.
Shatton,J.B., Halver,J.E., and Weinhouse,S. (1971). J.Biol.Chem. 246,4878.
Shaw,C.R. (1966). Science 153,1013.
Sies,H., and Brauser,B. (1970). Eur.J.Biochem. 15,531.
Smith,E.L. (1970). In "The Enzymes" (P.D.Boyer, ed.), I,267. Academic Press, New York.
Srivastava,S.K., and Beutler,E. (1969). J.Biol.Chem.244,6377.
Srivastava,S.K., Blume,K.G., Beutler,E., and Yoshida,A. (1972). Nature 238,240.
Stegeman,J.J., and Goldberg,E. (1971). Biochem.Genet. 5,579.
Stegeman,J.J., and Goldberg,E. (1972). Comp.Biochem.Physiol. 43B, 241.
Takahashi,T., and Hori,S.H. (1978). Biochim.Biophys.Acta 524,262.
Takahashi,T., Tanahashi,K., Kodama,T., and Hori,S.H. (1979). J.Fac.Sci.Hokkaido Univ.Ser.6, Zool. 22,12.
Takahashi,T., and Hori,S.H. (1981). Biochim.Biophys.Acta 672,165.
Tanahashi,K., and Hori,S.H. (1980). J.Histochem.Cytochem. 28,1175.
Thurman,R.G., Marazzo,D.P., Jones,L.S., and Kauffman,F.C. (1977). J. Pharmacol.Exp.Ther.201,498.
Van Cantfort,J., DeGraeve,J., and Gielen,J.E. (1977). Biochem. Biophys.Res.Commun. 79,505.
Walsh,K.A., and Neurath,H. (1964). Proc.Nat.Acad.Sci.U.S.A. 52, 884.
Watanabe,A., and Taketa,K. (1973). Arch.Biochem.Biophys. 158,43.

Wattenberg,L.W., and Leong,J.L. (1962). J.Histochem.Cytochem. 10, 412.

Weigl,K., and Sies,H. (1977). Eur.J.Biochem. 77,401.

Whitlock,J.P., Cooper,H.L., and Gelboin,H.V. (1972). Science 177,618.

Yamada,K., and Shimazono,N. (1961). Biochim.Biophys.Acta 54,205.

Yoshida,A., Baur,E., and Motulsky,A.G. (1970). Blood 35,506.

Yoshida,A., and Hoagland,V.D. (1970). Biochem.Biophys.Res.Commun. 40,1167.

Zaheer,N., Tewari,K.K., and Krishnan,P.S. (1967). Arch.Biochem. Biophys. 120,22.

SIGNALS REGULATING GLUCOSE-6-P DEHYDROGENASE LEVELS IN RAT LIVER

Patricia Manos, Nancy Taylor, Diana Rudack-Garcia, Norihiko Morikawa, Roderick Nakayama and Darold Holten[1]

Department of Biochemistry
University of California
Riverside, California, USA

From the pioneering work of Glock and Mclean (1), the Teppermans (2), Fitch and Chaikoff (3), and others, we know that liver glucose-6-P dehydrogenase (G6PD) becomes induced when rats are fed a high carbohydrate diet. My laboratory has been using G6PD as a model to study how diet and hormones regulate the levels of enzymes involved in the synthesis and secretion of triglyceride in liver. In the rat, liver makes about 30% of the triglyceride and lipogenesis increases when they are fed a high carbohydrate diet (4). Under these conditions, many of the enzymes along the pathway for fatty acid synthesis become induced and some increase more than 50-fold (5). Figure 1 illustrates this by marking with an asterisk some of the lipogenic enzymes which become induced under these conditions. Inducible enzymes along the direct pathway providing the carbon atoms for fatty acids include pyruvate kinase, citrate cleavage enzyme (CCE), acetyl CoA carboxylase (AcCC) and fatty acid

[1] *Supported in part by Research Grant AM 13324 from the United States Public Health Services.*

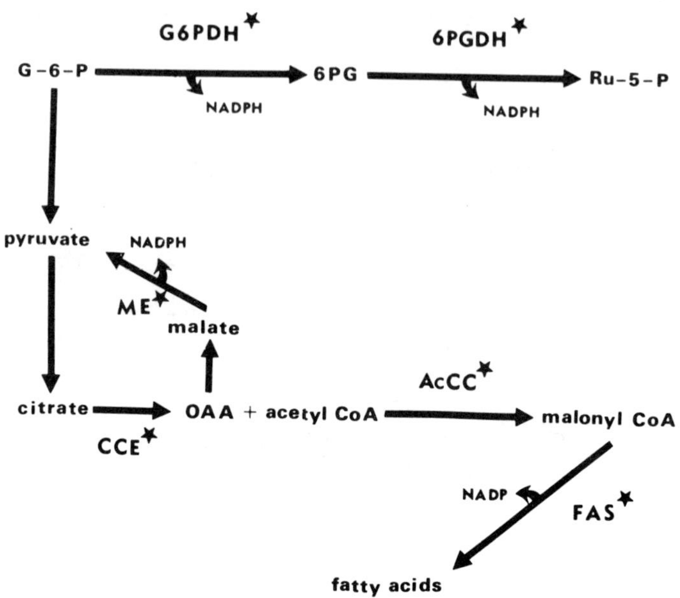

Figure 1. Asterisks indicate enzymes which become induced during increased lipogenesis.

synthase (FAS). The three enzymes which provide the reducing equivalents for fatty acid synthesis also become induced coordinately with the other lipogenic enzymes. These include the two pentose phosphate pathway dehydrogenases and malic enzyme.

Since the induction of G6PD is coordinated with the induction of lipogenic enzymes and with increased rates of fatty acid synthesis one might anticipate that the metabolic and hormonal signals which regulate fatty acid synthesis also regulate G6PD induction.

In rat liver, fatty acid synthesis is increased by insulin or by feeding rats a high carbohydrate diet (which increases blood insulin levels) and is decreased by glucagon or cAMP (6,7). The rate limiting and regulated step in fatty acid synthesis appears to be acetyl CoA carboxylase. Its activity is increased by poymerization of the enzyme when the citrate concentration (a fatty acid precursor) increases and decreased when it is phosphorylated by a cAMP-independent protein kinase and in some species by cAMP-dependent protein kinase (8,9). This suggested G6PD levels might be induced by insulin and possibly by some intermediate in fatty acid synthesis and that G6PD levels might be decreased by cAMP. In addition, polyunsaturated fatty acids have been shown to decrease the rate of fatty acid synthesis (6).

The work in my laboratory has focused on identifying those signals which regulate G6PD induction and the mechanisms by which those signals act. When rats are fasted for two days and refed a 60% glucose-fat free diet, G6PD activity increases to a new steady-state in about three days (Figure 2). The K calories of carbohydrate eaten/day/100 g rat is quite constant from days 3-7. The equation describes the time course for the increase in G6PD activity (P) from the initial fasting state (P_0) to the new

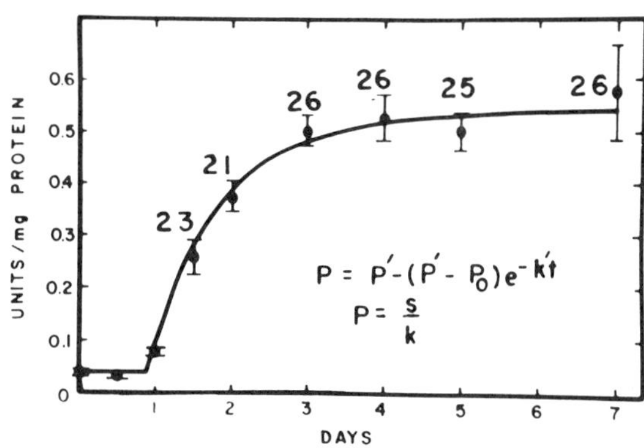

Figure 2. Time course for the induction of G6PD in rats fasted for two days and refed a 60% glucose-fat free diet for seven days. Numbers above the line indicate the K calories of carbohydrate consumed/100 g rat/day. The equation describes the time course for the change in G6PD (P) at any time t in transition between the non-induced state (P_0) and the induced steady-state (P'). The solid line was calculated using a half-life of 15 hours for G6PD degradation in induced rats (10).

induced state (P'). The solid line was calculated using this equation and a half-life of 15 hours for G6PD in induced rats.

We did a number of similar experiments where rats were fed differing amounts or types of carbohydrates with and without insulin or with various amounts of polyunsaturated fatty acid (PUFA) (10,11). The results were always consistent with a half-life of 15 hours for G6PD in rats fed these synthetic diets so we used the simple relationship $K_s = P' \times K_d$ to estimate the rate of G6PD synthesis in these animals. When the estimated rate of G6PD synthesis was plotted versus the amount of carbohydrate

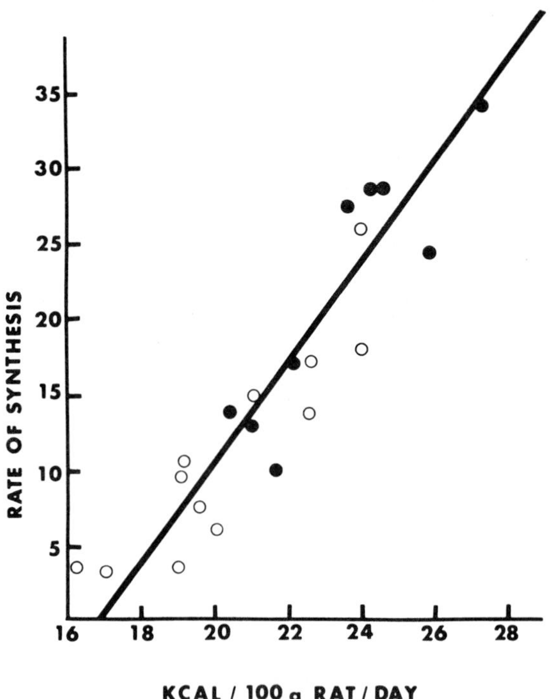

Figure 3. Correlation between G6PD synthesis and the amount of glucose consumed in the diet. Closed circles are diets with differing amounts of carbohydrate and open circles are diets containing differing amounts of polyunsaturated fatty acids and/or carbohydrate (11).

consumed, roughly a straight line relationship was obtained. This suggested that, at least within this range, G6PD synthesis was directly related to the amount of consumed carbohydrate. The open and closed circles represent diets with and without PUFA, respectively. Since all points were equally close to the line, this suggested that specific effects of PUFA to decrease G6PD synthesis may be masked when these large amounts of carbohydrate are consumed. Work by others who pair-fed constant and smaller amounts of carbohydrate to rats concluded that PUFA can decrease G6PD levels in rats (6).

The correlation between the amount of carbohydrate consumed and the rate of G6PD synthesis suggests that some carbohydrate or lipogenic precursor may regulate G6PD levels, that insulin released in response to the carbohydrate may regulate G6PD levels, or that some combination of the two may be involved. If rats are fed an inducing diet to produce a high level of G6PD and then injected with insulin for two days, G6PD is clearly increased by insulin (Table I). However, insulin injections

Table I

Effect of Insulin on the Specific Activities of the Hexose Monophosphate Shunt Dehydrogenases in Rats Previously Adapted to Two Different Steady State Enzyme Levels.

Treatment	Pretreatment	G6PD Specific Activity
		(units/mg protein)
Pellet	fed	0.048 ± 0.005
Sucrose	fed	0.41 ± 0.02
Sucrose + insulin	fed	0.75 ± 0.00
Sucrose	fasted	0.60 ± 0.03
Sucrose + insulin	fasted	0.78 ± 0.04

Rats were fed a pellet diet or a 60% sucrose diet for 8 days or rats were fasted for 48 hours and then refed the sucrose diet for 6 days. Insulin was injected at a dose of 10 units per 100 g rat per day during the final 48 hours of the experiment.

also increased consumption of the carbohydrate diet and the data could not be used to prove that insulin had an ability to independently regulate G6PD levels.

Glucagon, however, does specifically decrease the induction of G6PD (Figure 4). When fasted rats were fed the high carbohydrate diet and were injected with glucagon throughout the experiment, the induction of G6PD was decreased by about 50%. In this experiment, there was no decrease in the amount of carbohydrate consumed by glucagon-injected rats. Clearly, in the live rat, glucagon could inhibit G6PD induction (12).

In an attempt to determine how rapidly glucagon acts, we fed rats a 60% fructose diet until G6PD reached an induced steady-state and then injected them with dibutyryl cAMP and theophylline (Figure 5). In this experiment the rate of G6PD synthesis was measured directly by immunoprecipitating G6PD with a specific antiserum and determining the cpm in the G6PD after SDS PAGE

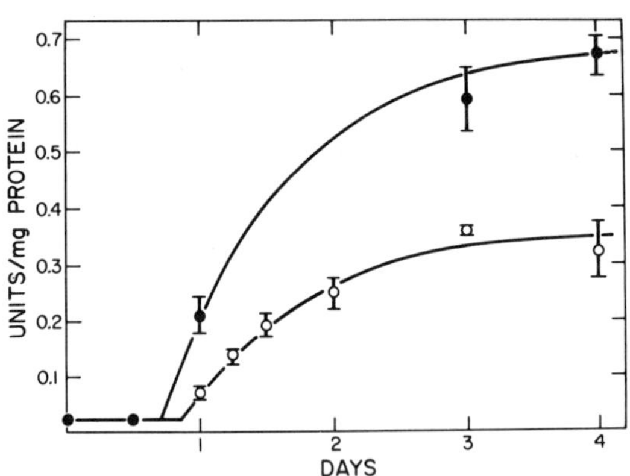

Figure 4. The effect of glucagon (0.2 mg/100 g every 6 hours) on the time-course for the induction of G6PD in rats fasted for two days and refed a 60% glucose-fat free diet. Open and closed circles represent animals treated with and without glucagon, respectively. The solid lines were calculated using a half-life of 15 hours for G6PD degradation.

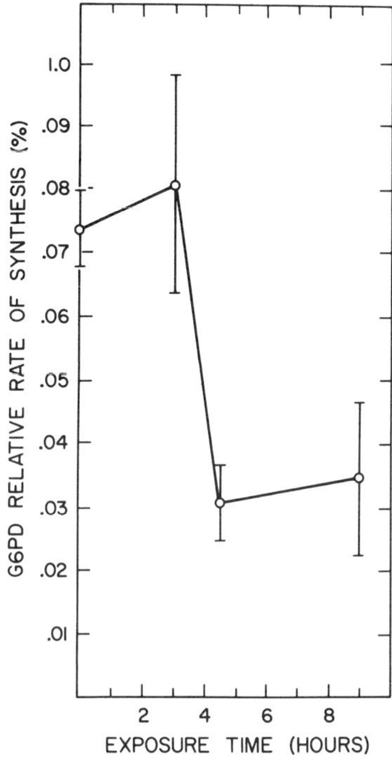

Figure 5. Time course for the decrease in G6PD synthesis in Bt_2cAMP-treated animals. All animals were injected with a mixture of 50 mg Bt_2cAMP and 20 mg theophylline/kg rat. Two injections were administered on the following schedule. For the 9-hour treatment, injection one was at zero time and injection two at 6.5 hours. For the 4.5-hour treatment, injection one was at zero time, and injection two at 1 hour. Untreated animals were injected with saline only. Hepatocytes were isolated and used to determine the rate of G6PD synthesis. This is plotted as a function of total exposure time including the time required to isolate hepatocytes and label protein. Each point represents an average of 3-5 animals plotted as the mean $\pm$ S.E.M.

electrophoresis (13). Dibutyryl cAMP decreased G6PD synthesis about two-fold by four and one-half hours. RNA was isolated from these animals, translated in a reticulocyte lysate, and the G6PD was immunopreciptated in order to quantitate the levels of G6PD mRNA by methods we had published previously (14). We found that G6PD mRNA decreased about as much as G6PD synthesis.

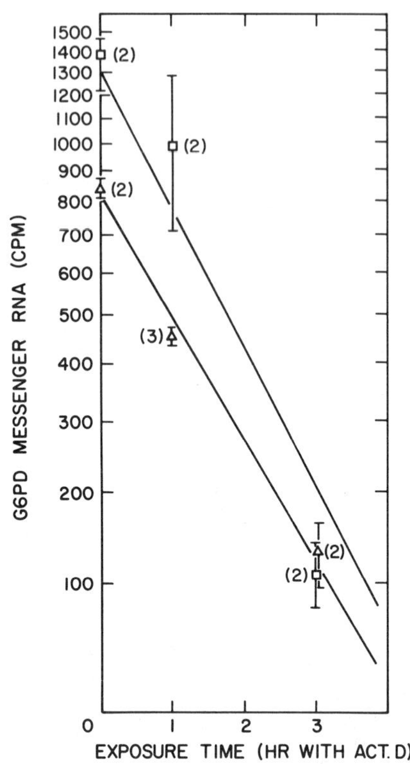

Figure 6. The effect of Bt_2cAMP on the half-life of G6PD mRNA. Animals induced with a 60% fructose diet were treated with 0.66 mg Actinomycin D/kg rat at zero time, one, and three hours. Two sets of animals were examined to determine the effect of cAMP: those treated with 50 mg Bt_2cAMP + 20 mg theophylline/kg rat for 4.5 hours and those injected with saline only. [Saline or Bt_2cAMP injections were started 4.5 hours before zero time in the figure.] G6PD mRNA was measured at each time point after translation in endonuclease treated rabbit reticulocyte lysate as described previously (14). All measurements included 2-3 animals per time point as shown in parentheses. Average values are shown as the mean $\pm$ S.E.M.

In the next experiment, we tried to determine if cAMP had any effect on the half-life of G6PD mRNA (Figure 6). Rats with high levels of G6PD were treated with dibutyryl cAMP for 4.5 hours before being injected with actinomycin D to inhibit mRNA synthesis. RNA was isolated for the next three hours, translated, and cpm in G6PD were determined. In animals treated with cAMP (bottom line) or without cAMP (top line), G6PD mRNA declined with a half-life of 1-2 hours. These data, although limited, suggest that cAMP does not decrease G6PD mRNA by increasing its rate of degradation. Presumably the cAMP acts by decreasing the synthesis of G6PD mRNA. The zero time point on the graph, which is 4.5 hours after cAMP treatment, confirmed that the cAMP decreased G6PD mRNA by about 1.6-fold. Although cAMP appears to specifically decrease G6PD synthesis, it clearly cannot completely account for the much larger changes in G6PD synthesis we have observed.

Recently, we have used more sensitive methods to determine how large these changes are in rats which are fasted for two days and then fed a 60% glucose-fat free diet for four days (15). Our standard procedure is to incubate hepatocytes (prepared from rats in the desired nutritional state) with [^{3}H]-leucine and to use an antiserum to immunoprecipitate G6PD. We have found that subjecting the immunoprecipitate to 2D gel electrophoresis gives a much lower background and higher signal to noise ratio. The lane with the G6PD spot (Figure 7) is excised, sliced and counted to determine the cpm in G6PD. Figure 8 illustrates the typical signal to noise ratio for immunoprecipitated G6PD after electrophoresis on 2D gels. The G6PD from chow-fed rats was immunoprecipitated from 3 ml of hepatocyte supernatant, while for induced rats we used only 0.5 ml of hepatocyte supernatant. Table II shows the relative rate of G6PD synthesis and specific activity in fasted and induced rats. In this table, specific activity is expressed relative to DNA content so it can be compared to results with cultured hepatocytes. Identical changes were observed if G6PD activity was expressed relative to protein. We now measure G6PD activity at the pH optimum for G6PD (pH 9) rather than at pH 8 as in the assay of Bottomley _et al_. (16). This gives a more sensitive assay

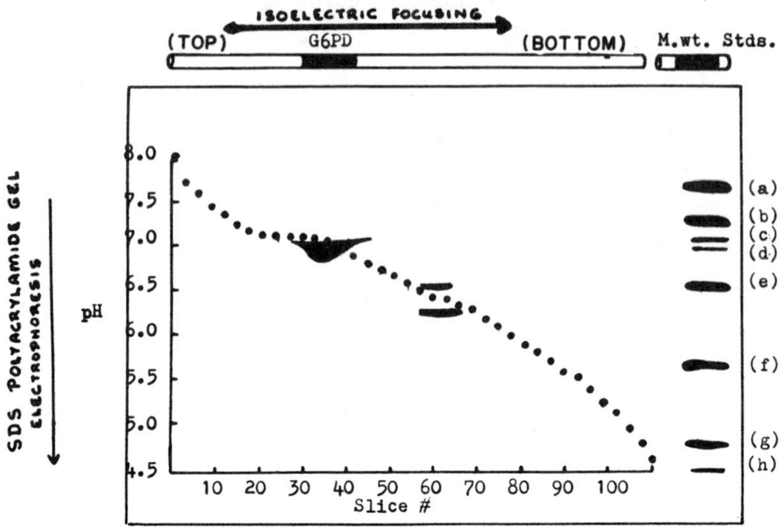

Figure 7. Two dimensional gel of immunoprecipitated G6PD stained with Coomassie Blue.

Table II

Changes in G6PD Activity and Synthesis

	G6PD Sp. Activity (mU/µg DNA)	Relative Rate of Synthesis (% of Total)
Fasted	0.20 ± 0.04	0.0015 ± 0.0002
	63X	73X
Fasted-refed	12.5 ± 2.4	0.11 ± 0.02

for G6PD in fasted rats where G6PD is much less than 6-phosphogluconate dehydrogenase. Using these more sensitive assays, we find that both G6PD activity and synthesis increase 60 to 70-fold when fasted rats are refed a 60% glucose-fat free diet.

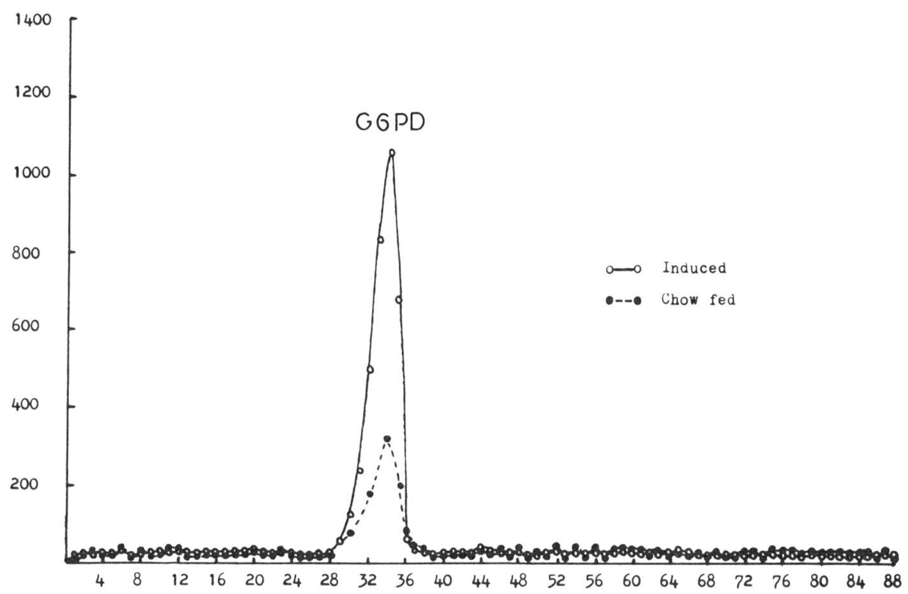

Figure 8. Radioactivity in G6PD immunoprecipitated from 0.5 ml or 3 ml of radiolabeled liver supernatant fraction from induced and fasted rats, respectively. The immunoprecipitated G6PD was subjected to 2 D gel electrophoresis as in Figure 7 and the gel lane containing the G6PD spot was excised, sliced, counted and cpm (ordinate) were plotted versus slice number.

Our work with live rats suggested that we would need an isolated system where we could independently control hormones and carbohydrate before we could hope to determine which signals regulate G6PD induction. We chose isolated primary hepatocytes cultured on collagen coated nylon discs. Patricia Manos has developed a system which plates and cultures hepatocytes from fasted rats completely in the absence of serum and in defined media. This system is very responsive to added hormones and supports an induction of G6PD which, even in the absence of serum, approaches the 60 to 70-fold increase seen in live rats. Figure 9 illustrates the time course for the induction of G6PD when hepatocytes from a two-day fasted rat are cultured in L-15 medium which is devoid of any carbohydrate. Dexamethasone and insulin were present at 10^{-6} and 10^{-7} M, respectively. For the first six days, the induction of G6PD in the absence of carbohydrate and serum (solid circles) proceeds almost as well as it does in the presence of serum (open circles). After six days, G6PD activity decreases in the absence of serum. In the presence of serum, G6PD induction is maximum after eight days in culture.

Figure 10 illustrates some preliminary results on the effect of adding 5 mM fructose or insulin to cells cultured in the absence of serum. Fructose alone has a small effect on G6PD induction. Insulin, in the absence of any carbohydrate in the medium, supported a 45-fold increase in G6PD. Insulin plus fructose supported a 70-fold increase in G6PD. Thus, in hepatocytes from fasted rats cultured in the absence of serum, the combination of insulin, dexamethasone, and a utilizable carbohydrate supports an induction of G6PD which equals that seen in live rats fasted for two days and refed a high carbohydrate-fat free diet. Recent results indicate that induction by insulin in the absence of carbohydrate does not require dexamethasone. Clearly, insulin alone is a major factor regulating the induction of G6PD. In addition, there appears to be some interaction between fructose and insulin which results in an induction equaling that seen in live rats. We are currently investigating if dexamethasone is required for this interaction.

In summary, the 60 to 70-fold induction of G6PD synthesis and activity which occurs in live rats can also be produced by insulin, carbohydrate, and dexamethasone in primary hepatocytes cultured in the absence of serum. Of these three signals,

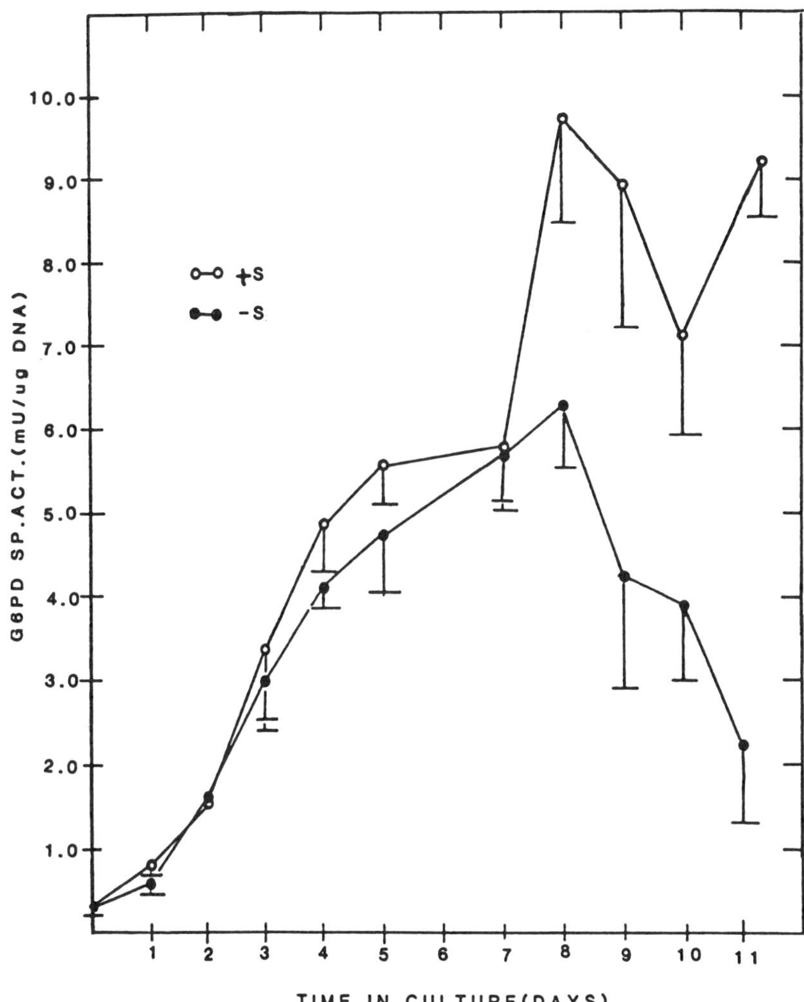

Figure 9. The effect of serum on G6PD activity in primary cultures of hepatocytes. Hepatocytes were isolated from a two-day fasted rat by collagenase perfusion of the liver, and cultured on floating collagen gels embedded on a nylon mesh. The medium, Leibovitz's L-15, was modified by omitting galactose and pyruvate, by replacing arginine with ornithine, and was supplemented with Hepes (25 mM), $NaHCO_3$ (1.18 g/L), BSA (2 mg/ml), dexamethasone (10^{-6} M) and insulin (10^{-7} M). The serum condition was supplemented with 5% Newborn Calf Serum (+S). The medium was changed 4 hours after the isolation procedure, and every 24 hours thereafter. Each value represents the mean ± standard deviation for three plates.

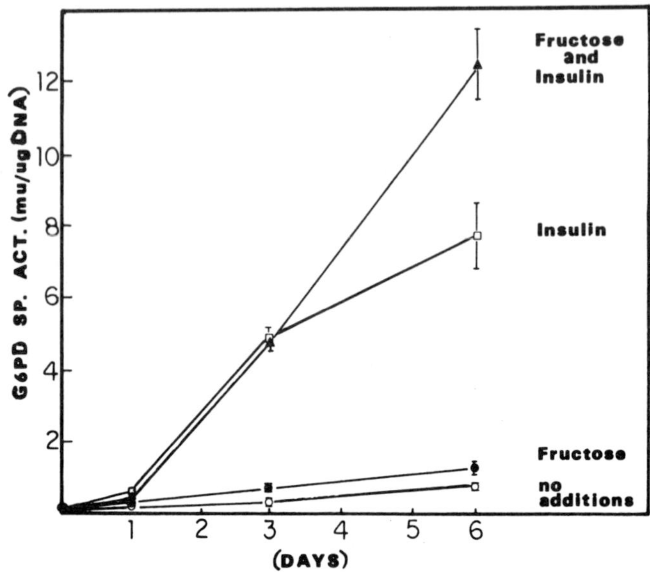

Figure 10. The effect of insulin, dexamethasone, and fructose on the activity of G6PD. The experimental procedure and medium were as described in the legend to Figure 9. Dexamethasone (10^{-6} M) was present in all conditions. Where indicated, fructose was present at a concentration of 5mM, and insulin was present at a concentration of 10^{-7} M. The values represent the mean $\pm$ S.D. of three plates.

insulin seems to have the largest effect. Glucagon acting via cAMP can repress G6PD synthesis about two-fold and may also play a role in regulating G6PD synthesis. In 1978 (14), we used a translation assay to show that G6PD mRNA only increased enough to partially account for the increased rate of G6PD synthesis. We now know that since G6PD levels and synthesis are so low in fasted rats, those methods almost certainly underestimated the increase in G6PD mRNA. Our current evidence, therefore, suggests that those signals regulating G6PD synthesis are likely to also cause large changes in G6PD mRNA.

REFERENCES

1. Glock, G. E., and McLean, P. (1955). Biochem. J. 61, 390-397.
2. Tepperman, H. M., and Tepperman, J. (1958). Diabetes 7, 478-485.
3. Fitch, W. M., and Chaikoff, I. L. (1960). J. Biol. Chem. 235, 554-557.
4. Bouillon, D. J., and Berdanier, C. D. (1980). J. Nutr. 110, 286-297.
5. Muto, Y., and Gibson, D. (1970). Biochem. Biophys. Res. Commun. 38, 9-15.
6. Herzberg, G. (1983). Adv. Nutr. Res. 5, 221-223.
7. Witters, L., Kowaloff, E., and Avruch, J. C. (1979). J. Biol. Chem. 254, 245-248.
8. Lee, K., and Kim, K. (1977) Ann. Rev. Biochem. 52, 1748-1751.
9. Wakil, S. J., Stoops, J. K., and Joshi, V. C. (1983). Ann. Rev. Biochem. 52, 537-579.
10. Rudack, D., Chisholm, E., and Holten, D. (1971). J. Biol. Chem. 246, 1249-1254.
11. Gozukara, E., Frolich, M., and Holten, D. (1972). Biochim. Biophys. Acta 286, 155-163.
12. Rudack-Garcia, D., and Holten, D. (1975). J. Biol. Chem. 250, 3960-3965.
13. Winberry, L., and Holten, D. (1977). J. Biol. Chem. 252, 7796-7801.
14. Sun, J., and Holten, D. (1978). J. Biol. Chem. 253, 6832-6836.
15. Morikawa, N., Nakayama, N., and Holten, D. (1984). Biochem. Biophys. Res. Commun. 120, 1022-1029.
16. Bottomley, R. H., Pitot, H. C., Potter, V. R., and Morris, H.P. (1963). Cancer Res. 23, 400-409.

HEPATIC GLUCOSE-6-PHOSPHATE DEHYDROGENASE: NUTRITIONAL AND HORMONAL REGULATION OF mRNA LEVELS

Rolf F. Kletzien[1], R. Scott Fritz, Christopher R. Prostko, Evan A. Jones, and Kevin L. Dreher

Department of Biochemistry, West Virginia University School of Medicine, Morgantown, West Virginia, U.S.A.

I. INTRODUCTION

Thirty years have elapsed since it was first recognized that hepatic glucose-6-phosphate dehydrogenase (G6PD; EC 1.1.1.49) is subject to nutritional regulation (1). Subsequently, numerous investigators have demonstrated that feeding a high carbohydrate diet to previously fasted rats results in a sharp increase in hepatic G6PD activity (2-5). Studies have been conducted to delineate the factors responsible for the dietary induction of the enzyme. It has been demonstrated that refeeding fasted rats a carbohydrate deficient diet containing high protein (6) or high fat (5) fails to elicit the characteristic induction of G6PD activity. Johnson and Sassoon (5) also demonstrated that fructose or sucrose could substitute for

---

[1]This work was supported by Grant AA 06728 from the National Institute on Alcohol Abuse and Alcoholism. R.K. is the recipient of Career Development Award AM-00733 from the National Institutes of Health.

glucose as the carbohydrate portion of the diet responsible for dietary induction. Thus, it is clear that in the intact animal, high carbohydrate in the diet plays a fundamental role in the induction of heaptic G6PD.

The potential involvement of the endocrine system in the dietary induction of G6PD has been examined in some detail. The role of insulin has been assessed since a high carbohydrate diet would be expected to elicit the release of this hormone. When diabetic rats were fasted and then refed a high carbohydrate diet, the induction of G6PD was not observed (7-9) unless insulin was administered concurrently (9,10). Therefore, the studies employing animals strongly suggested the involvement of insulin in the dietary induction but did not establish that insulin had a direct effect of liver.

The involvement of the adrenal glucocorticoids in the induction of hepatic G6PD was clearly demonstrated by Berdanier and colleagues (11). Adrenalectomized rats fed a high carbohydrate diet failed to exhibit as great an increase in G6PD activity compared to controls unless the animals were treated with cortisol. Berdanier and Shubeck (12) analyzed the interaction of the glucocorticoids and insulin and demonstrated, using adrenalectomized, diabetic, and adrenalectomized-diabetic rats, that upon refeeding a high carbohydrate diet to fasted animals there was no increase in G6PD activity unless glucocorticoid, insulin, and glucocorticoid plus insulin, respectively, were administered. Thus, it was shown that both insulin and the glucocorticoids were involved in the dietary induction of G6PD although a direct effect on liver could not be established.

The cellular regulatory mechanisms within the hepatocyte responsible for the increase in G6PD activity have been a focus of attention and the subject of some controversy. The dietary induction of G6PD activity seems most likely to be the result of increased rate of synthesis of the enzyme (13,14) but data has been presented which suggests that enzyme protein stays constant and that activity must be modulated through an activation-inactivation mechanism (15). Further controversy existed as to the mechanism responsible for the increased rate of synthesis. Sun and Holten (16) have suggested that the increased synthesis results from an increase in translational efficiency in addition to a small increase in the concentration of G6PD mRNA. Miksicek and Towle (14) attributed the increased rate of synthesis entirely to an increase in mRNA. Both of these studies used indirect measurements of mRNA since the assay system employed _in vitro_ translation of functional mRNA.

For several years our laboratory has been studying the interaction of the glucocorticoids and insulin in the regulation of hepatic carbohydrate metabolism. We have currently focused our studies on two aspects of the hormonal and dietary induction of hepatic G6PD. First, since studies with intact animals cannot differentiate between direct and indirect effects of hormones, we have examined the effect of the glucocorticoids and insulin on G6PD in primary cultures of hepatocytes maintained in a chemically defined medium. Second, we have carried out the molecular cloning of DNA sequences complementary to hepatic G6PD mRNA and have employed these molecular probes to directly measure G6PD mRNA levels. The results of our studies have established that the glucocorticoids and insulin directly modulate hepatic G6PD activity and relative rate of synthesis (17,18). In addition, molecular hybridization studies using cDNA probes have established that dietary induction of G6PD (activity) can be accounted for by a commensurate increase in G6PD mRNA and that both insulin and the glucocorticoids can modulate the cellular level of G6PD mRNA (19).

II. EXPERIMENTAL PROCEDURES

The methods used to generate the data presented here have been published elsewhere (17-20) with the exception of the use of ultraviolet light to cross-link RNA to nylon membranes for dot blot analysis (21).

III. RESULTS

A. Structure and Organization of G6PD cDNA

We have previously described the molecular cloning of DNA sequences for G6PD mRNA (19). The organization and size of the cDNA sequences from the initial report (19) and others that have been obtained subsequently are shown in Figure 1. The first cDNA inserts obtained (pGDC-4, pGDF-34, pGDF-23, pGDD-5) are characterized by an internal Pst I site and represented at most 630 nucleotides of the 2300 bases found in the mature G6PD mRNA (19). In the construction of other duplex cDNAs, we employed the method of Gubler and Hoffman (22) for the synthesis of the second strand of DNA. This resulted in the construction of a plasmid (pGDY-38) with a 830 base insert which contained substantially more bases toward the 3' end of the mRNA. Single stranded primers were prepared from the 340 base Pst I fragment of pGDC-4 and were used to prime first strand synthesis of cDNA. Following synthesis of the second strand, homopolymer tailing and annealing with G-tailed pBR322, the preparation was used to transform *E. coli* MC 1061. The recombinant plasmid, pGDP-2, was selected from this library which extended the sequence approximately 150 bases toward the 5' end of the mRNA. The recombinant plasmid pGDS-33 was selected in our laboratory from a cDNA library constructed in the laboratory of Dr. Darold Holten. Interestingly, this cDNA insert which is the largest found in this cDNA library complementary to G6PD mRNA, is approximately the same size as pGDY-38, the largest cDNA found in libraries constructed in our laboratory. This may indicate that a region of secondary structure exists at this point in the mRNA which causes reverse transcriptase to stutter. The recombinant plasmid, pGDY-28, is from a human fibroblast cDNA library which was screened with the 340 base fragment of pGDC-4. Preliminary restriction enzyme analysis of this insert has revealed some similarity with the rat cDNA inserts but no sequence information is available at this time.

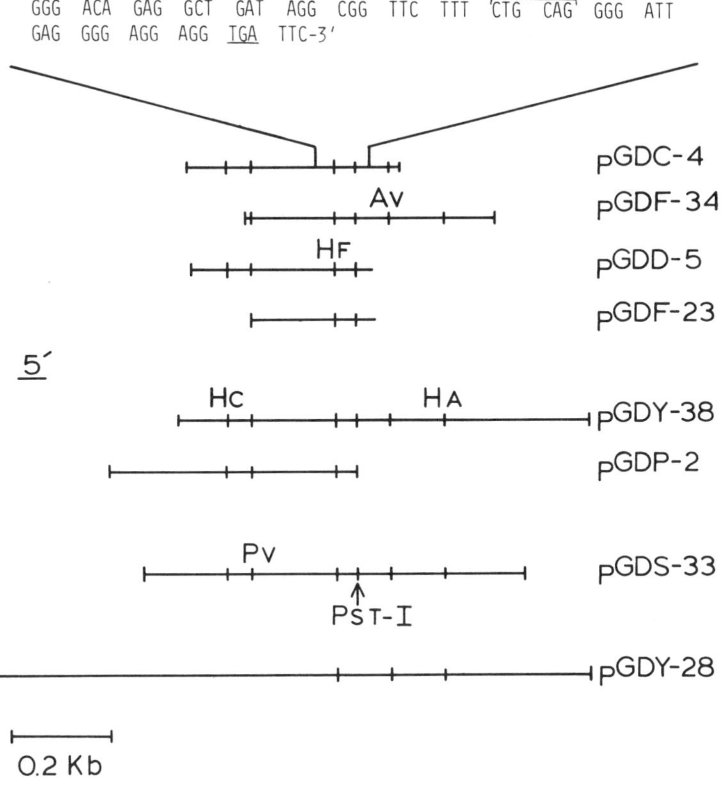

Figure 1. The size and organization of G6PD cDNA. The nucleotide sequence of the indicated portion of pGDC-4 is shown. Restriction enzyme abbreviations used are: Av, Ava II; Hf, Hinf I; Hc, Hinc II; Ha, Hae III; Pv, Pvu II.

Fragments of the hepatic G6PD cDNA were subcloned into M-13 mp 18 and 19 phages. Sequencing of the fragments by the dideoxyribonucleotide chain termination method (23) has revealed upon computer analysis of the sequence data that an open reading frame exists from the far 5' sequences we have cloned through 20 bases to the 3' side of the internal Pst I fragment. Therefore, we have currently cloned and sequenced approximately 550 bases of the putative coding region of the mRNA.

B. Hormonal Regulation of Hepatic G6PD

Previous studies from this laboratory have demonstrated that dietary induction of G6PD activity can be accounted for by a commensurate increase in functional G6PD mRNA (19,20). In addition, we have shown directly that treatment of primary cultures of hepatocytes with glucocorticoids and insulin increases G6PD activity, relative rate of G6PD synthesis and functional G6PD mRNA (17,18,20).

The development of cDNA probes for G6PD mRNA (19) has permitted us to directly assess the effect of dietary induction and hormones on G6PD mRNA levels. In the experiment shown in Figure 2 we have analyzed relative G6PD mRNA abundance by dot blot analysis. The results establish that dietary induction of G6PD activity can be accounted for by an increase in G6PD mRNA. Treatment of primary cultures of hepatocytes with insulin and the glucocorticoids, individually and together, also resulted in an increase in G6PD mRNA sequences. In each case the increase in G6PD mRNA is sufficient to account for the increase in G6PD activity. However, with cultures treated with only glucocorticoids there is an increase in mRNA without any increase in enzyme activity. A similar result was obtained in earlier studies in which functional mRNA levels were determined

(20). The increase in G6PD mRNA and activity elicited by hormone treatment of the primary cultures could not completely account for the increases observed in intact liver for dietary induction. We have considered the possibility that T_3 is involved in the dietary induction since other investigators have examined its potential involvement. However, in our studies employing intact animals we have demonstrated that T_3 does not modulate dietary induction of G6PD mRNA although G6PD activity is increased, suggesting an involvement of this hormone at a translational level (R. Scott Fritz, data not presented).

IV. CONCLUSIONS

The use of primary cultures of hepatocytes maintained in a chemically defined medium has permitted us to directly assess the individual and combined effects of insulin, the adrenal glucocorticoids and glucose on G6PD activity, relative rate of synthesis, and mRNA levels (17-20, Fig. 2). The conclusions that we can draw from these studies are that the hormones and glucose directly modulate G6PD activity (17-20). Insulin and the glucocorticoids increase both the relative rate of G6PD synthesis (18) and G6PD mRNA as measured by either indirect (20) or direct (Fig. 2) methods. In the intact rat, the dietary induction of the enzyme activity can be accounted for by an increase in the mRNA as measured by direct assay of the G6PD mRNA sequences (19, Fig. 2). The increase in G6PD mRNA observed in the intact animal cannot as of yet be duplicated by insulin and glucocorticoid treatment of primary cultures of hepatocytes. The reason for this is not clear but could involve some other unidentified hormone or nutritional factor. In addition, it is possible that our model system does not completely reflect the *in vivo* status of the hepatocyte and thus, the hepatocyte's responses to physiological stimulus may be somewhat attenuated.

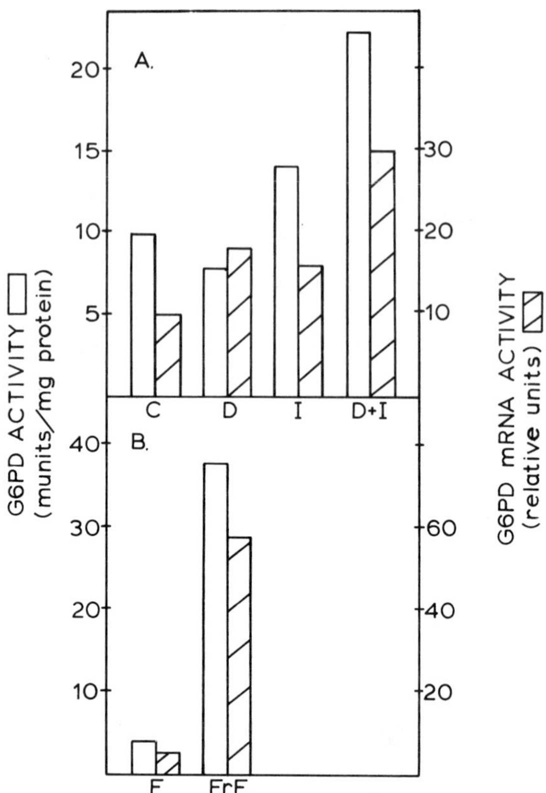

Figure 2. The effect of diet and hormones on G6PD activity and mRNA. Primary cultures of hepatocytes were prepared and G6PD activity determined as described previously (20). Cultures were also extracted for RNA and poly A^+ RNA was isolated as described (20). Rats were fasted for 48 hours or refed a high carbohydrate diet (20) for 48 hours following a 48 hour fast. The livers were removed, homogenized, and assayed for G6PD activity. RNA was extracted and poly A^+ RNA was isolated as described (20). Dot blot analysis of the poly A^+ RNA was carried out as described previously (19) with the following modification. The RNA was cross-linked to nylon filters by exposure to ultraviolet light (23). The units of mRNA activity are based on cpm's of ^{32}P-cDNA probe (280 base Pst I fragment of pGDF-34) hybridized per μg of input RNA. Each determination represents the mean of at least three determinations. Panel A: C is control; D is dexamethasone (1 μM) treatment for 48 hours; I is insulin (45 munits/ml). Treatment for 48 hours; D+I is Treatment with both hormones for 48 hours. Panel B: F represents rats fasted for 48 hours; FrF represents rats fasted for 48 hours and then refed a high carbohydrate diet for 48 hours.

The future direction of research in this laboratory will be focused on the mechanism responsible for the increased G6PD mRNA. Our current working hypothesis is that the hormones and/or nutritional factors stimulate transcription of the G6PD gene. Thus, work in this laboratory is directed toward *in vitro* transcription analysis and isolation of the genomic sequence with the adjacent 5'-flanking control region.

REFERENCES

1. Glock, G.E., and McLean, P. (1955). Biochem. J. 61, 390.
2. Tepperman, J., and Tepperman, H.M. (1958). Am. J. Physiol. 193, 55.
3. Tepperman, H.M., and Tepperman, J. (1958). Diabetes 7, 478.
4. Tepperman, H.M., and Tepperman, J. (1963). Adv. Enz. Reg. 1, 121.
5. Johnson, B.C., and Sassoon, H.F. (1967). Adv. Enz. Reg. 5, 93.
6. Potter, V.R., and Ono, T. (1961). Cold Spring Harbor Symp. Quant. Biol. 26, 355.
7. Rudack, D., Chisholm, E.M., and Holten, D. (1971). J. Biol. Chem. 246, 1249.
8. Benevenga, N.J., Stielau, W.J., and Freedland, R.A. (1964). J. Nutr. 84, 345.
9. Weber, G., and Convery, H.J.H. (1966). Life Sci. 5, 1139.
10. Nepokroeff, C.M., Lakshmanan, M.R., Ness, G.C., Muesing, R.A., Kleinsek, D.A., and Porter, J.W. (1974). Arch. Biochem. Biophys. 162, 340.
11. Berdanier, C.D., Wurdeman, R., and Tobin, R.B. (1976). J. Nutr. 106, 1791.
12. Berdanier, C.D., and Shubeck, D. (1979). J. Nutr. 109, 1766.
13. Winberry, L., and Holten, D. (1977). J. Biol. Chem. 252, 7796.
14. Miksicek, R.J., and Towle, H.C. (1982). J. Biol. Chem. 257, 11,829.
15. Kelley, D.S., Watson, J.J., Mack, D.O., and Johnson, B.C. (1975). Nutr. Rep. Int. 12, 121.
16. Sun, J.D., and Holten, D. (1978). J. Biol. Chem. 253, 6832.
17. Kelley, D.S., and Kletzien, R.F. (1984). Biochem. J. 217, 543.

18. Stumpo, D.J., and Kletzien, R.F. (1985). Biochem. J. 226, 123.
19. Kletzien, R.F., Prostko, C.R., Stumpo, D.J., McClung, J.K., and Dreher, K.L. (1985). J. Biol. Chem. 260, 5621.
20. Stumpo, D.J., and Kletzien, R.F. (1984). Eur. J. Biochem. 144, 497.
21. Church, G.M., and Gilbert, W. (1984). Proc. Natl. Acad. Sci. 81, 1991.
22. Gubler, U., and Hoffman, B.J. (1983). Gene 25, 263.
23. Sanger, F., Nicklenz, S., and Coulson, A.R. (1977). Proc. Natl. Acad. Sci. 74, 5463.

TRANSCRIPTIONAL AND POST-TRANSCRIPTIONAL REGULATION OF UTERINE GLUCOSE-6-PHOSPHATE DEHYDROGENASE BY ESTRADIOL[1]

Kenneth L. Barker, Kim-Yen T. Pham and Ward H. Lutz

Department of Biochemsitry, School of Medicine,
Texas Tech University Health Sciences Center
Lubbock, Texas, U.S.A.

I. INTRODUCTION

Estradiol induces uterine hypertrophy when given to either immature or ovariectomized mature female rats. Following hypertrophy, the various cell types characteristic of the myometrium, stroma and endometrium of this organ undergo hyperplasia (1). In the normally cycling female these responses, which characterize the estrogen induced proliferative phase of the estrous or menstrual cycle, are followed by a progesterone induced transformation of the endometrium into a secretory or glandular epithelial layer into which a developing blastocyst can implant and be nurtured during early pregnancy. The hypertrophy response by the uterus to estrogen consists of increases in uterine hyperemia and permeability to a variety of carbohydrate

[1]*This work supported in part by grants from the National Institutes of Health (HD16236 and HD16235).*

and amino acid substrates; increases in the rate of synthesis of RNA, lipid, protein and glycogen; increases in the activity of a variety of metabolic pathways, including the hexose monophosphate (HMP) or pentose phosphate pathway; and increases in the activities of a variety of enzymes required to support and/or initiate these hormone induced effects (2-3). Most striking are the very early sequential increases that occur in the rates of RNA, fatty acid and phospholipid synthesis, the HMP pathway activity and then the total tissue levels of glucose-6-phosphate dehydrogenase (G6PD) activity.

The uterus of the ovariectomized mature rat is a tissue in which the mechanisms of regulation of G6PD can be uniquely studied under in vivo conditions because: 1) the activities of the enzyme are acutely and precisely regulated by intravenous administration of estradiol (4,5), 2) the tissue is directly accessible for the experimental introduction of inhibitors, cofactors and radiolabeled precursors of macromolecule synthesis, etc., via a non-surgical transcervical intrauterine application method which minimizes the systemic effects and/or dilution of these compounds (6), 3) the uterine G6PD cross-reacts with antiserum from rabbits immunized with rat liver G6PD, which makes it possible to study G6PD protein synthesis, processing and degradation using this reagent to isolate G6PD from uterine cytosols (7) and 4) the rate of protein synthesis in the estrogen-deprived rat uterus is slow relative to the more frequently studied tissues, such as the liver and reticulocyte, making it easier to dissect the overall sequence of events involved in the in vivo synthesis and processing of this protein (8,9). The main disadvantage of studies of G6PD regulation in this tissue are: 1) the small amount of tissue available for study (100-300 mg wet weight per uterus), 2) the relatively low amounts of G6PD activity in the tissue (0.1 to 1 unit per uterus), 3) the presence of multiple cell types, all of which respond to estradiol, in the organ (4), 4) the uterotropic response to estradiol results in the simultaneous stimulation of synthesis of many proteins, in addition to G6PD, which make measurements of changes in specific activities (units per mg protein) and changes in G6PD synthesis relative to total uterine protein synthesis (i.e., dpm of ^{3}H-amino acid incorporated into G6PD relative to dpm incorporated into total protein) meaningless parameters with regard to evaluating the effects of the hormone on rates of G6PD synthesis (4,5,7), and 5) the uterotropic responses by *in vitro* addition of estradiol to uterine tissue in organ culture is incomplete and slower with lower magnitudes of change than can be induced by the hormone *in vivo* (10,11).

II. REGULATORY SEQUENCES INVOLVING G6PD AND HMP PATHWAY ACTIVITIES IN THE ESTROGEN-INDUCED UTERUS

Using this model system we have determined several features regarding the sequence of events involved in regulation of G6PD activity, and its participation in HMP pathway activity, which could provide some insight into the more compressed sequence of events which occur in metabolically more active tissues such as the liver. *First*, HMP pathway activity in uterine tissue is clearly elevated prior to any change in the levels of G6PD activity in the tissue (5,12). We have observed that uterine G6PD activity remains constant for a period of 6 hours following estradiol administration after which it increases dramatically. By 6 hours, however, the relative rate of HMP utilization, as estimated by the ratio of formation of $^{14}CO_2$ from the C-1/C-6 positions of glucose by uterine strips *in vitro*, is increased by 75%. The increase in HMP pathway activity is highly correlated with the rate of lipid synthesis ($\underline{r}$ = 0.83). This suggests that the activity of the HMP pathway is mainly controlled by factors other than rate limiting tissue levels of G6PD. Logical candidates are the levels of glucose-6-phosphate and the $NADP^+/NADPH$ ratios which are also found to increase during the first 6 hours of the uterine response to estradiol (2,13,14). Second, HMP pathway utilization appears to provide some regulatory signal which triggers the synthesis of G6PD. Temporally, HMP pathway utilization in the uterus is elevated prior to the induction of G6PD activity following estradiol administration suggesting a cause and effect relationship between these two parameters (5). In addition, the rate of glucose utilization by the HMP pathway relative to glycolysis can be significantly increased by the intrauterine administration of fluoride which is a strong inhibitor of enolase activity in the glycolytic pathway (15). Again, 4-6 hours after increasing HMP pathway activity by this non-hormonal method the levels of G6PD activity begin to increase. Uterine G6PD activity can also be increased by the intrauterine administration of $NADP^+$ (6,7). While the ability of this nucleotide to cross uterine cell plasma membranes may be open to question, we have found that uterine strips from tissues treated in this manner *in vivo* exhibit an elevated rate of glucose utilization via the HMP pathway as determined by the ratio of $^{14}CO_2$ formation from the C-1/C-6 positions of glucose under *in vitro* conditions. All three inducers; estradiol, NaF and $NADP^+$, increase the rate of G6PD synthesis as measured by the rate of incorporation of radiolabeled amino acids into immunoprecipitable G6PD protein (4,15). Interestingly, the induction of G6PD activity by estradiol, but not NaF and $NADP^+$, is inhibited by prior treatment of the uterus with actinomycin D, while the induction of G6PD activity by all three inducers is blocked by prior treatment with cycloheximide (6). This suggests that the induction of G6PD synthesis by factors associated with HMP pathway utilization occur at a post-transcriptional level while estradiol induction of G6PD synthesis is initiated by responses which also require RNA synthesis. Third, uterine G6PD synthesis is preceded by 2 hours by an increase in total uterine $NADP^+$ + NADPH levels (14). This suggests that synthesis of this apoenzyme and coenzyme system may be coupled. We determined that the estradiol-induced synthesis of $NADP^+$ occurs at two levels, the conversion dNMN to dNAD and the conversion of NAD to NADP (16). The amount of NAD kinase activity, measured under optimal conditions, was not increased. Since NADPH is a known inhibitor of NAD kinase, we have tentatively concluded that $NADP^+$ synthesis is regulated by the early decrease in NADPH levels which result from estrogen-induced fatty acid synthesis (5,17). Together these results suggest that estradiol may induce uterine G6PD synthesis by two mechanisms, directly by inducing the synthesis of the mRNA for G6PD and indirectly by stimulating reductive biosynthetic reactions (probably fatty acid synthesis) which results in: 1) an increase in the $NADP^+$/NADPH ratio, 2) an increase in the rate of HMP pathway activity, 3) the synthesis of additional $NADP^+$ and 4) the accumulation of some HMP pathway linked factor (perhaps $NADP^+$ itself) which facilitates G6PD synthesis. The indirect effect appears to occur at a post-transcriptional level and probably increases the rate of translation of either preexisting basal levels of G6PD mRNA or estrogen-induced levels of G6PD mRNA.

III. MOLECULAR EVENTS THROUGH WHICH THE AMOUNT OF UTERINE G6PD APOPROTEIN IS REGULATED BY ESTRADIOL

To directly assess the effect of estradiol on uterine G6PD synthesis, an antiserum to purified rat liver G6PD was prepared and procedures were developed to use this antiserum to immunoprecipitate G6PD from uterine cytosol preparation (7). By immunotitration of G6PD activity from uteri of control and estradiol-treated ovariectomized mature rats, it was found that the end-point of immunotitration of G6PD activity was identical for G6PD from control and estradiol-induced uteri and they were also equivalent to values obtained for rat liver G6PD (7). These results suggest that changes in uterine G6PD activity reflect changes in the amount of G6PD protein in the uterine cytosol preparations rather than changes in specific enzyme activity of the protein. The G6PD antiserum and immunoglobulin fractions obtained from it have been used to quantify the effects of estradiol on the rates of various reactions in the uterus which relate to G6PD synthesis, processing and degradation. A summary of the effects of 12 and 24 hours of estradiol-treatment on several parameters related to G6PD synthesis, processing and degradation in the uterus of the ovariectomized mature rat is given in Table I.

To evaluate the effect of estradiol on G6PD synthesis rates, groups of identically-treated ovariectomized mature rats were given ^{14}C-amino acids by the intrauterine route 4 hours before sacrifice with the pulse labeling period beginning at various times (from $^-$4 to 72 hours) after the intravenous administration of estradiol. Uterine cytosols were prepared and the amount of radioactivity incorporated into immunoprecipitable G6PD from these cytosols (corrected for differences in specific activities of the free amino acid pool for each sample) was determined. From these we determined that the hormone induced an 18-fold and

TABLE I. Molecular events through which the amount of uterine G6PD is regulated by estradiol

Parameter Evaluated	Hours After Estradiol Treatment			Ref.
	0	12	24	
G6PD Synthesis Rates (dpm ^{14}C-amino acids into G6PD uterus $^{-1} \cdot 4h^{-1}$)	10	180 (18x)[a]	110 (11x)	7,9
G6PD Translatable mRNA Levels (dpm ^{35}S-methionine into G6PD·250 ng^{-1} uterine RNA) x 10^{-3}	4	26 (6.5x)	24 (6x)	18,19
Ribosome Transit Times on G6PD-mRNA (min between initiation and release of G6PD from uterine ribosomes)	4.9	2.0 (2.5x)	3.4 (1.4x)	9,18
G6PD N-Terminus Processing Time (min between release of G6PD from ribosomes and formation of N-terminal pyroglutamate)	9	2.5 (3.6x)	4 (2.3x)	9,12,18
G6PD Degradation ($t_{\frac{1}{2}}$ hours)	50	-	>100	11,20,21

[a]Numbers in parentheses are the fold increase relative to zero time control rates

an 11-fold increase in the rate of uterine G6PD synthesis during the 4 hour pulse labeling interval beginning at 12 and 24 hours after hormone administration, respectively (7).

Translatable G6PD mRNA levels were determined by translating total uterine RNA in a mRNA-dependent reticulocyte lysate translation system followed by immunoprecipitation of the translation products with the anti-G6PD immunoglobulin and determining the mRNA-dependent rate of incorporation of

^{35}S-methionine into G6PD. We observed that G6PD mRNA levels were increased 6.5-fold and 5.9-fold, 12 and 24 hours after estradiol administration, respectively (18,19). From these observations, we concluded that estradiol-induction of uterine G6PD synthesis was not solely controlled by effects of the steroid hormone on transcription.

To directly evaluate the effects of estradiol on the time required to translate the G6PD mRNA, the ribosome transit times were estimated by determining the differences between the time required to obtain initial linear rates of incorporation of either ^{3}H-leucine or ^{3}H-glutamate into total uterine protein and the time required to observe initial linear rates of incorporation of these amino acids into G6PD in the released or "post-ribosomal" fraction of the uterine cytosol after intrauterine administration of the labeled substrates. This difference has been shown to be a statistical estimate of one-half the time interval between initiation and release of a specific protein from ribosomes and is independent of the rate of initiation which would be controlled by the availability of mRNA and other factors required to form the 80S ribosomal initiation complex (22,23). We observed ribosome transit times of 4.9, 2.0 and 3.4 min for G6PD mRNA in uteri from rats at 0, 12 and 24 hours after estradiol administration, respectively (9). Thus estradiol caused a 2.5-fold and a 1.4-fold increase in the rate of G6PD mRNA translation, respectively. Together with the increase in G6PD mRNA pool sizes, these two parameters nearly account for the observed increase in G6PD synthesis rates (i.e. 6.5-fold increase in mRNA x a 2.5-fold increase in the rate of translation equals a 16.3-fold theoretical increase in rate of synthesis which is comparable to the observed 18-fold increase at 12 hours after estradiol-treatment and a 5.9-fold increase in mRNA x a 1.4-fold increase in rate of translation equals an 8.3-fold theoretical increase in rate of synthesis which is comparable to the observed 11-fold increase seen at 24 hours after estradiol-treatment).

The amino-terminus of G6PD purified from erythrocytes, bovine adrenal, rat liver and rat uterus has been found to be pyroglutamate (24-26). Since all eukaryotic cells are considered to initiate protein synthesis with methionine, the presence of pyroglutamate instead of methonine on the amino-terminus of G6PD infers that modification of the amino-terminus of a precursor form of G6PD must occur. As part of the study to estimate the ribosomal transit times for G6PD mRNA, ^{3}H-glutamate was used as the labeled precursor. Aliquots of the immunoprecipitated G6PD were used to estimate transit times and the remainder was used to evaluate the presence or absence of ^{3}H-pyroglutamate on the amino-terminus of the labeled protein. Labeled pyroglutamate was first found in G6PD at 16, 4 and 8 min after administration of the ^{3}H-glutamate to tissues from 0, 12 and 24 hour estradiol-treated rats, respectively. Theoretically, labeled amino-termini should have first appeared in the released G6PD fraction one transit-time after the onset of initial labeling of the total protein. Subtracting these theoretical values from the observed time of labeling of amino-terminal pyroglutamate in the immunoprecipitates revealed that processing times of 9, 2.5 and 4 min after the release of G6PD from uterine ribosomes were required for the newly synthesized enzyme at 0, 12 and 24 hours after estradiol-treatment, respectively. These observations suggest that pre-G6PD is released from ribosomes before it is processed and that the processing of its amino-terminus is also an estrogen-dependent event (6).

Since G6PD functions intracellularly, it was considered that intracellular degradation of the protein in the rat uterus may also be controlled by estradiol. Based on the kinetics of the rate of loss of enzyme activity after its induction by either estradiol or $NADP^+$, we determined that the apparent half-life of the enzyme in the estrogen deprived animal *in vivo* was 50 hours (20). Because elevated steady state levels reflecting estrogen-induced changes in the rates of synthesis and degradation could not be maintained by *in vivo* daily administration of estradiol for longer than a few hours, a kinetic analysis of the effects of estradiol on enzyme half-life could not be evaluated. To estimate the effects of estradiol on the G6PD degradation *in vivo*, total uterine proteins, including G6PD were pulse-labeled by the intrauterine administration of 1-[^{14}C]glutamate during the interval of 12-16 hours after estradiol-treatment. Within one hour after the last injection of this rapidly degraded precursor, less than 0.1% remained, which effectively eliminated contributions of enzyme synthesis on subsequent estimates of radioactivity remaining in the pulse-labeled G6PD. We observed that from 16 to 36 hours after injection of estradiol, the half-life of G6PD was greater than 100 hours. After 36 hours it was degraded with a half-life of 24 hours as the tissue underwent involution. A second injection of estradiol at the 48th hour after the first (in the middle of the period of estrogen-withdrawal) totally blocked G6PD degradation for an additional 24 hours. Paradoxically, intrauterine administration of $NADP^+$ at the 48th hour actually increased the rate of loss of G6PD from uterine cytosols resulting in a half-life of 9-12 hours, while enzyme activity was increased relative to the levels present at the time of $NADP^+$ administration (21). We have conducted a similar experiment using uterine strips in organ culture in which conventional pulse-chase experiments could be conducted by addition and then removal of ^{3}H-glutamate from the culture medium. Under these conditions G6PD labeling could be done in control tissues followed by addition of estradiol (2×10^{-8} M) under *in vitro* conditions. Under these totally *in vitro* conditions the half-life of G6PD was 92 hours for control tissues and degradation was again totally blocked by addition of estradiol (11). Thus under both *in vivo* and *in vitro* conditions estradiol blocks the degradation of G6PD thus indicating another mechanism of intracellular regulation of this enzyme in the uterus.

IV. MECHANISMS BY WHICH UTERINE G6PD SYNTHESIS AND MODIFICATIONS MAY BE REGULATED BY ESTRADIOL

The estradiol-regulated mechanisms involved in controlling G6PD levels in the uterus are presently being pursued. Clearly, based on a wealth of literature regarding estradiol-induced gene activation (1), it is not surprising that estradiol induces a 6 to 7-fold increase in the amount of translatable G6PD mRNA. We are anticipating the availability of cDNA probes for G6PD mRNA to evaluate the effects of estradiol on the amount of G6PD mRNA sequences, the rate of G6PD mRNA synthesis and the rate of processing of pre-G6PD mRNA in the rat uterus. Although we doubt that estradiol and its receptor will interact directly with the regulatory sequences on the uterine G6PD genome, we will hopefully be able to evaluate this possibility as well as the possibility that a regulatory intermediate may be produced in the uterus in response to estradiol, that can in turn activate the more general process of transcription of multiple genes which are characteristic of the uterine hypertrophy response (1).

We have made two observations which may explain the mechanism by which estradiol increases the rate of G6PD translation. First, during short pulse labeling studies, we found incorporation of radioactive leucine and glutamate initially in a ribosome-associated anti-G6PD immunoprecipitable protein of a M_r of 42,000 d. A comparison of ^{35}S-methionine labeled peptides derived from the 42,000 d protein and from authentic M_r 57,000 d

^{35}S-methionine labeled G6PD, after complete papain digestion revealed extensive sequence homology between these proteins. This supported our suggestion that the 42,000 d protein is a nascent chain of G6PD (9,27). In vivo pulse labeling studies under conditions which gave identical tissue precursor specific activities, revealed that 12 hours of estradiol-treatment expanded the pool size of the nascent G6PD peptide by 8-fold which approximates the effect of estradiol on the G6PD mRNA pool (9). The relative rate of labeling of this nascent peptide and the completed G6PD suggested that the rate of flux of labeled amino acid through the nascent peptide was increased by estradiol (9,27). Accumulation of a nascent chain of a uniform size suggests that the G6PD mRNA may contain an inherent "pause-site" which slows translation resulting in a pile-up of this relatively uniform nascent peptide. A "pause site" could occur due to the secondary structure of the mRNA or to the presence of rate-limiting amounts of an aminoacyl-tRNA required for a codon or codons that are present at this location in the mRNA. Second, we have examined the specific amino acid acceptor activity (pmoles of amino acid incorporated into aminoacyl-tRNA per nmole of tRNA) for uterine tRNA isolated from control and 14 hour estradiol-treated rats and have found an overall increase in acceptor activity that is increased by 2.6-fold for a total of 18 amino acids (all except asparagine and cysteine). The specific acceptor activity for three amino acids; proline, glycine and leucine, were increased by 3.5 to 4.5-fold. Pretreatment of the tRNA with tRNA-nucleotidyltransferase, which repairs the defective CCA amino acid acceptor terminus on the 3'-end of tRNA, caused the tRNA acceptor activity of the tRNA from control uteri to be raised to the level seen for tRNA from estradiol-treated rats. Uterine tRNA from estradiol-treated rats was only slightly increased by this tRNA repair enzyme (28). Together, these observations suggest that the rate of translation of uterine G6PD mRNA may be regulated by the repair and/or synthesis of tRNA's required to translate unique codons that are found in this region of the G6PD mRNA corresponding to the site found at a pause site which gives rise to the accumulated nascent peptide. This effect appears analogous to the estrogen-induced synthesis of the serine-rich protein, phosvitin, in the chick liver which is accompanied by an increase in the amount of active $tRNA^{ser}$ that is present in the tissue (29). It has also been shown that tRNA from estrogen-dominated hen oviducts facilitates the *in vitro* synthesis of the estrogen-induced protein ovalbumin (30).

The intracellular processing of G6PD to a form which contains pyroglutamate on its amino-terminus may be a feature that occurs with an unknown overall effect on the enzyme activity of the molecule. We have found a minor anti-G6PD immunoprecipitable protein with a M_r of 64,000 d. This protein is also labeled during both *in vivo* and *in vitro* synthesis prior to the predominant M_r 57,000 d protein and the two proteins contain common papain produced methionine containing peptides (27). The uterine M_r 64,000 d immunoprecipitable protein appears to be a precursor of the M_r 57,000 G6PD and it appears to have its amino-terminus trimmed to produce the predominant form of the enzyme. Preliminary results suggest that the clip site may be between lys(arg)-glx residues in a lys(arg)-lys(arg)-glx-gly sequence that is present at the amino-terminal region of the final molecule and that the formation of amino-terminal pyroglutamate may occur spontaneously by cyclization of a terminal glutamate or glutamine residue thus produced. The functional significance of such processing of the protein relative to enzyme activity may be nil but such processing may give rise to electrophoretic variants of the enzyme. Such variants could be due to changes in G6PD clip site sequences or to changes in "processing enzyme activity" between tissues and/or individuals. The latter could give rise to non X-chromosome-linked variants in the G6PD molecule.

The mechanism of degradation of uterine G6PD and its control by estradiol is uncertain. It is noteworthy that degradation of total cytosolic uterine proteins, including G6PD, is inhibited by estradiol (25). Lysosomes are generally considered to be the site of intracellular degradation of cytosolic proteins and since G6PD is either a cytosolic protein or a protein that is "loosely" associated with intracellular particles it is likely degraded by a lysosomal mechanism (31-33). The stabilization of cytosolic uterine proteins by estradiol may be related to the estrogen-induced migration of lysosomes from the cytosol to or near the nucleus which has been observed in several estrogen target tissues (34,35). Cytosolic depletion of lysosomes may stabilize newly synthesized estrogen-induced proteins in the cytosol, such as G6PD, to extend their period of service to the cell during estrogen-induced uterine hypertrophy.

V. CONCLUSIONS

Observations summarized above for the regulation of uterine G6PD synthesis, processing and degradation by estradiol may have relevance as a model system to identify basic regulatory mechanisms by which the levels of this enzyme may be regulated in other tissues. Specifically, the stimulation of uterine lipid synthesis by estradiol and the subsequent induction of G6PD activity in this tissue may have certain analogies to the events which occur in the mammary gland at the onset of lactation and in the liver when starved animals are refed a high carbohydrate diet. Of particular relevance may be our observations that estradiol and $NADP^+$ are able to induce post-transcriptional effects on uterine G6PD synthesis and that translational control of G6PD synthesis may occur through regulation of the availability of selected aminoacyl tRNAs. Others have reported that both transcriptional and translational control of G6PD synthesis may occur in the rat liver after refeeding a high carbohydrate diet to starved animals (36-38). The pronounced translational effect we see in uterus during a brief period of G6PD synthesis "overshoot" 12 hours after estradiol treatment may be particularly relevant to the classically observed "overshoot" of G6PD activity (and probably G6PD synthesis) which was initially observed in livers of refed rats by Tepperman and Tepperman (39).

REFERENCES

1. Knowler, J.T. and Beaumont, J.M. (1985) Essays in Biochemistry 20, 1.
2. Barker, K.L., Nielson, M.H. and Warren, J.C. (1966) Endocrinology 79, 1069.
3. Baquer, N.Z. and McLean, P. (1972) Biochem. Biophys. Res. Commun. 48, 729.
4. Scott, D.B.M. and Lisi, A.G. (1960) Biochem. J. 77, 52.
5. Barker, K.L. and Warren, J.C. (1966) Endocrinology 78, 1205.
6. Barker, K.L. (1967) Endocrinology 81, 791.
7. Smith, E.R. and Barker, K.L. (1974) J. Biol. Chem. 249, 6541.
8. Whelly, S.M. and Barker, K.L. (1974) Biochemistry 13, 341.
9. Donohue, T.M., Jr. and Barker, K.L. (1983) Biochim. Biophys. Acta. 739, 148.
10. Keran, E.E. and Barker, K.L. (1976) Endocrinology 99, 1386.
11. Pham, K-Y.T. (1981) University of Nebraska, Ph.D. Dissertation.
12. Barker, K.L. (1980). In "The Endometrium" (K.T. Kirton and F.T. Kimball, eds.) p. 227. Spectrum Press, Inc.
13. Smith, D.E. and Gorski, J. (1968) J. Biol. Chem. 243, 4169.
14. O'Dorisio, M.S. and Barker, K.L. (1970) Endocrinology 86, 1118.
15. Smith, E.R. and Barker, K.L. (1976) Biochim. Biophys. Acta 451, 223.
16. O'Dorisio, M.S. and Barker, K.L. (1976) Biol. Reprod. 15, 504.
17. Aizawa, Y. and Mueller, G.C. (1961) J. Biol. Chem. 236, 381.
18. Barker, K.L., Adams, D.J. and Donohue, T.M., Jr. (1981) In "Cellular and Molecular Aspects of Implantation" (S.R. Glasser and D.W. Bullock, eds.) p. 269. Plenum Publishing Co., New York.

19. Adams, D.J. (1979) University of Nebraska, Ph.D. Dissertation.
20. Moulton, B.C. and Barker, K.L. (1971) Endocrinology 89, 1131.
21. Smith, E.R. and Barker, K.L. (1977) J. Biol. Chem. 252, 3709.
22. Fan, H. and Penman, S. (1970) J. Mol. Biol. 50, 655.
23. Nielson, P.J. and McConkey, E.H. (1980) J. Cell. Physiol. 104, 269.
24. Yoshida, A. (1972) Anal. Biochem. 49, 320.
25. Singh, D. and Squire, P.G. (1975) Int. J. Pept. Protein Res. 7, 185.
26. Donohue, T.M., Jr., Mahowald, T.A., Adams, D.J. and Barker, K.L. (1981) Biochim. Biophys. Acta 658, 356.
27. Cummings, A.C. and Barker, K.L. (1986) Biochim Biophys. Acta in press.
28. Lutz, W.H. and Barker, K.L. (unpublished observation).
29. Lizardi, P.M., Vahdavi, V., Shields, D. and Candelas, G. (1979) Proc. Natl. Acad. Sci. U.S.A. 76, 6211.
30. Sharma, O.K. and Borek, E. (1976) Nature 262, 62.
31. Baquer, N.Z. and McLean, P. (1972) Biochem. Biophys. Res. Commun. 46, 167.
32. Natori, Y. (1975) In "Intracelluton Protein Turnover" (R.T. Schimke and N. Katunuma, eds.) p. 237. Academic Press, New York.
33. Schimke, R.T. (1975) In "Intracellular Protein Turnover" (R.T. Schimke and N. Katunuma, eds.) p. 173. Academic Press, New York.
34. Szego, C.M. (1974) Rec. Progr. Horm. Res. 30, 171.
35. Szego, C.M. (1975) In "Lysosomes in Biology and Pathology" (J.T. Dingle and R.T. Dean, eds.) p. 385. American Elsevier Publishing Co., New York.
36. Sun, J.D. and Holten, D. (1978) J. Biol. Chem. 253, 6832.
37. Miksicek, R.J. and Trowle, H.C. (1982) J. Biol. Chem. 257, 11829.
38. Kletzien, R.F., Prostko, C.R., Stumpo, D.J., McClung, J.K. and Dreher, K.L. (1985) J. Biol. Chem. 260, 5621.
39. Tepperman, J. and Tepperman, H.M. (1958) Am. J. Physiol. 193, 55.

EXPRESSION OF Gd LOCUS

G6PD AS A TOOL AND A TARGET FOR STUDIES ON X–CHROMOSOME LINKAGE[1]

Antoniettina Rinaldi
Istituto di Biologia Generale dell'Universita'
Cagliari, Italy

Michele Purrello
Memorial Sloan-Kettering Cancer Center, NY, NY, USA

Giorgio Filippi
Cattedra di Genetica Medica dell'Universita'
Ospedale per l'Infanzia, Trieste, Italy

Marcello Siniscalco[2]
Memorial Sloan-Kettering Cancer Center, NY, NY, USA

---

1. The linkage studies carried out by the authors were supported by grants of the Italian National Research Council to A.R. and to G.F., the New York Community Fund to M.P. and the NIH (HD-16742), the National Foundation (#782) and the Muscular Dystrophy Association to M.S.

2. Symposium's speaker

The discovery of common genetic variation affecting the G6PD enzyme in the late Fifties was a bonus for students of mammalian genetics. In the twentyfive years that followed there were practically no aspects of this discipline to which G6PD was not applied either as a tool or as the very target of genetic research: from Mendelian and population genetics to somatic cell genetics and molecular biology. This explains why scientists of the most diverse expertise are convened at this meeting. The present report is meant to summarize what is known to date about the use of G6PD as a tool and a target for X-linkage analysis.

First of all let us recall how G6PD was linked to sex. When Carson and colleagues (1) demonstrated that the primary genetic defect of primaquine sensitivity was a deficiency in G6PD activity, geneticists (2) had already realized that the severe gluthatione instability found in primaquine sensitive individuals and in patients with drug or food induced hemolytic anemias (among which clinical favism) was preponderant only among males of well defined ethnical groups (such as Black Americans, Italians, Greeks and Sephardic Jews) and was inheritable in a fashion compatible with X-linked inheritance. However, since at the time the family data available were scanty and the G6PD variation was studied only in a quantitative manner, the possibility that the enzyme deficiency could be the result of an autosomal mutation with a sex-limited expression could not be entirely ruled out. It was indeed through linkage studies that G6PD deficiency was first proven to be the primary consequence of an X-linked mutation in view of its segregation in a complete linkage association with colorblindness defects of Deutan or Protan type (3-5), the only examples of common X-linked polymorphism known at the time. But, the unquestionable proof that the structural gene for G6PD is X-linked was obtained only when electrophoretic variants of G6PD were discovered and shown to follow the same pattern of X-linked inheritance as the G6PD deficiency (6)

As it is well known and thoroughly discussed by other reports at this meeting, the G6PD variation provided also an ideal opportunity for confirming the Lyon hypothesis with the demonstration of the expected X-inactivation dependent cellular mosaicism in the somatic tissues of heterozygous carriers (7-8). In heterozygotes for the Mediterranean mutant, this mosaicism can be clearly identified in the mature red cells (9-10), as well as in other cells or tissues (11), including hair bulbs (12) and _in vitro_ cultured fibroblasts or myoblasts (13).

Studies on the distribution of G6PD deficiency spread to a world-wide scale thanks to the dye decoloration test of Motulsky and Campbell-Kraut (14), a simple screening procedure for G6PD activity that allowed the collection of population data in a fast and reliable manner. Such studies revealed a close association between the incidence of the G6PD deficient alleles and that of malarial morbidity (15-16). In particular, it was realized that the lowland populations of the Mediterranean basin (where malignant malaria was eradicated in relatively recent times) have a very high incidence of the G6PD Mediterranean mutant. Thus, it is no surprise that for a number of years most of the linkage studies between G6PD and other X-linked markers, and in particular between G6PD and rare X-linked diseases, were performed in Mediterranean countries.

The hopes towards a rapid construction of a genetic map of the human X were brightened by the discovery of the Xg(a) red cell antigen, which for a long time remained the best example of a common X-linked polymorphism known in Man. However, the first sets of published data on the linkage relationship between G6PD and Xg(a) were puzzling, since they suggested the existence of measurable linkage, with very high Lod scores, in Israeli (17) and Greek pedigrees (18), but a complete absence of linkage among Sardinians. The finding prompted speculations about the existence of more than one G6PD locus, possibly resulting from a chromosomal inversion that might have further increased the adaptiveness of the G6PD deficient mutant in malarial environments (19). Soon after, it became clear that this alleged difference in the G6PD-Xg linkage was a statistical artifact, since the estimates of meiotic recombination had been largely derived from "phase known" pedigrees in the Sardinian study, but rarely so in the other two studies (20). When the same approach was used also for the Israeli and the Greek series, the absence of measurable linkage between the two loci was confirmed (21-22). It may be redundant to refer to these early problems, now that we know that the loci for G6PD and Xg are physically located at the opposite subtelomeric regions of the X-chromosome, but it is certainly a good story for reminding newcomers to the field of human linkage that even large Lod scores are not enough to prove or disprove a linkage relationship when the segregant haplotype combinations are classified as recombinants or non-recombinants in a statistical manner rather than on the basis of the known phase of the parental genotypes.

During the the first twenty years which followed the discovery of the G6PD polymorphism, the linkage relationship of G6PD with numerous normal X-linked traits and with X-linked diseases continued to be established through classical pedigree and population studies. It thus became clear that the locus for G6PD is part of a cluster of closely linked loci that include the genes for Deutan and Protan colorblindness, Factor VIII deficiency (hemophilia A) and adrenoleukodystrophy (23-24). To date, no recombinants have been found between G6PD and hemophilia A out of 58 scorable sibs (25). A total of 3 recombinants out of 238 scorable sibs have been reported in the world literature for the linkage G6PD/Deutan and respectively 1 out of 51 for G6PD/Protan (26). The occurrence of recombinants between the two loci for colorblindness (27-28), though infrequent, is estimated to be less rare than that between either one of them and G6PD, thus supporting the conclusion that the latter locus must be intermediate to the other two (27-29). The very tight linkage between G6PD deficiency and the loci for colorblindness was confirmed in Sardinia by the finding of a linkage disequilibrium of different haplotype combinations in a fashion that suggests the location of the G6PD locus nearer to the Protan than to the Deutan locus (30).

With the advent of somatic cell genetics, G6PD variation proved to be a very suitable tool for complementation studies in intraspecific (31-33) and interspecific somatic cell hybrids (34). The unquestionable proof in favor of X-linkage of the structural gene for hypoxanthine-phospho-ribosyl-transferase (HPRT) was given by demonstrating the concordant segregation of human G6PD and HPRT in unstable mouse-human somatic cell hybrids (34). Conversely, the physical location of the G6PD locus to the subterminal region of the X-chromosome long arm was established on the basis of its co-segregation with the HPRT locus in rodent-human somatic cell hybrids derived from the fusion of HPRT deficient rodent cells with human cells carrying inborn X-autosomal translocations (35-36).

The co-segregation of genetic markers of the same chromosome in somatic cell hybrids has been termed "synteny" to distinguish it from the true genetic linkage that implies a significant deviation from the random assortment of gametic combinations at meiosis. In the early days of cell hybridization, markers of the same chromosomes were supposed to behave always syntenically, thus permitting their assignment to the same linkage group independently of their physical distance. However, it was soon realized (once again using G6PD as a cell marker) that a certain degree of mitotic separation (37) does occur between markers of the same chromosome and that this phenomenon can be experimentally enhanced, by irradiating the human parental cells prior to fusion (38), and exploited as a parasexual approach to the measurement of physical distances between loci (39).

One of the most surprising news of the last few years in the field of X-chromosome pathology has been the discovery that about half of the males affected by an X-linked form of mental retardation carry an inducible fragile site in the subtelomeric region of the X-chromosome long arm at the interface between bands Xq27 and Xq28 (40-42). This cytological marker is inherited in an almost complete association with all other phenotypical features that characterize this X-linked condition, which is known as the fragile-X syndrome. These reports prompted us to search for families segregating for X-linked mental retardation of all types and G6PD deficiency of Mediterranean type in Sardinia. The first set of families studied suggested a close linkage relationship between G6PD deficiency and the fragile-X syndrome, but absence of measurable linkage between G6PD deficiency and other forms of X-linked mental retardation, such as the Renpenning syndrome (43). Taken at its face value, the close linkage association between the fragile-X syndrome and the G6PD cluster supports the current belief that the inducible fragile site at Xq28 is the phenotypical expression of a mutant gene localized at the same chromosomal site. Likewise, the absence of measurable linkage between G6PD and the Renpenning syndrome favors the hypothesis that the latter type (and perhaps all other types) of X-linked mental retardation must be the consequence of mutation(s) at a different locus or loci. However, as it will be discussed later on, it is also possible that the mutation(s) responsible for the fragile-X syndrome may itself (themselves) affect the frequency of meiotic recombination between the flanking loci.

As it is well known, the application of DNA restriction analysis to the study of genetic variation in Man has changed dramatically the perspectives and the methodologies for the mapping of human chromosomes (44-46). At the last workshop on human gene mapping (Helsinki, August, 1985), the number of DNA polymorphic sites assigned to the human X-chromosome has been updated to a total of 68. Among these, only five have been detected through the use of cloned genes and the remainder with arbitrary X-specific DNA probes (47). This harvest of polymorphic X-linked loci has contributed already significant advances towards a total genetic map of the human X-chromosome (48) and to establish indirectly the genetic distance between several loci of medical importance. However, in spite of these tremendous advances, it has not yet been possible to determine the relative position of the G6PD cluster with respect to its neighboring loci, let alone the order of genes within the cluster.

To answer the first question, our group has recently carried out a series of in situ hybridization studies (49-50) using preparations of metaphase chromosomes from patients with the fragile-X syndrome and the cloned genes for G6PD, Factor IX and Factor VIII as labeled molecular probes. These experiments have clearly shown that the locus for Factor IX is located at the telomeric end of band Xq27, proximally to the fragile site, whereas the loci for Factor VIII and G6PD are both located distally to it, within the band Xq28. This is to say that the physical distance separating these two sets of loci cannot be greater than eight million DNA base pairs and, probably, it is much less (50). This estimate, evaluated in conjunction with the well documented absence of measurable linkage between Factor IX deficiency and at least three genes of the G6PD cluster (49), is at variance with the generally held contention that 1% of genetic recombination corresponds to a physical distance of one million base pairs (44), unless one makes the additional assumption that the narrow chromosomal region, which overlaps with the location of the fragile site at Xq27-Xq28 includes a hot spot of recombination. In turn, the existence of close linkage between all the genes located distally to this site of frequent recombination could be regarded as the effect of interference. However, the situation has been further complicated by the recent report (51) that a TaqI restriction fragment length polymorphism (RFLP), detected with a molecular probe for FIX, segregates in a close linkage relationship with the fragile-X syndrome, just like the G6PD cluster does. This is to say that the genetic distance between the G6PD cluster and FIX appears to be non-measurable (i.e. with a frequency of recombination not significantly different from 50%) when derived from direct linkage comparisons in pedigrees without the fragile-X mutation, but it turns out to be very close if based on the sum of the two separate estimates of recombination derived from the linkage comparisons fragile-X/FIX and fragile-X/G6PD (49). To account for this paradox we have hypothesized that the high proneness to recombination typical for the region may be suppressed in the heterozgotes for the fragile-X mutation(s) (49-50), as if these were themselves the consequence of irregular crossing over events leading to chromosomal alterations capable of influencing to a variable degree the type and the frequency of meiotic exchanges. To evaluate this hypothesis, we are presently examining the segregation of several of the newly discovered DNA polymorphic markers of the region Xq27-Xqter in an extensive set of Sardinian pedigrees that segregate already for the fragile-X syndrome and/or at loci of the G6PD cluster.

The linear order of genes within the G6PD cluster cannot be resolved through the mere accumulation of random pedigree data, because this approach would obviously require tens of thousands of informative meioses to detect significant differences (if they exist) in the frequency with which the rare recombination events occur between the individual loci. On the other hand, when rare pedigrees with a recombination event within the cluster are found, they may become very precious tools for fine genetic mapping. During the course of our long-term population survey in Sardinia we found three of these rare "recombinant" pedigrees out of a total of several hundred informative families: two with a recombination event within the G6PD cluster (27,19) and a third between G6PD and the fragile-X syndrome (43). A fourth pedigree with the exceptional finding of two apparent recombinants between G6PD and Deutan within the same sibship, turned out to include a triplo-X mother and no recombinants (26). Needless to say that after this experience we have established the rule of checking the maternal karyotype in all pedigrees that suggest the occurrence of rare recombination events between closely X-linked loci.

The three recombinant families mentioned above are now being restudied with respect to the newly discovered higly polymorphic DNA restriction sites of the region Xq26-Xqter. Thus far, we found that the pedigree including the rare recombinant within the G6PD cluster and between the two loci for colorblindness, segregates also for the BamHI polymorphic site detected through the HPRT molecular probe (52). The analysis of this pedigree suggests the sequence HPRT-Deutan-G6PD-Protan as the most likely one within the cluster, in close agreement with the genetic map of the human X recently derived by Keats (53) from the overall computer evaluation of all Lod score data thus far published on human X-linkage.

We have mentioned already the usefulness of the population approach for estimating genetic distances through the hitch-hiking effect that the adaptive G6PD mutant allele must have had on its closest neighboring loci (30). We intend to extend this approach to the series of the new polymorphic loci now known in the region of the G6PD cluster and in the one proximal to the fragile site. If it is true that the chromosomal site between Xq27 and Xq28 includes a hot spot of meiotic recombination, it should follow that X-linked RFLPs located distally to this site could be in linkage disequilibrium with the adaptive G6PD deficient mutant, but those proximally to it should not. On the other hand, if the fragile-X mutations suppress meiotic crossing-over, the distribution of the haplotype combinations of the same set of loci in the hemizygous fragile-X mutants should be considerably different from that found among the normal males of the same population.

Last but not least, the contribution of G6PD variation to the classification of X-chromosome non-disjunction should be mentioned for the potential application that these non-disjunction families may have in X-chromosome mapping (45). The recent report of a X-specific human satellite DNA probe (54), detecting centromeric heterozygosity in practically every human female (47), and the profusion of X-specific RFLPs now available are expected to permit the classification of all XXY individuals with respect to the parental origin of the supernumerary X-chromosome and to the time of the non-disjunctional event. This type of information can be eventually used for identifying the distribution of crossing-over events along the X-chromosome and for mapping the still unassigned DNA polymorphic DNA sites in a fashion that is essentially similar to the tetrad analysis of Neurospora genetics.

In contrast to all other genes thus far cloned, the availability of molecular probes for G6PD has not led as yet to the discovery of common RFLPs, in spite of the extensive number of restriction sites thus far screened within and around the gene (55). This is a puzzling result in view of the existence of over 200 examples of genetic variants known at the level of the gene end product (56). There is no doubt that the picture will change when more will be known about the primary structure of the G6PD mutations and the right DNA sequences will be used to screen for restriction site variation around them. Thus, for the time being, geneticists from the Mediterannean basin can continue to enjoy their exclusivity in the performance of linkage and population studies on G6PD.

REFERENCES

1. Carson, P.E., Flanagan, C.L., Ickes, C.E., and Alving, A.S. (1956) Science 124, 484.

2. Childs, B., Zinkham, W.H., Browne, E.A., Kimbro, E.L., and Torbert, J.V. (1958) Johns Hopkins Med. J. 102, 21.

3. Siniscalco, M., Motulsky, A.G., Latte, B., and Bernini, L. (1960). Rend. Acad. Naz. Lincei 28, 903.

4. Adam, A. (1961). Nature 189, 686.

5. Porter, I.H., Schulze, J., and McKusick, V.A. (1962). Nature 193, 506.

6. Boyer, S.H., Porter, I.H., and Weilbacher, R.G. (1962). Proc. Natl. Acad. Sci. USA 48, 1868.

7. Beutler,E., Yeh, M., and Fairbanks, V.F. (1962) Proc. Natl. Acad. Sci. USA 48, 9.

8. Davidson, R.G., Nitowski, H.M., and Childs, B. (1963). Proc. Natl. Acad. Sci. USA 50, 481.

9. Sansone, G., Rasore-Quartino, A., and Veneziano, G. (1964). Lancet, i, 329.

10. Rinaldi, A.,Filippi, G., and Siniscalco, M. (1976). Am. J. Hum. Genet. 28, 496.

11. Linder, D., and Gartler, S.M. (1965). Am. J. Hum. Genet. 17, 212.

12. Gartler, S.M., Gandini, E., Angioni, G., and Argiolas, N.. (1969). Am. J. Hum. Genet. 33, 171.

13. Webster, C., Filippi, G., Rinaldi, A., Mastropaolo, C., Tondi, M., Siniscalco, M., and Blau, H.M. (1985). In "Molecular Biology of Muscle Development. UCLA Symp. on Mol. and Cell Biology. (C. Emerson et al., eds.), Alan Liss, NY (in press).

14. Motulsky, A.G. and Campbell-Kraut, J.M. (1961). In "Proc. Conf. on Genetic Polymorphisms and Geographic Variations in Disease" (B. Blumberg, ed.), p. 159. Grune & Stratton, NY.

15. Motulsky, A.G. (1960). Hum. Biol. 32, 28.

16. Siniscalco, M., Bernini, L., Latte, B., and Motulsky, A.G. (1961). Nature 190, 1179.

17. Adam, A., Sheba, C., Sanger, R., Race, R.R., Tippett, P., Hamper, J., Gavin, J.,and Finney, D.J. (1963). Ann. Hum. Genet. Lond. 26, 187.

18. Fraser, Defranas, B., Kattamis, C.A., Race, R.R., Sanger, R. and Stamatoyamopoulos, G. (1964). Ann. Hum. Genet. Lond. 27, 395.

19. Siniscalco, M. (1964). In "Genetics Today" Proc. XIth Int. Cong. Genetics, The Hague, The Netherlands, 1963, p. 851. Pergamon Press.

20. Siniscalco, M., Filippi, G., Latte, B., Piomelli, S., Rattazzi, M., Gavin, J., Sanger, R., and Race, R.R. (1966). Ann. Hum. Genet. Lond. 29, 231.

21. Adam, A., Tippett, P., Gavin, J., Noades, J., Sanger, R., and Race, R.R. (1967) Ann. Hum. Genet. Lond. 30, 211.

22. Stamatoyannopoulos, G., Sofroniadou, C., Akrivakis, A., and Fraser, G.R. (1968). Am. J. Hum. Genet. 20, 528.

23. Boyer, S.H., and Graham, J.B. (1965). Am. J. Hum. Genet. 17, 320.

24. Migeon, B.R., Moser, H.W., Axelman, J.A., Sillence, D., and Norum, R.A. (1982). Cytogen. Cell Genet. 32, 298.

25. Filippi, G., Mannucci, P.M., Coppola, R., Farris, A., Rinaldi A., and Siniscalco, M. (1984). Am. J. Hum. Genet. 36, 44.

26. Rinaldi, A., Velivasakis, M., Latte, B., Filippi, G., and Siniscalco, M. (1978). Am. J. Hum. Genet. 30, 339.

27. Siniscalco, M., Filippi, G., and Latte, B. (1964). Nature 204, 1062.

28. Arias, S., and Rodriguez, A. (1972). Humangenetik 14, 264.

29. Fraser, G.R. (1969). Am. J. Hum. Genet. 21, 593.

30. Filippi, G., Rinaldi, A., Palmarino, R., Seravalli, E., and Siniscalco, M. (1977). Genetics 86, 192.

31. Siniscalco, M., Klinger, H., Eagle, H., Koprowski, H., Fujimoto, W., and Seegmiller, E. (1969). Proc. Natl. Acad. Sci. USA 62, 793.

32. Silagi, S., Darlington, G., and Bruce, S. (1969) Proc. Natl. Acad. Sci. USA 62, 1085.

33. Migeon, B.R., Norum, R.A., and Corsaro, Ch.M. (1974). Proc. Natl. Acad. Sci. USA 71, 937.

34. Nabholz, M., Miggiano, V., and Bodmer, W. (1969) Nature 223, 358.

35. Siniscalco, M. (1970) In "Control Mechanisms in Expression of Cellular Phenotypes" (M. Padikula, ed.) p. 205, Acad. Press, NY.

36. Ricciuti, F.C., and Ruddle, F.H. (1973) Genetics 74, 661.

37. Miller, O.J., Cook, P.R., Meera Khan, P., Shin, S., and Siniscalco, M. (1971). Proc. Natl. Acad. Sci. USA 68, 116.

38. Pontecorvo, G. (1971). Nature 230, 367.

39. Goss, S.J., and Harris, H. (1977 a,b). J. Cell Sci. 25, 17 & 25.

40. Sutherland, G.R. and Ashforth, P.L.C. (1979). Hum. Genet. 48, 117.

41. Jacobs, P.A., Glover, R.W., Mayer, M., Fox, P, Gerrard, J.W., Dunn, H.G., and Herbst, D.S. (1980). Am. J. Hum. Genet. 7, 481.

42. Harrison, C.J., Jack, E.M., Allen, T.D., and Harris, R. (1983). J. Med. Genet. 20, 280.

43. Filippi, G., Rinaldi, A., Archidiacono, N., Rocchi, M., Balazs, I., and Siniscalco, M. (1983). Am. J. Med. Genet. 15, 113.

44. Botstein, D., White, R., Skolnick, M., and Davis, R. (1980). Am. J. Hum. Genet. 32, 314.

45. Siniscalco, M., Szabo, P., Filippi, G., and Rinaldi, A. (1982). In "Human Genetics, Part A: The Unfolding Genome" pp. 103-124. Alan R. Liss, Inc., NY, NY.

46. White, R. (1985). T I G 1, 177.

47. Willard, H., Skolnick, M., Pearson, P., and Mandel, J.L. (1986). In "Human Gene Mapping VIII", Cytogen. & Cell Genet. (suppl.), in press.

48. Drayna, D., Davies, K.E., Hartley, D.A., Williamson, R., and White, R. (1984). Proc. Natl. Acad. Sci. USA 81, 2836.

49. Szabo, P., Purrello, M., Rocchi, M., Archidiacono, N., Alhadeff, B., Filippi, G., Toniolo, D., Martini, G., Luzzatto, L., and Siniscalco, M. (1984). Proc. Natl. Acad. Sci. USA 81, 7855.

50. Purrello, M., Alhadeff, B., Esposito, D., Szabo, P., Rocchi, M., Truett, M., Masiarz, F., and Siniscalco, M. (1985). EMBO J. 4, 725.

51. Camerino, G., Mattei, M.G., Mattei, J.F., Jayne, M., and Mandel, J.L. (1983). Nature 306, 701.

52. Rinaldi, A., Filippi, G., Traccis, S., Latte, B., and Siniscalco, M. (1984). Hum. Genet. 65, 295.

53. Keats, B. (1983). Hum. Genet. 64, 28.

54. Willard, H.F. (1985). Am. J. Hum. Genet. 37, 524.

55. Luzzatto, L. (1985). This symposium.

56. Beutler, (1978) In " The metabolic basis of inhereted diseases" (J. Stanbury, J.B. Wyngaarden, D.S. Fredrickson, eds.) p.1430, McGraw-Hill, NY, NY.

X-INACTIVATION IN FEMALES HETEROZYGOUS FOR G-6-PD VARIANTS[1,2]

Ernest Beutler

Scripps Clinic & Research Foundation

Department of Basic & Clinical Research

10666 North Torrey Pines Road

La Jolla, California 92037

The topic of this chapter dealing with X-inactivation in females heterozygous for G-6-PD variants, seems especially appropriate for this conference. The X-inactivation hypothesis was originally developed at the City of Hope, a co-sponsor of this meeting, and it was the expression of the G-6-PD deficient allele in heterozygous females which led to its formulation.

[1]This is publication number 4197 BCR from the Research Institute of Scripps Clinic.

[2]This work was supported in part by Grant # HL 25552 from the National Institutes of Health.

I. THE ORIGIN OF THE CONCEPT OF X-INACTIVATION

In 1959 Dr. Susumo Ohno at the City of Hope was exploring the cytological origin of the Barr chromatin body. This condensed chromatin had been discovered in the nuclei of female mammals (1) but was not found in males. Interestingly, the number of chromatin bodies was always one less than the number of X-chromosomes. Until Ohno's studies it had generally been believed that the Barr chromatin body was comprised of fragments derived from both X-chromosomes, but he was able to show that in reality it represented a single heterochromatic condensed X-chromosome (2). Since it was well known that heterochromatin was genetically inactive (3,4), Ohno's findings clearly implied that only one of the two X-chromosomes were active in each cell.

At the time that Ohno was making these seminal discoveries, studies of G-6-PD deficient individuals posed questions that cried out for solutions. Browne, Childs, et al (5,6) had demonstrated that G-6-PD deficiency was sex-linked. Using the glutathione stability test they had demonstrated that transmission of G-6-PD deficiency occurred in a mother-to-son pattern. Strangely, some of the females known to be heterozygotes by a pedigree analysis manifested the effect in very severe form while others seemed to be normal. This marked variability of expression in heterozygous females was reminiscent of the variability in the expression of the phenotype of other sex-linked genes like those of hemophilia in heterozygous females, but was even more striking. Another puzzle was posed by the fact that females having two X-chromosomes had no more G-6-PD activity than did males.

It was Ohno's observation that suggested to us an answer to this puzzle. If only one of the two X-chromosomes were active in each cell of the female then the activity in normal females would be the same as in normal males (7). Ohno had suggested the possibility that it was always the maternally-derived or paternally-derived X which might be the inactive one. However, this state of affairs would be inconsistent with the known manner of inheritance of G-6-PD deficiency. If that had been the case, expression of the defect in females would be the same as that in males: some would be severely deficient and others would be normal. But the latter state of affairs was the exception rather than the rule. Most heterozygous females had intermediate G-6-PD activity. This lead us to suggest inactivation of the X-chromosomes in females was random with the paternally-derived X being inactivated in some cells and the maternally-derived X in others (7-9).

In the twenty-five years since we first proposed that this state of affairs existed, ample evidence has accumulated that X-inactivation involving not only the G-6-PD gene but most other genes on the X-chromosomes does indeed occur.

II. EXPRESSION OF G-6-PD DEFICIENCY IN THE RED CELLS OF HETEROZYGOUS FEMALES

The original observations of Browne, Childs, et al (5,6) on the variability of expression in heterozygous females has been amply confirmed. We now recognize that in a population of females heterozygous for G-6-PD deficiency, expression of the two alleles is on the average generally the same, but that among individual women expression may vary from exclusive expression of one allele to exclusive expression of the other. An interesting exception to this rule is found in women heterozygous for G-6-PD variants that produce hemolytic anemia. In such women it is not uncommon for enzyme activity to be normal or nearly so. The explanation for this state of affairs lies in the life span of the cells that are produced. Cells which are produced by progenitors with the G-6-PD deficient allele have a short intravascular life span, while the progeny of the G-6-PD normal cells survive normally. Thus, in sampling the peripheral blood the normal cells are over-represented.

III. DIAGNOSIS OF THE HETEROZYGOUS DEFICIENT STATE

The diagnosis of heterozygotes of G-6-PD deficiency poses special problems.

Given the biologic variation of "normal" G-6-PD activity it is not surprising that mixtures of normal and deficient cells will fall into the normal range even when a substantial proportion of deficient cells are present. The solution to the problem of heterozygote detection lies, then, in measuring individual red cell G-6-PD levels and this has been technically a difficult task.

The first approach to detecting small populations of deficient cells in heterozygous individuals consisted of measuring the course of methemoglobin reduction in the red cells of putative heterozygotes in the presence of methylene blue (10). In the course of these investigations it was appreciated that there was considerable cell-cell interaction and that this was mediated by the passage of leukomethylene blue from one cell to another. Another dye, Nile blue sulfate could, however, be used (11). Although a proportion of deficient cells too small to detect by enzyme assay could be appreciated using this technique, its sensitivity in heterozygote detection was limited. Combination of this approach with the measurement of individual red cell methemoglobin represented a more sensitive approach. Use of the methemoglobin elution technique (12), however, had limitations imposed by a "threshold" effect which could give rise to "pseudomosaicism" (13). Employing a tetrazolium dye as an electron receptor (14,15) seems to provide the most sensitive method of detecting heterozygotes. Even this approach is limited, however, to the detection of 5% or more of deficient cells (16).

IV. TISSUE DISTRIBUTION OF GENE PRODUCTS IN HETEROZYGOTES

The representation of cells with the paternally-derived and of the maternally-derived X-chromosomes active may vary greatly from tissue to tissue. There are two reasons why this may be the case. One of these factors is selection. The predominance of normal phenotypes in the mothers of children with non-spherocytic hemolytic anemia due to G-6-PD deficiency, to which I have eluded, is an example. Here selection is based upon the relative survival of the cells in which one or the other allele of interest is active. A similar, and even more striking example of this type of selection is observed in heterozygotes for hypoxanthine-guanine-phosphoribosyl transferase deficiency. Here selection is strongly in favor of the $HGPRT^+$ allele, and thus peripheral blood erythrocytes of heterozygotes generally manifest normal enzyme activity (17). But it is important to recognize that selection may also be based upon other alleles on the X-chromosome since inactivation also affects them. Since the importance of various genes varies from organ to organ, the representation of cells with one or the other X-chromosome active might easily differ in different tissues (18,19). One could conceive, for example, of a gene which was absolutely essential for the development of hepatocytes which was not even actively transcribed in hematopoietic stem cells. A defective copy of such a gene would have a profound effect on the representation of the two chromosomes in the liver, but none in the red blood cells.

Relatively few studies of tissue specific expression of G-6-PD alleles have been performed. Gandini et al (20) found that four heterozygotes for G-6-PD deficiency who expressed virtually no enzyme activity in their red cells had granulocyte activity which was also in the hemizygous deficient range. In contrast, two of the four heterozygotes had lymphocyte activity which was higher than the hemizygous deficient range and the activity of skin extracts and of subcutaneous fat was intermediate or normal in those heterozygotes that were tested. We (21) found that the liver of a heterozygote of G-6-PD A and B manifested only G-6-PD B in normal liver while red cells and other organs studied showed the A-B phenotype.

V. CLINICAL MANIFESTATIONS IN HETEROZYGOTES

The number of heterozygotes for deficient variants of G-6-PD who have a high proportion of deficient cells in their circulation is sufficiently great so that clinical expression in heterozygotes is not at all unusual. From this point of view, enzyme assays performed on hemolysates are a reasonably satisfactory guide to the clinical significance of G-6-PD deficiency in females. If the enzyme activity is not detectably reduced then presumably less than 30 - 40% of the red cells are G-6-PD deficient. Such individuals will not develop severe hemolytic anemia from exposure to hemolysis-precipitating agents. When a sufficiently high proportion of red cells are involved to produce reduction of enzyme activity to half normal or less, however, and particularly when enzyme activity is in the hemizygous range, the risk of untoward clinical consequences is the same as in males.

REFERENCES

1. Barr ML, Bertram EG: A morphological distribution between satellite during accelerated nuclear protein synthesis. Nature 163:676-677, 1949.

2. Ohno S, Kaplan WD, Kinosita R: Formation of the sex chromatin by a single X-chromosome in liver cells of rattus norvegicus. Exp Cell Res 18:415-418, 1959.

3. Brown SW: Heterochromatin. Science 151:417-425, 1966.

4. Brown SW, Nur U: Heterochromatic chromosomes in the coccids. Science 145:130-136, 1964.

5. Childs B, Zinkham W, Browne EA, Kimbro EL, Torbert JV: A genetic study of a defect in glutathione metabolism of the erythrocyte. Johns Hopkins Med J 102:21-37, 1958.

6. Browne EA: The inheritance of an intrinsic abnormality of the red blood cell predisposing to drug-induced hemolytic anemia. Johns Hopkins Med J 101:115, 1957.

7. Beutler E: Biochemical abnormalities associated with hemolytic states. In: Mechanisms of Anemia in Man. Weinstein IM, Beutler E (eds.): Chapter 7. pp. 195-236. McGraw-Hill, Inc., New York 1962.

8. Beutler E, Yeh M, Fairbanks VF: The normal human female as a mosaic of X-chromosome activity: Studies using the gene for G-6-PD deficiency as a marker. Proc Natl Acad Sci USA 48:9-16, 1962.

9. Lyon MF: Sex chromatin and gene action in the mammalian X-chromosome. Am J Hum Genet 14:135-148, 1962.

10. Beutler E, Baluda MC: Methemoglobin reduction. Studies of the interaction between cell populations and of the role of methylene blue. Blood 22:323-333, 1963.

11. Beutler E, Dern RJ, Baluda MC: A new technique for the ascertainment of heterozygotes for G-6-PD deficiency. In: Proc.9th Cong. Europ. Soc. Haematol. pp. 675-684. S. Karger, Lisbon 1963.

12. Gall JC Jr, Brewer GJ, Dern RJ: Studies of glucose-6-phosphate dehydrogenase activity of individual erythrocytes: The methemoglobin-elution test for identification of females heterozygous for G6PD deficiency. Am J Hum Genet 17:359-368, 1965.

13. Papayannopoulou T, Stamatoyannopoulos G: Pseudo-mosaicism in males with mild glucose-6-phosphate dehydrogenase deficiency. Lancet 2:1215-1217, 1964.

14. Fairbanks VF, Lampe LT: A tetrazolium-linked cytochemical method for estimation of glucose-6-phosphate dehydrogenase activity in individual erythrocytes: Applications in the study of heterozygotes for glucose-6-phosphate dehydrogenase deficiency. Blood 31:589-603, 1968.

15. van Noorden CJF, Tas J, Vogels IMC: Cytophotometry of glucose-6-phosphate dehydrogenase activity in individual cells. Histochem J 15:583-599, 1983.

16. Fairbanks VF, Fernandez MN: The identification of metabolic errors associated with hemolytic anemia. JAMA 208:316-320, 1969.

17. McDonald JA, Kelley WN: Lesch-Nyhan syndrome: absence of the mutant enzyme in erythrocytes of a heterozygote for both normal and mutant hypoxanthine-guanine phosphoribosyl transferase. Biochem Genet 6:21-26, 1972.

18. Gartler SM, Linder D: Developmental and evolutionary implications of the mosaic nature of the G-6-PD system. Cold Spr Har Symp Quant Biol 29:253-260, 1964.

19. Beutler E: The distribution of gene products among populations of cells in heterozygous humans. Cold Spr Har Symp Quant Biol 29:261-271, 1964.

20. Gandini E, Gartler SM, Angione G, Argiolas N, Dell Acqua G: Developmental implications of multiple tissue studies in glucose- 6-phosphate dehydrogenase-deficient heterozygotes. Proc Natl Acad Sci USA 61:945-948, 1968.

21. Beutler E, Collins Z, Irwin LE: Value of genetic variants of glucose-6-phosphate dehydrogenase in tracing the origin of malignant tumors. N Engl J Med 276:389-391, 1967.

MECHANISM OF MAMMALIAN X-CHROMOSOME INACTIVATION[1]

T. Mohandas
Division of Medical Genetics
Harbor-UCLA Medical Center
Torrance, California, U.S.A.

[1]Research work from the author's laboratory cited in this article was supported by a research grant (HD 15193) and a research career development award (HD 00378) from the National Institutes of Health.

I. INTRODUCTION

The phenomenon of X chromosome inactivation is one of the most interesting and important observations in the study of mammalian genetics. The basic concepts of X-inactivation as a mechanism to explain dosage compensation for X-linked genes were put forth more than 20 years ago (1). Since the formulation of the X-inactivation hypothesis biochemical and cytogenetic investigations have resulted in detailed descriptions of many aspects of the developmental biology of X-chromosome inactivation. These results have been reviewed in detail previously (2-5), and therefore will not be presented here. The molecular mechanism of X chromosome inactivation has always been of much interest and recently some progress has been made in this area, which will be discussed in this paper. A consideration of the mechanism of X-inactivation must deal with the distinct steps of initiation, spreading and maintenance of this process. Initiation step must also account for the mechanism that leaves a single X chromosome in the active state per diploid set of autosomes. Recent studies provide strong support for DNA modification, specifically DNA methylation, as a mechanism for maintenance of X chromosome inactivation.

II. DNA METHYLATION AND CONTROL OF GENE ACTIVITY

Several years ago, it was hypothesized that DNA methylation plays a key role in the control of gene expression in vertebrates (6-9). About three percent of the cytosine residues in mammalian DNA is converted to 5-methylcytosine following DNA replication by transferring a methyl group from S-adenosylmethionine to the 5-carbon in the pyrimidine ring of cytosine (9). Ninety percent of the methylated cytosine residues are found in the dinucleotide sequence CG and 50-70% of the CG sites are methylated (9). The essential features of the DNA methylation model for control of gene activity are the following: 1) DNA is methylated symmetrically at CG residues, and therefore replication of DNA will result in two hemimethylated DNA molecules; and 2) a maintenance methylase present in somatic cells which acts preferentially on hemimethylated sites will then restore the original symmetrical pattern of DNA methylation. This will ensure that patterns of DNA methylation are maintained through cycles of DNA replication and cell division. DNA sequence-specific proteins could alter the pattern of DNA methylation at specific stages in development. Such proteins, termed determinator proteins (10) could be *de novo* methylases, demethylases, or inhibitors of the maintenance methylase. Altered protein DNA interaction following DNA methylation (8) or altered confirmation of methylated DNA in turn affecting protein-DNA interactions (11) could explain the resulting transcriptional silencing of genes. In the last five years numerous studies have been reported which support a role for DNA methylation in control of gene activity and several of the predictions of DNA methylation model have now been verified. These have been reviewed recently and will not be discussed in detail here (10,12-14). In brief, there is good experimental evidence to support symmetrical methylation of DNA (15,16), and the presence of maintenance methylase in somatic cells (17,18). *In vitro* methylation of cloned genes is inversely correlated with their expression in somatic cells following gene transfer (19-21). Importantly, these studies also show that methylation at CG sites is maintained in somatic cells. Recent studies demonstrate *de novo* methylation of DNA in preimplantation embryos (22,23), supporting the presence of determinator proteins at specific stages in development.

Two other lines of evidence support a role for DNA methylation in control of gene activity. One of these involves studies using inhibitors of DNA methylation. In a series of important papers, Jones and colleagues showed that the base analog 5-azacytidine (5AC) can induce stable changes in cellular phenotypes (24-25). A major cellular effect of 5AC following its incorporation into DNA is inhibition of DNA methylation (26,27). Since the original observations of Jones and colleagues, several reports have appeared showing activation of previously silent genes by 5AC (10). Second line of evidence comes from studies using cloned probes for specific genes in which more direct evidence for the role of DNA methylation in control of gene activity is sought. In these studies methylation status of CG residues in the vicinity of a gene is examined in tissues that express the gene and compared to the methylation status of the same sites in tissues that do not express the gene. These studies employ restriction enzymes that are sensitive to methylation at their recognition sites. Particularly useful are the enzymes Msp I and Hpa II, both of which recognize CCGG sites; however, Hpa II does not cleave this site when the internal C is methylated whereas Msp I is not sensitive to methylation of this internal C residue. For many of the genes examined using the Msp I/Hpa II restriction enzyme assay, sites in the vicinity of the gene, particularly at the 5' flanking region are more methylated in non-expressing tissues (10). However, it has not been possible in most cases to demonstrate the presence of sites that show a strict correlation with gene activity. Also, there are genes which show no correlation between their activity and methylation status. These observations have raised questions about the importance of the role of DNA methylation in control of gene activity. There are several possible explanations for these findings. First, it is important to note that the Msp I/Hpa II assay examines only a fraction of the CG sites in the DNA and it may be that the critical sites are not being evaluated with this methodology. Alternatively, methylated sites in a given domain may act cooperatively to alter DNA confirmation or DNA-protein interactions, such that methylation of any given site may not show a strict correlation with gene activity. There is also no need to expect that all genes will show a correlation between methylation and activity, as DNA methylation is likely to be one of several mechanisms controlling gene activity. The available evidence strongly suggests that DNA methylation acts as a mechanism which locks in states of gene activity; thus methylation appears to be a necessary but not sufficient mechanism for control of gene activity.

III. DNA METHYLATION AND MAINTENANCE OF X-CHROMOSOME INACTIVATION

Riggs in 1975 (8) proposed that DNA modification, specifically methylation, could explain many aspects of X-chromosome inactivation. In particular, DNA methylation could explain the remarkable stability of inactivation in somatic cells. The stability of inactivation in mammalian somatic cells has been appreciated for some time (28). Mouse-human cell hybrids containing partial complements of human chromosomes including the inactivated human X have been isolated (29-31). These inactive X chromosomes maintain their state in the hybrid cells in the absence of the majority of other human chromosomes. Based on this, it was proposed that the control for maintenance of X-inactivation must reside on the X chromosome itself (32). More direct evidence for the role of DNA modification in the maintenance of X-inactivation has come from DNA-mediated gene transfer studies and reactivation of genes from the inactive human X chromosome.

A. DNA-Mediated Gene Transfer Studies

In these studies deproteinized, high molecular weight DNA is prepared from desired donor cells and presented to recipient cells as DNA-calcium phosphate precipitate (33). The recipient cells internalize the DNA which is ultimately integrated into their chromosomes and expressed. The presence of the selectable marker hypoxamthine phosphoribosyltransferase (HPRT) on the mammalian X chromosome, has made the gene transfer assay particularly useful in studying the mechanism of maintenance of X-inactivation. Liskay and Evans in 1980 (34) prepared DNA from two independent clones isolated from an established mouse cell line that were deficient in HPRT. The HPRT deficient clones as well as the wild type parental line had two structurally normal X chromosomes, one of which was late replicating and presumably inactive. The DNA isolated from the mutant clones was unable to transform HPRT deficient Chinese hamster cells, whereas DNA from the wild type parental line transformed the Chinese hamster cells giving rise to clones able to grow in HAT medium and expressing mouse form of HPRT. These results indicate that the wild type HPRT allele residing on the inactive X chromosome, which is present in the mutant cells, was incompetent in the transformation assay. These findings suggest that there is a difference between the DNA, per se, of the active and inactive X chromosome at or near the HPRT gene and that this difference could account for maintaining the HPRT gene in the inactive state. In an elegant study, Chapman et al (35) confirmed these findings using female mice heterozygous for isolectric focussing variants of HPRT and carrying a balanced X/16 translocation (Searle's translocation). In these mice, the normal X chromosome (but not the translocated X) is usually inactivated in the somatic cells. DNA isolated from the somatic cells of these mice was used to transform HPRT deficient Chinese hamster cells and the HPRT phenotype of the transformants were determined. These studies showed that only a single transformant from a total of 58 expressed the HPRT allele (HPRT-A) located on the structurally normal inactive X chromosome. On the other hand, when DNA isolated from the somatic cells of XX females heterozygous for HPRT was used as the donor, 9 or 35 HAT resistant clones expressed mouse HPRT-A. In the somatic cells of the XX females, the inactivation is, of course, random. Thus the HPRT allele on the inactive X is highly inefficient, if not incompetent altogether in DNA-mediated gene transfer. The single transformation event observed could be due to a small proportion of cells in the translocation carrier mice in which the normal X is active and a translocated X is inactive. Inactivation of a translocated X, although rarely seen in adult somatic cells, has been observed in these mice carrying Searle's translocation in early stages of development (36). Similar results were also observed by Venolia and Gartler using DNA from cloned fibroblasts of a human female heterozygous for HPRT deficiency (37). Clones with a normal HPRT allele on the active or inactive X chromosome were isolated by plating the fibroblasts in hypoxanthine-aminopterin-thymidine (HAT) medium or 6-thioguanine medium, respectively. DNA isolated from these clones was used to transform HPRT deficient mouse cells. Again, HPRT gene on the inactive X chromosome was highly inefficient in producing transformants compared to the HPRT gene on an active X chromosome. A report by de Jonge et al., on the other hand, showed that DNA from HPRT positive and negative clones obtained from SV40 transformed human fibroblasts heterozygous for HPRT deficiency was equally efficient in the transformation assay (38). Venolia and Gartler were unable to reproduce these results (37). Thus, the weight of the evidence from transformation studies clearly favors a DNA modification mechanism for maintenance of mammalian X-inactivation. It is known that the inactive X chromosome can undergo reactivation _in vivo_. Primordial female germ cells apparently undergo X-inactivation as do somatic cells, but the inactive X is reactivated at or about the time these cells enter meiosis (39-42). Given that the inactive X can be reactivated _in vivo_, it is highly unlikely that inactivation is controlled by DNA modification which involves major genomic rearrangements. However, DNA methylation is compatible with reactivation of the X as demethylation of DNA could be brought about by determinator proteins, as noted earlier.

B. Reactivation of Genes from the Inactive X Chromosome

Further evidence supporting a role for DNA methylation in the maintenance of X-inactivation comes from reactivation of genes from the inactive X-chromosome using 5-azacytidine. In a mouse-human somatic cell hybrid clone derived from the fusion of mouse HPRT deficient cells and containing only an inactive human X and no other human X chromosome material, we were able to show that 5AC induces the expression of human HPRT at remarkable frequencies (43). In untreated cells the frequency of reactivation of the HPRT gene from the inactive X, monitored by plating efficiency of the hybrid clone in HAT medium is approximately $1x10^{-6}$. In cells treated with 5AC, under optimal conditions, as much as 2% of the cells surviving 5AC treatment were able to grow in HAT medium and expressed human form of HPRT (44). These results were confirmed in independently derived rodent-human hybrid clones retaining inactive human X chromosomes (45,46) and also in an established line derived from an interspecific mouse hybrid embryo (47). Reactivation is not limited to HPRT, as other X-linked enzyme loci are also reactivated from the inactive X, following 5AC treatment (44-47). The following points need to be emphasized with regard to the action of 5AC on the inactive X. First, reactivation of genes induced by 5AC is a permanent and heritable change even in the absence of selection for the reactivated gene. It should also be noted that cells are not grown continuously in the presence of the analog after the initial treatment, which is usually for the duration of a cell cycle time. Secondly, the entire X chromosome is not reactivated following 5AC treatment. Therefore, the control for maintenance of X-inactivation appears to be localized, whereas the initiation of inactivation must be a chromosomal event.

Transformation studies using DNA from these mouse-human hybrid clones retaining either active, inactive or partially reactivated (for HPRT) human X chromosome, further support the premise that 5AC acts at the DNA level (45, 48). It was observed that DNA from hybrid clones containing an active or reactivated human X, transforms HPRT deficient rodent lines whereas DNA from the hybrid clone retaining the inactive human X is unable to confer HAT resistance to these same rodent cells. DNA from the hybrid with the inactive X, was able to transform thymidine kinase (TK) deficient cells or adenine phosphoribosyltransferase (APRT) deficient cells, suggesting that the inability to transform HPRT deficient cells is not due to a poor DNA preparation (45, 48). These results show that 5AC has induced a permanent, heritable change at the DNA level. Since the major effect of 5AC on DNA is hypomethylation, reactivation studies strongly support a key role for DNA methylation in the maintenance of X-inactivation.

Studies using 5AC have also provided other interesting findings which have some bearing on the mechanism of X-inactivation. Following treatment of cells with 5AC, reactivation of human HPRT is maximal, if the cells are allowed to recover for two cell cycles before plating in HAT medium (49). This is consistent with the idea that 5AC acts by demethylation of DNA, since two cell cycles will be required for symmetrically demethylating DNA. Another interesting finding is that, following reactivation the level of expression of the reactivated genes is variable (45, 50, 51). Normally, in a mouse-human hybrid clone retaining an active human X chromosome, electrophoretic evaluation of glucose-6-phosphate dehydrogenase (G6PD) expression reveals three prominent bands representing human, human-mouse, and mouse enzymes. Following reactivation of human G6PD, clones have been isolated in which the presence of the human enzyme band is minimal or even absent. These clones do show a prominent mouse band and a less prominent hybrid band, the latter indicating unequivocally that the human gene is in fact expressed. Cytogenetic analysis has shown that this phenotype cannot be attributed to a loss of the human X chromosome from most of the hybrid cells. Furthermore, single-cell clones isolated from such a hybrid continue to express the G6PD phenotype of the parental clone. Thus, 5AC has induced a heritable change such that partial expression of G6PD is possible from the X chromosome. Expression of other X-linked enzyme loci, phosphoglycerate kinase (PGK) and HPRT, are similarly affected, with the activity of HPRT in some clones being higher than that observed in a hybrid with an active X (45, 51). A low level of expression of G6PD has also been observed in a case of spontaneous reactivation of this gene in a cloned human fibroblast (52). It is also of note that the activity of steroid sulfatase (STS), a locus which escapes X-inactivation on the human X chromosome is reduced when expressed from an inactive X (53). In the case of 5AC, it appears as though the extent of demethylation induced by this base analog affects protein-DNA interactions differently which in turn affects the level of transcription of the gene. Independent reactivation events very likely result in different levels of demethylation, and the induced change is clonally heritable.

Another interesting observation is that it has not been possible to reactivate genes from the inactive X chromosome in diploid human fibroblasts using 5AC (54). The reason for this is not clear, although it may be indicative of additional complexities in the maintenance of the inactive state. For example, it has been suggested that organization of chromosomes in the interphase nucleus may play a role in maintaining X-inactivation (Gartler, S.M., personal communication). Such organization is naturally disrupted when cell hybrids are formed and therefore the inactive human X in a cell hybrid may be more susceptible to the action of 5AC.

IV. MOLECULAR STUDIES ON THE MECHANISM OF X-INACTIVATION

With the availability of cloned probes from the X chromosome, restriction analysis of DNA from active, inactive and reactivated human X chromosomes have now been carried out, using enzymes that are sensitive to methylation at their recognition sites. Using a probe that codes for an apparently transcribed gene from the long arm of the X chromosome, Wolf and Migeon (54) did not observe any consistent changes in DNA methylation between male and female cells. These results clearly ruled out the possibility that the inactive X is ubiquitously methylated. Lack of differences in DNA methylation between the active and inactive X chromosomes, around a second random sequence from the X has recently been reported (55). However, using cloned probes for specific X-linked genes which undergo inactivation differences in methylation patterns between active and inactive human chromosomes have now been demonstrated. Yen et al. (56) using genomic clones of human HPRT gene and compared the DNA methylation patterns in male DNA, female DNA, and DNA isolated from mouse-human hybrids containing an active human X, inactive human X, or an inactive X containing a reactivated HPRT gene following 5AC treatment. These studies showed that the CG sites at the 5' end of this gene are generally undermethylated on active and reactivated X chromosomes compared to the same sites on inactive X chromosomes. Specifically, two sites in the first intron of the gene showed very good correlation with gene activity, such that they were methylated on inactive X's and unmethylated on active or reactivated X chromosomes. Surprisingly, a site in the middle of the gene showed a reverse correlation in that it was methylated on active X chromosomes and undermethylated on inactive X's. These results also showed that treatment with 5AC did not result in stable demethylation of long stretches of DNA. Independently, Wolf et al. (57) also studied the methylation pattern of HPRT and showed that the CG sites clustered in the 5' end of HPRT are hypomethylated on active X chromosomes compared to inactive X chromosomes. However, sites at the 3' end of the gene showed a reverse correlation between methylation and activity of the gene. Wolf et al. (57) also observed that the methylation pattern on the active X chromosome was more stable than on the inactive X chromosome and based on this they suggested that the mechanism of inactivation may be directed at the active X rather than the inactive X chromosome. Studies using a probe for the 3' region of human G6PD, have shown that most of the CG sites in this region are methylated on active and inactive X chromosomes (58). However, specific sites were seen which were unmethylated on active X's, but were partly methylated on inactive X's (58).

The molecular studies using probes for HPRT and G6PD, while showing consistent differences between active and inactive X chromosomes do not support a straightforward correlation between X-inactivation and DNA methylation that would be predicted based on studies using 5-azacytidine. As observed for the autosomal genes there are sites on the X chromosome which show undermethylation correlated with gene activity. However, there are also sites whose methylation is correlated with activity. It may be that activity of X-linked genes is controlled by a given pattern of methylation and unmethylation of sites in the vicinity of the gene. It is also possible that the critical sites are not being evaluated, as these studies are limited to sites recognized by the restriction enzymes. Limitations are also imposed by distribution of restriction sites and the location of single-copy restriction fragments that can be used as probes. Thus it is not realistic to evaluate the methylation status of all sites using the restriction enzyme methodology. Alternative strategies to evaluate methylation in genomic DNA may provide more definitive evidence on the role of DNA methylation in maintenance of X-inactivation. The recently introduced technique of genomic sequencing may prove useful in this regard (59).

Further studies using the Msp I/Hpa II assay are also in order. It would be important to evaluate the methylation status of an X-linked gene showing tissue specific expression; so far these studies have been limited to X-linked housekeeping genes. Also of importance is the need to evaluate the methylation status of genes that escape the usual X-inactivation. Recent studies have shown that the gene loci for STS and 12E7 (a cell surface antigen) escape X-inactivation (60, 61). A molecular probe MIC2 (which codes for 12E7) has recently been isolated (62, 63) and it should be of particular interest to evaluate the methylation status of this gene that escapes X-inactivation and is expressed from both the X and Y chromosomes. The phosphoglycerate kinase (PGK) gene has an autosomal counterpart showing tissue specific expression and comparison of the methylation status of these genes will be of much interest. Probes for the X-linked and autosomal PGK genes have been isolated (64-67). Autosomal genes attached to the X chromosome can undergo X-inactivation (68, 69), suggesting that there may be nothing unique about the X-linked genes in their ability to undergo inactivation. It should be of considerable interest to compare the methylation status of an inactivated autosomal gene with its normal counterpart on an autosome. With the increasing availability of cloned gene probes it should be possible to undertake these studies in the near future and they are likely to provide more definitive information on the role of DNA methylation in control of X chromosome activity.

ACKNOWLEDGEMENTS

I am most grateful to N. Hitt for assistance in preparation of this manuscript.

REFERENCES

1. Lyon, M. F. (1961). Nature 190, 372.
2. Lyon, M. F. (1972). Biol. Rev. 47, 1.
3. Gartler, S. M., and Andina, R. J. (1976). Adv. Hum. Genet. 7, 99.
4. Mohandas, T. and Shapiro, L. J. (1983). in "Cytogenetics of the Mammalian X Chromosome, Part A, Basic Mechanisms of X Chromosome Behaviour" (A.A. Sandberg, ed.), p. 271. Alan R. Liss, Inc., New York.
5. Gartler, S. M. and Riggs, A. D. (1983). Ann. Rev. Genet. 17, 155.
6. Scarano, E. (1971). Adv. Cytopharmacol. 1, 13.
7. Holliday, R., and Pugh, J. E. (1975). Science 187, 226.
8. Riggs, A. D. (1975). Cytogenet. Cell Genet. 14, 9.
9. Razin, A., and Riggs, A. D. (1980). Science 210, 604.
10. Riggs, A.D. and Jones, P.A. (1983). Adv. Cancer Res. 40, 1.
11. Felsenfeld, G., and McGhee, J. (1982). Nature 296, 602.
12. Doeffler, W. (1983). Ann. Rev. Biochem. 52, 93.
13. Bird, A. P. (1984). Nature 307, 503.
14. Razin, A., Cedar, H., and Riggs, A. D. (1984). "DNA Methylation, Biochemistry and Biological Significance". Springer-Verlag, New York.
15. Bird, A. P. (1978). Phil. Trans. Royal Soc. Lond. B. 283, 325.
16. Cedar, H., Solage, A., Glaser, G., and Razin, A. (1979). Nucleic Acid Res. 6, 2125.
17. Jones, P. A., and Taylor, S. M. (1981) Nucleic Acid. Res. 9, 2933.
18. Gruenbaum, Y., Cedar, H., and Razin, A. (1982) Nature 295, 620.
19. Wigler, M., Levy, D., and Perucho, M. (1981). Cell 24, 33.
20. Stein, R., Razin, A., and Cedar, H. (1982). Proc. Natl. Acad. Sci. USA 79, 3418.
21. Busslinger, M., Hurst, J. and Flavell, R. A. (1983) Cell 34, 197.
22. Jahner, D., Stuhlmann, H., Stewart, C. L., Harbers, K., Loehler, J., Simon, I., and Jaenisch, R. (1982). Nature 298, 623.
23. Stewart, C. L., Stuhlmann, H., Jahner, D., and Jaenisch, R. (1982). Proc. Natl. Acad. Sci. USA 79, 4098.
24. Constantinides, P. G., Taylor, S. M., and Jones, P.A. (1977). Nature 267, 364.
25. Taylor, S. M., and Jones, P. A. (1979). Cell 17, 771.
26. Jones, P. A., and Taylor, S. M. (1980). Cell 20, 85.
27. Creuso, F., Acs, G., and Christman, J. K. (1982). J. Biol. Chem. 257, 2041.
28. Migeon, B. R. (1972). Nature 239, 87.
29. Kahan, B., and DeMars, R. (1975). Proc. Natl. Acad. Sci. USA 72, 1510.

30. Hellkuhl, B., and Grzeschik, K. H. (1978). Cytogenet. Cell Genet. 22, 527.
31. Mohandas, T., Sparkes, R. S., Hellkuhl, B., Grzeschik, K. H., and Shapiro, L. J. (1980). Proc. Natl. Acad. Sci. USA 77, 6759.
32. Kahan, B., and DeMars, R. (1980). Som. Cell Genet. 6, 309.
33. Pellicer, A., Robins, D., Wold, B., Sweet, R., Jackson, J., Lowy, I., Roberts, J. M., Sim, G. K., Silverstein, S., and Axel, R. (1980). Science 209, 1414.
34. Liskay, R. M., and Evans, R. J. (1980). Proc. Natl. Acad. Sci. USA 77, 4895.
35. Chapman, V. M., Kratzer, P. G., Siracusa, L. D., Quarantillo, B. A., Evans, R., and Liskay, R. M. (1982). Proc. Natl. Acad. Sci. USA 79, 5357.
36. Takagi, N. (1980). Chromosoma 81, 439.
37. Venolia, L., and Gartler, S. M. (1983). Nature 302, 82.
38. de Jonge, A. J. R., Abrahams, P. J., Westerveld, A., and Bootsma, D. (1982). Nature 295, 624.
39. Gartler, S. M., Andina, R., and Gant, N. (1975). Exptl. Cell Res. 91, 454.
40. Kratzer, P. G., and Chaptman, V. M. (1981). Proc. Natl. Acad. Sci. USA 78, 3093.
41. McMahon, A., Foster, M., and Monk, M. (1981). J. Embryol. Exptl. Morphol. 64, 251.
42. Monk, M., and McLaren, A. (1981). J. Embryol. Exptl. Morphol. 63, 75.
43. Mohandas, T., Sparkes, R. S., and Shapiro, L. J. (1981). Science 211, 393.
44. Shapiro, L. J., and Mohandas, T. (1983). Cold Spring Harbor Symp. Quant. Biol. 37, 631.
45. Lester, S. C., Korn, N. J., and DeMars, R. (1982). Somat. Cell Genet. 8, 265.
46. Hors-Cayla, M. C., Heuertz, S., and Frezal, J. (1983). Somat. Cell Genet. 9, 645.
47. Graves, J. A. M. (1982). Exptl. Cell Res. 141, 99.
48. Venolia, L., Gartler, S. M., Wassman, E. M., Yen, P., Mohandas, T., and Shapiro, L. J. (1982). Proc. Natl. Acad. Sci. USA 79, 2352.
49. Jones, P. A., Taylor, S. M., Mohandas, T., and Shapiro, L. J. (1982). Proc. Natl. Acad. Sci. USA 79, 1215.
50. Mohandas, T., Sparkes, R. S., Bishop, D., Desnick, R. J., and Shapiro, L. J. (1984). Amer. J. Hum. Genet. 36, 916.
51. Migeon, B. R., Johnson, G. G., Wolf, S. F., Axelman, J., Schmidt, M. (1985). Amer. J. Hum. Genet. 37, 608.
52. Migeon, B. R., Wolf, S. F., Mareni, C., and Axelman, J. (1982). Cell 29, 595.
53. Migeon, B. R., Shapiro, L. J., Norum, R. A., Mohandas, T., Axelman, J., and Dabora, R. L. (1982). Nature 299, 838.
54. Wolf, S. F., and Migeon, B. R. (1982). Nature 295, 667.

55. Lindsay, S., Monk, M., Holliday, R., Husehtscha, L., Davies, K. E., Riggs, A. D., and Flavell, R. A. (1985). Ann. Hum. Genet. 49, 115.
56. Yen, P., Patel, P., Chinault, A. C., Mohandas, T., and Shapiro, L. J. (1984). Proc. Natl. Acad. Sci. USA 81, 1759.
57. Wolf, S. F., Jolly, D. J., Lunnen, K. D., Friedmann, T., and Migeon, B. R. (1984). Proc. Natl. Acad. Sci. USA 81, 2806.
58. Toniolo, D., D'Urso, M., Martini, G., Persico, M., Tufano, V., Battistuzzi, G., and Luzzatto, L. (1984). EMBO Journal 3, 1987.
59. Church, G. M., and Gilbert, W. (1984). Proc. Natl. Acad. Sci. USA 81, 1991.
60. Shapiro, L. J., Mohandas, T., Weiss, R., and Romeo, G. (1979). Science 204, 1224.
61. Goodfellow, P., Pym, B., Mohandas, T., and Shapiro, L. J. (1984). Amer. J. Hum. Genet. 36, 777.
62. Darling, S., Banting, G., Pym, B., Wolfe, J., and Goodfellow, P. N. (1985). Proc. Natl. Acad. Sci. USA (in press).
63. Buckle, V., Mondello, C., Darling, S., Craig, I. W., and Goodfellow, P. N. (1985). Nature 317, 739.
64. Michelson, A. M., Markham, A. F., and Orkin, S. H. (1983). Proc. Natl. Acad. Sci. USA 80, 472.
65. Singer-Sam, J., Simmer, R. L., Keith, D. H., Shively, L., Teplitz, M., Itakura, K., Gartler, S. M., Riggs, A. D. (1983). Proc. Natl. Acad. Sci. USA 80, 802.
66. Michelson, A. M., Bruns, G. A. P., Morton, C. C., and Orkin, S. H. (1985). J. Biol. Chem. 260, 6982.
67. Gartler, S. M., Riley, D., Eddy, R., and Shows, T. B. (1985). Amer. J. Hum. Genet. 37, A230.
68. Couturier, J., Dutrillaux, B., Garber, P., Raoul, O., Croquette, M. F., Fourlinnie, J. C., and Maillard, E. (1979). Hum. Genet. 49, 319.
69. Mohandas, T., Sparkes, R. S., and Shapiro, L. J. (1982). Amer. J. Hum. Genet. 34, 811.

INSIGHTS INTO G6PD REGULATION FROM STUDIES OF X DOSAGE COMPENSATION

Barbara R. Migeon

Department of Pediatrics
The Johns Hopkins University School of Medicine
Baltimore, Maryland 21205 USA

In mammals, the sex difference in dosage of X chromosomes is compensated by silencing all but one X chromosome in female somatic cells (Reviewed in Reference 1). Dosage compensation is not established simultaneously in all tissues (2), but is programmed along with other tissue specific functions. At the time of tissue differentiation, only one X chromosome achieves potential transcriptional activity; the others become inactive, condense at interphase, and replicate asynchronously (later than the autosomes and the active X).

The locus for glucose 6 phosphate dehydrogenase (G6PD) has provided an extraordinary marker for studies of X chromosome dosage compensation. The availability of common enzyme variants differing in electrophoretic mobility has made it possible to discriminate between active and inactive chromosomes in cells from females who are heterozygous for the variant (See Figure 1).

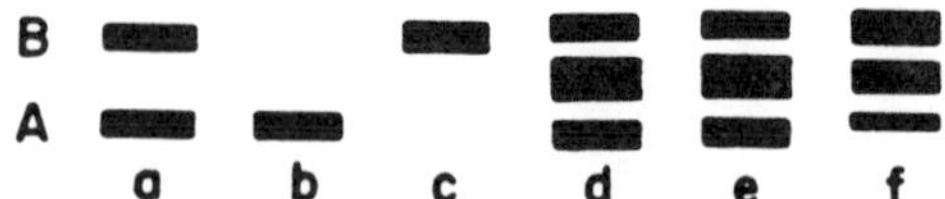

Figure 1. Diagram showing G6PD isozyme pattern in heterozygous cells from (a), uncloned fibroblasts; (b), (c), derivative clones; (d), oocytes; (e), 69, XXY triploid fibroblasts, and (f), reactivant with derepressed G6PD A allele on inactive X, or diploid clone from chorionic villi with partial expression of G6PD A on inactive X.

The expression of only one of the two enzymes in clonal populations from these heterozygotes (3) provided the first compelling evidence that there is only a single active X in female somatic cells, and that the inactivation is transmitted through cell division. Furthermore, the enzyme is a dimer, so that heterodimers are generated if both genes function in the same cell. The presence of heterodimers has revealed that the two X chromosomes are expressed in oocytes (4,5), in triploid cells (6), and in cells whose locus on the inactive X has been reactivated (7-9).

The availability of DNA probes for the G6PD locus provides further opportunities to explore the molecular basis of X inactivation. Our studies of the organization of the G6PD locus on human X chromosomes have been carried out in collaboration with Persico and Toniolo using their probes pGD3 and pGD1.4 (10). These genomic DNA sequences are located 45 kb 3' to those that code for the G6PD protein (See chapter). Our studies of human and mouse cells, and somatic cell hybrids having only relevant portions of the human X, indicate that there is a single locus in both species; situated at the long arm telomere of the human X, but closer to the centromere (in the A region) of the mouse X, reflecting intrachromosomal transposition of the locus (11). Furthermore, no difference in the arrangement of DNA sequences of active and inactive X loci has been observed.

On the other hand, there are significant differences in the methylation of the G6PD on active and inactive X (10,12). The differences involve two remarkable clusters of CpG dinucleotides in the 3' region of the gene (10). In CpG clusters from the silent gene, the cytosines are modified by methyl groups but cytosines are unmethylated in clusters from the active gene (12). Furthermore, the correlation between the state of methylation of the clusters and the expression of this locus is absolute; these clusters become demethylated in genes on the inactive X that are derepressed, either spontaneously (or) induced with cytidine analogues (like 5 azacytidine) that inhibit DNA methylation (12).

Whereas it is clear that differential methylation of these clusters helps maintain silence of the G6PD locus on the inactive X, DNA methylation may not have a general role in X dosage compensation (12,13). Consistent differences in DNA methylation of homologous sequences seems to occur only in X linked housekeeping genes like G6PD, hypoxanthine phosphoribosyl transferase (HPRT)

(14,15), and phosphoglycerate kinase (PGK) (16).

Studies of random DNA sequences from the X chromosome have revealed no consistent differences in their methylation (17,16) and preliminary studies of factor IX, an X linked clotting factor with tissue specific expression, have failed to reveal such generalized sex differences (18). The CpG clusters are not special features of X linked genes as they are absent in the factor IX gene (18), but are general features of housekeeping genes on all chromosomes - being present in the promoter regions of autosomal housekeeping genes (dihydrofolate reductase (19), adenine phosphoribosyl transferase (19), and HMG CoA reductase (20)), in lieu of the usual TATA and CAT sequences. Clustered CpG dinucleotides are also found in other regions of genes - as in the 3' region of the G6PD locus. Recent studies suggest that there are as many as 30,000 clusters in our genome (21).

That these CpG clusters are important regulatory features of housekeeping genes like G6PD is clear from studies of their chromatin (13). When the clusters are unmethylated, the relevant chromatin is hypersensitive to nucleases like DNAseI and SI, the hallmark of chromatin from other regulatory sequences. In contrast the homologous chromatin from the inactive X is not nuclease sensitive when the cluster is methylated; however, the demethylation of these clusters associated with derepression of the gene on the allocyclic X induces nuclease hypersensitivity identical to that which characterizes the homologous locus on the active X (13). The demethylation of the G6PD cluster is specifically associated with reexpression of G6PD, as reactivation events affecting other X loci, but not G6PD, do not affect methylation of the G6PD clusters (12).

Sex differences in DNA methylation also have been noted in the body of X housekeeping genes. Based on studies of reactivants (12,13), this differential methylation is not important for gene activity. Paradoxically, the active gene is highly methylated in this region, whereas the inactive gene is less methylated. However, in contrast to the consistent pattern for the active genes, methylation of the inactive gene is variable, and tissue specific differences are prominent, supporting the hypothesis that differential methylation is not the primary mechanism for maintaining inactivity of the silent X.

Because the pattern of DNA methylation of the active X at G6PD and HPRT loci is so constant, it can be used to identify the presence of the active X in any cell, and has been useful in distinguishing active from inactive X in hybrid cells. Furthermore, combined with the presence of common restriction fragment length polymorphisms (RFLP), the pattern of DNA methylation of X housekeeping genes can be used as a marker for cell population homogeneity (clonality) of malignant (or normal) tissues when protein variants are not available (22). For instance, in DNA digested by HhaI plus BamHI from females heterozygous for the 12, 24 kb BamHI RFLP at the HPRT locus, the active X yields

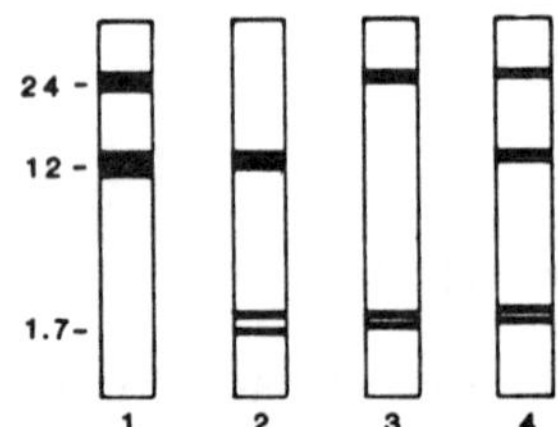

Figure 2. Diagram showing methylation patterns as indicators of clonal populations; restriction fragments from the HPRT gene in a female heterozygous for the 12-24 kb RFLP. (1) Southern blots of BamHI digested DNA showing equal presence of both alleles, (2-4) blots of DNA digested with both BamHI and HhaI. (2) clone with 12 kb allele on active X, (3) clone with 24 kb allele on active X, and (4) uncloned cells from the heterozygote (See text). Adapted from Vogelstein, et. al. Science, 227, 642-645, 1985.

0.8 kb and 24 kb (or 12 kb if polymorphic) fragments whereas the inactive X yields fragments of variable size, but none the same size as those from the active X (See Figure 2). In HhaI/BamHI digests the presence of only one intact allele, (either 12 or 24 kb) indicates that the DNA came from the clonal progeny of a single cell. This kind of analysis should be possible for G6PD when RFLPs are identified.

The pattern of methylation of active G6PD and HPRT genes is so consistent in all tissues that it has been useful in identifying tissues in which X dosage compensation is incomplete. Because of a paucity of restriction enzyme fragments originating from the inactive X (less than expected on the basis of equal numbers of active and inactive chromosomes per cell), we suspected that some genes on the inactive X might be expressed in DNA from chorionic villi of human placenta (23). From studies carried out on the villi cultured from fetal specimens heterozygous for G6PD variants, it is clear that some cells express only a single allele, whereas others express both of them (24). The patterns of G6PD enzymes observed in clonal cell populations (See Figure 1f), consisting of two major bands (one parental homodimer and the heterodimer) and a minor band (the other homodimer), indicate that both loci are expressed in the same cell; yet, the prominence of one homodimer in each clone means that one X is expressed more than the other. Unlike mouse cells of trophoblast derivation where the paternal X is not expressed (25), the choice of X in the analogous human cells is random with respect to parental origin of the active X (23).

The partial expression of the inactive X is not associated with any detectable change in the replication of the chromosome and a late replicating allocyclic X is seen in each cell (23). However, this late replicating X chromosome that is partially expressed in chorionic villi begins to replicate synchronously and appears to undergo complete reactivation when introduced into the milieu of the interspecies hybrid cell. The examination of hybrids derived from a **clonal** population of cells from villi heterozygous for G6PD A (46 XX, with allocyclic X carrying G6PD A) and mouse A9 cells, indicate that at least 3 loci (G6PD, HPRT and PGK) are fully functional on the human inactive X; furthermore, the human X chromosomes carrying either G6PD A or B replicate synchronously with each other and with murine chromosomes (24). The X with G6PD A was identified as the original late replicating X, because methylation of HPRT on this chromosome differed from the consensus pattern for the locus on active X chromosomes (14); although CpG clusters were unmethylated in HPRT on both human chromosomes, only those with the G6PD A allele were hypomethylated in the body of the gene.

These results indicate that X inactivation is completely reversible in cells of trophoblast origin; induction of full transcriptional activity is accompanied by acquisition of isocyclic replication, showing an intimate relationship between these processes. The molecular events responsible for this reversal may be similar to those responsible for reversal of inactivation that occurs during maturation of oocytes (26). Chorionic villi and derivative hybrids provide _in vitro_ models for exploring early events that program the single active X.

It is clear that studies of the G6PD locus have revealed many insights into X dosage compensation; on the other hand, studies of X inactivation provide a unique opportunity to learn about the regulatory features of this gene. Comparison of G6PD loci on active and inactive X have revealed the importance of CpG clusters in transcription of the gene. It is likely that we will understand the relationship of the 3' CpG clusters probed by pGD3 to the sequences that code for the enzyme protein from studies of the organization of this locus not only on the human

X chromosome, but also on the X chromosome in other mammals. In any event, the ability to compare active with inactive loci within the same cell should discriminate those regulatory features that are related to function from the red herrings that are not.

ACKNOWLEDGEMENT

I gratefully acknowledge the contributions of Malgorzata Schmidt, Joyce Axelman, Catherine Ruta Cullen, David Kaslow and especially Dr. Stanley F. Wolf, who recognized these CpG clusters as regulatory features of housekeeping genes. The studies described in this paper were supported by NIH Grant HD 05465.

REFERENCES

1. Lyon, M.F. (1972). Biol. Rev. 47, 1-35.
2. Monk, M. and Harper, M.I. (1979). Nature 281, 311-313.
3. Davidson, R.G., Nitowski, H.M. and Childs, B. (1963). Proc. Natl. Acad. Sci. U.S.A. 50, 481-483.
4. Gartler, S.M., Andina, R. and Gant, N. (1975). Exp. Cell Res. 92, 454-457.
5. Migeon, B.R. and Jelalian, K. (1977). Nature 269, 242.
6. Migeon, B.R., Sprenkle, J.A. and Do, T.T. (1979). Cell 18, 637-641.
7. Migeon, B.R., Wolf, S.F., Mareni, C. and Axelman, J. (1982). Cell 29, 595-600.
8. Mohandas, T., Sparkes, R.S. and Shapiro, L.J. (1981). Science 211, 393-396.
9. Lester, S.C., Korn N.J. and DeMars, R. (1982). Somat. Cell Genet. 8, 265-284.
10. Toniolo, D., D'Urso, M., Martin, G., Persico, M.G., Tufano, V., Battistuzzi, G. and Luzzatto, L. (1984). Embo J. 3, 1987-1995.
11. Martin-DeLeon, P.A., Wolf, S.F., Persico, G., Toniolo, D. and Migeon, B.R. (1985). Cytogenet. Cell Genet. 39, 87-92.

12. Wolf, S.F., Dintzis, S., Toniolo, D., Persico, G., Lunnen, K., Axelman, J. and Migeon, B.R. (1984). Nucl. Acids Res. 12 (24), 9333-9348.
13. Wolf, S.F. and Migeon, B.R. (1985). Nature 314, 467-469.
14. Wolf, S.F., Jolly, D.J., Lunnen, K.D., Friedmann, T. and Migeon, B.R. (1984). Proc. Natl. Acad. Sci. U.S.A. 81, 2806-2810.
15. Yen, P.H., Patel, P., Chinault, A.C., Mohandas, T. and Shapiro, L.J. (1964). Proc. Natl. Acad. Sci. U.S.A. 81, 1759-1763.
16. Lindsay, S., Monk, M., Holliday, R., Huschtscha, L., Davies, K.E., Riggs, A.D. and Flavell, R.A. (1985). Ann. Hum. Genet. 49, 115-127.
17. Wolf, S.F. and Migeon, B.R. (1982). Nature 295, 667-671.
18. Ruta Cullen, C., Hubberman, P., Kaslow, D.C., Brownlee, G. and Migeon, B.R. DNA Methylation of the Human Factor IX Locus: Implications for Transcription and X Chromosome Inactivation. In Preparation.
19. Stein, R., Sciaky-Galilli, N., Razin, A. and Cedar, H. (1983). Proc. Natl. Acad. Sci. U.S.A. 80, 2422-2426.
20. Reynolds, G.A., Basu, S.K., Osborne, T.F., Chin, D.J., Gil, G., Brown, M.S., Goldstein, J.L. and Luskey, K.L. (1984). Cell 38, 275-285.
21. Bird, A., Taggart, M., Frommer, M., Miller, O.J. and Macleod, D. (1985). Cell 40, 91-99.
22. Vogelstein, B. Fearon, E., Hamilton, S.R. and Feinberg, A.P. (1985). Science 227, 642-645.
23. Migeon, B.R., Wolf, S.F., Axelman, J., Kaslow, D.C. and Schmidt, M. (1985). Proc. Natl. Acad. Sci. U.S.A. 82, 3390-3394.
24. Migeon, B.R., Schmidt, M., Axelman, J., Cullen, C.R. (1986). Proc. Natl. Acad. Sci. U.S.A. In Press.
25. West, J.D., Frels, W.I., Chapman, V.M. and Papaioannou, V.E. (1977). Cell 12, 873-882.
26. Kratzer, P.G. and Chapman, V.M. (1981). Proc. Natl. Acad. Sci. U.S.A. 78, 3093-3097.

POLYCLONAL TUMORS

David Linder

Department of Pathology, University of Oregon Medical School, Portland, OR

Stanley M. Gartler

Departments of Medicine and Genetics and the Center for Inherited Diseases, University of Washington, Seattle, WA

INTRODUCTION

The question of tumor cell origin is one of long standing with theories ranging from the so-called field theory or critical mass to single cells altered by mutation. One way of analyzing such a developmental problem is to utilize cell mosaics as a cell tracer method (1). X-inactivation-derived mosaics are especially useful for such applications (2). An individual heterozygous for an X-linked gene expresses only one allele in any one somatic cell. Either the maternal or paternal allele can be expressed in a cell, and the pattern of expression is fixed within the cell's somatic heredity. Consequently, any somatic event that follows the embryological initiation of X-inactivation may be traced through the mosaicism. For example, a tumor originating in a single cell of an individual heterozygous for the X-linked electrophoretic glucose-6-phosphate dehydrogenase (G6PD) variation would exhibit a single G6PD phenotype (monoclonal), whereas a tumor starting from a large number of cells could have a mixed G6PD phenotype (polyclonal). Over the last 20 years a number of different tumors have been studied by these methods and a general picture of tumor monoclonality has emerged (2-5). Considering the mechanisms that are presumed to play a major role in tumor initiation and progression (eg. mutations, oncogene activation), monoclonality is the expected result. However, some polyclonal tumors have been reported during this period (6, 7), and the purpose of this paper is to draw attention to these cases and consider the possible significance of polyclonal growth in the early stages of tumorigenesis.

POLYCLONAL GROWTH AS AN INITIAL STAGE IN HEREDITARY TUMORS

The first demonstration of a polyclonal tumor using cell mosaics was for hereditary trichoepitheloma (6). Susceptibility to development of this neoplasm is transmitted as a simple dominant. A family segregating for this tumor and the Mediterranean variant of G6PD deficiency was investigated, and it was shown that the mosaic composition of the tumor was the same as that of adjoining normal tissue; that is, both cell types of the mosaic were identified. Later, another hereditary tumor, multiple neurofibromatosis, was studied in G6PD electrophoretic variant heterozygotes and again the mosaic composition of the tumors was found to be the same as the adjacent nontumorous tissue (7). More recently, a third dominantly transmitted susceptibility to tumors, Gardner's Syndrome, has been studied in G6PD electrophoretic variant heterozygotes and again the tumors were all polyclonal (8). Since these are all benign tumors, the polyclonal growth may represent an initial or early stage in the development of these hereditary tumors. Furthermore, it is possible that malignant progression in these cases could be of a monoclonal nature, and in one case of hereditary neurofibromatosis, a neurofibrosarcoma was monoclonal while the benign fibromas were polyclonal (9). A fourth hereditary tumor, inherited medullary carcinoma, has also been studied in subjects heterozygous for G6PD electrophoretic variants. Several individual carcinomas were shown to have single G6PD phenotypes; however, a single portion of a large tumor was shown to be polyclonal (10, 11). The authors of this work interpret the mixed phenotype tumor as representing different single cell carcinomas. However, the mixed phenotype tumor can just as well be interpreted as the remnants of an initial polyclonal growth. Furthermore, there is evidence in medullary thyroid carcinoma for an early c-cell hyperplasia preceding the appearance of carcinomas (12). In either event it is possible that a polyclonal hyperplasia may precede the appearance of malignant tumors.

Although these hereditary tumors may form a consistent group illustrating an initiating polyclonal phase in tumor progression, it does not follow that all hereditary tumors will exhibit this pattern. Retinoblastoma represents a hereditary tumor which has been studied very intensively in recent years. It seems clear that rare chromosomal events underlie the appearance of malignant tumors, and though not proven by definitive methods, it seems almost certain that these tumors will be monoclonal (13). It is conceivable that some form of early hyperplastic growth in the form of retinomas (14) may precede the tumors, but for the present, retinoblastoma must be considered as an example of a hereditary tumor which initiates as a monoclonal malignant form.

The presence of an initial polyclonal stage in hereditary tumors may not appear surprising as every cell in the body carries the susceptibility mutation. One can imagine an altered sensitivity to a growth-promoting substance; however, what determines the focal nature of the tumor is not at all apparent.

POLYCLONAL MALIGNANT TUMORS

Polyclonal tumors are not only found as initiating benign forms but also occur in malignant cases. The first such case reported was a G6PD heterozygote with primary carcinoma of the colon and a number of metastases (15). The primary tumor exhibited a double G6PD phenotype while the various metastatic nodules exhibited single G6PD phenotypes, either A or B. Carcinoma of the breast is also polyclonal as well as chronic lymphocytic leukemia (16). A most surprising apparent polyclonal tumor involves a leiomyosarcoma of the uterus (4). The primary tumor in the uterus could not be analyzed but several metastatic sites were studied, and these (pancreas, chest, pelvis) showed both A and B cells while a fourth metastasis in the lung showed only B cells. Since leiomyomas are clonal these results suggest that leiomyosarcomas do not progress in a direct way from a leiomyoma. It is of interest to point out that necrotic leiomyomas are usually polyclonal and it is conceivable that they represent incipient leiomyosarcomas.

When known carcinogenic agents have been used and when host susceptibility is not critical, it might be expected that the titer of the carcinogen would determine whether the tumor was of single cell or multicellular origin. In natural viral tumors like Burkitt's the tumors are monoclonal (18), whereas in experimentally induced tumors like herpesvirus saimiri or EBV lymphomas the tumor may be multiclonal (19). Chemically induced tumors may also be polyclonal as shown with X inactivation mosaic systems in experimental mice, and in these cases the polyclonality may remain over a period of several tumor transfers (20).

However, the polyclonal tumors we have discussed are not induced by any obvious agent and their origins are not necessarily the same. A "field effect" may be operative in causing hereditary neoplasm such as neurofibroma, and, consequently, it is not likely that any infrequent event like oncogene activation, somatic mutation or chromosome loss could be directly involved in this kind of mass change. Also surprising is the maintenance of the polyclonal state over many cell generations of malignant tumor cells as one would expect tumor evolution to always lead to a monoclonal state. It seems possible that cell recruitment could play an important role in the development of some of these malignant polyclonal tumors. Another possibility is that neoplastic growth is enhanced because of its mixed cell population by a process similar to complementation.

LEIOMYOMAS OF THE UTERUS

As mentioned earlier most tumors show a monoclonal pattern; however, it is not clear in most cases whether the monoclonality represents an initiating stage or a late stage in tumor evolution. One of the few cases in which a monoclonal pattern appeared to clearly involve an incipient event was the leiomyoma of the uterus (2). Leiomyomas are benign tumors that usually occur multiply in the uterus. They are easily and clearly separated from the surrounding uterine wall and vary in size from a few millimeters to several centimeters in diameter. Several hundred of these tumors from over 50 subjects heterozygous for the G6PD electrophoretic variant have now been studied by ourselves and others, and the vast majority of specimens show single G6PD phenotypes, with both phenotypes being present in a single subject (Table I). A small number of double phenotype leiomyomas have been seen (~4%) and in most

TABLE I. Summary of leiomyoma data in G6PD heterozygotes.

# Cases	# Tumors	G6PDA	G6PDB	G6PDAB
65	360	184	162	14*

*3 necrotic

instances they can be explained away by necrosis or inclusion of capsular or other stromal elements in tumor extracts. The necrotic tumors may have intermediate mosaic compositions (A≅B), but the nonnecrotic mixed leiomyomas always show one major and one trace G6PD component. These latter cases are most likely monoclonal tumors with a trace of stromal contamination. A number of the single phenotype tumors have been dissected into sections representing 1% of the tumor volume, and these sections then subjected to electrophoretic analysis; in all instances the same single phenotype characteristic of the parent tumor was detected in the subsections. In contrast to single tumor phenotypes, adjacent normal myometrium samples as small as 1 mm^3 nearly always showed double phenotypes.

What mechanisms, other than single cell origin, might account for a single-phenotype tumor? It might be argued that a single-phenotype tumor could start from multiple cells simply by chance sampling of all cells of the same G6PD type. The chance of such an event would depend on the fine structure of the tissue mosaicism; the greater the variegation, or larger the patch size, the greater the chance that a number of adjacent cells will be of the same mosaic type. We have carried out an analysis of the data gathered on leiomyomas and can statistically exclude the possibility that the more than 300 tumors studied could all be single-phenotype tumors simply due to chance sampling of two or more adjacent cells of like type (20).

Another possible explanation for such tumors could be that through selective overgrowth, one cell type has predominated in a population originally mixed. Although it is true that it would take only a slight growth rate advantage for one cell type in a tumor of several centimeters in diameter to account for 95% of the tumor (a minor component of 5% or less would not be detectable), several arguments can be made against this possibility. Selection would predict a nonlinear relationship of tumor G6PD type to the A/B ratio of the myometrium. The curve would be S shaped rather than linear; the data, however, appear quite linear (Fig. 1). Further, if selection were the reason for the single-phenotype leiomyomas, one would always expect one tumor type (i.e., GdA or GdB) to predominate in a subject since selection would be acting on X chromosomes which carry the only known genetic differences between the cell types. However, this is not the case. Finally, in those cases where there is a majority tumor type, that tumor type should be larger than the minority type. Again the results are contrary to this prediction.

A more complex explanation for single-phenotype tumors could be that during the growth of the tumor there have been repeated growth cycles; that is, growth followed by extensive cell death, and then reseeding from a smaller group of cells to start the growth cycle again. This process would be analogous to genetic drift and, given sufficient number of cycles, could account for a single-phenotype tumor. This idea predicts that a significant fraction of early tumors should show mixed phenotypes. We have examined this possibility by carrying out microdissection of a number of tumors, especially the core, to see if we can find any trace of a second G6PD component. All such tumors have only had a single G6PD phenotype, and we conclude that drift is not a likely explanantion of the single G6PD phenotypes of leiomyomas.

The leiomyoma of the uterus would appear to be a near perfect example of a tumor originating from a single cell. In

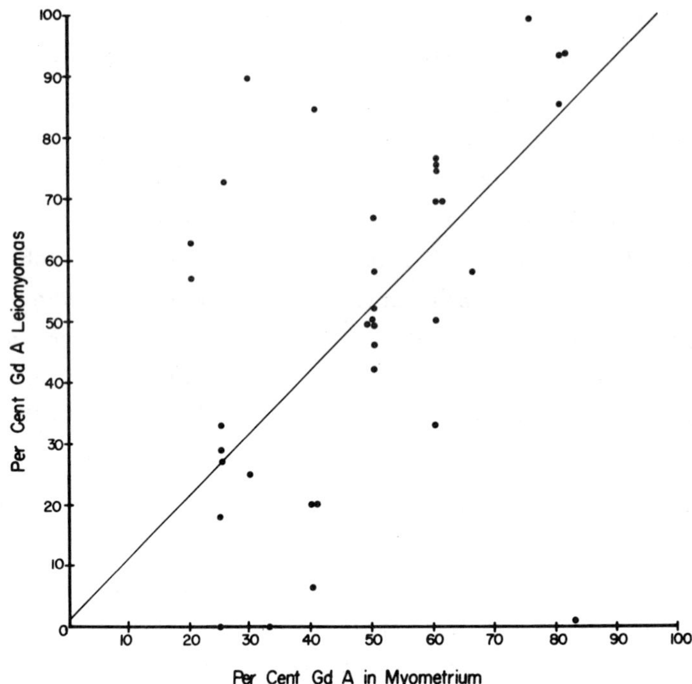

FIGURE 1. Comparison of the number of myomas with the percent of A myomas within the same uterus.

fact, with respect to the well defined structure of this tumor, we believe this to be the case. However, one aspect of the relationship between myometrial and tumor G6PD composition leads us to consider the possibility of myometrial hyperplasia preceding the initiation of single cell myomas.

There is a fairly linear relationship between the A:B cell ratio of the myometrium and the A:B cell ratio of the tumors (Fig. 1). The regression coefficient (b) of these values is 0.96 ($\bar{y} = a + b\ x = 0.09 + 0.96\ \bar{x}$) which suggests that the A:B tumor ratio in some way reflects the A:B cell ratio of the parent myometrial population. However, the tumor ratios tend to be more extreme than the myometrial values. This effect is shown graphically in Fig. 2 where the A:B tumor ratios and their absolute numbers are recorded. A peculiarly flat distribution is seen which is due to the excess of extreme

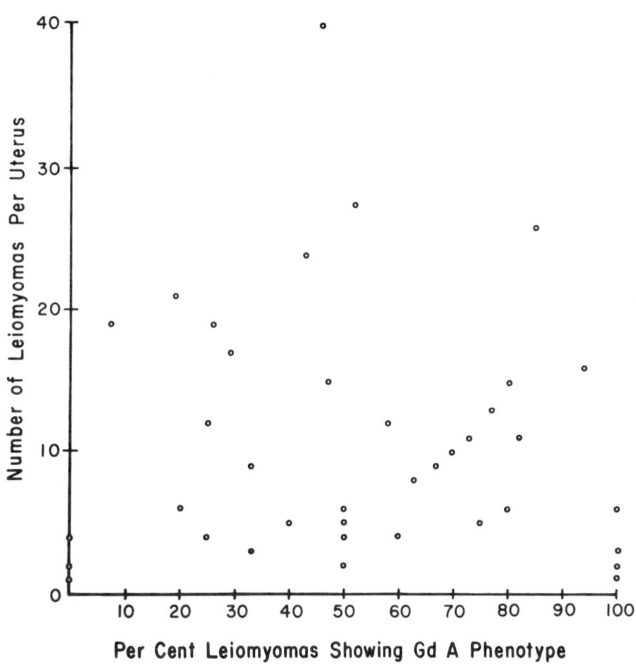

FIGURE 2. Comparison of the A:B cell ratio expressed as percent GdA of the myometrium with the percent of GdA myomas in the same uterus.

cases. By applying a cumulative chi square analysis to the data, the A:B tumor ratio and the A:B cell ratio of the myometrium were shown to be significantly different.

One possible explanation of the extreme tumor ratios is that a localized hyperplasia of the myometrium could precede the initiation of leiomyomas. In the case of equal A/B myometrial ratios, the tumor A/B ratios would closely approximate the myometrial composition. However, where the resultant myometrial

A/B ratio is extreme then the chance of a majority type tumor could be greater than the proportion of that G6PD type cell in the myometrium. In fact, the myometrial A/B distribution is somewhat broader than would be expected from observations of A/B distributions observed in other tissues and in newborn myometrium. If the myometrium or some other precursor cell in tumor cases does go through a hyperplastic stage, then it might be detectable as a target in tumor foci transfection studies.

SUMMARY

We have pointed out that polyclonal growth plays a definite role in the origin of benign hereditary tumors and various malignant tumors. We further point out that the classic monoclonal leiomyoma of the uterus may be preceded by a localized hyperplasia of the myometrium. Since monoclonal stages in tumor evolution are nearly always expected, it seems possible that an early polyclonal stage in tumorigenesis may be fairly general.

ACKNOWLEDGMENT

This work was supported in part by National Institutes of Health Grant GM15253. S.M.G. is the recipient of a NIH Research Career Award.

REFERENCE

1. Sturtevant, A. H. (1929). Z. Wiss. Zool. 135, 323.
2. Linder, D. and Gartler, S. M. (1965). Science 150, 67.
3. Fialkow, P. J., Gartler, S. M. and Yoshida, A. (1967). Proc. Natl. Acad. Sci. USA 58, 1468.

4. Linder, D. (1969). Proc. Natl. Acad. Sci. USA 63, 699.
5. Fialkow, P. J. (1980). Cold Spring Harbor Conf. Cell Prolif. 7, 1171.
6. Gartler, S. M., Ziprkowski, L., Krabowski, A., Ezra, R., Szeinberg, A. and Adam, A. (1966). Amer. J. Hum. Genet. 18, 282.
7. Fialkow, P. J., Sagebiel, R. W., Gartler, S. M. and Rimoin, D. L. (1971). New Eng. J. Med. 284, 298.
8. Hsu, S. H., Luk, G. D., Krush, A. J., Hamilton, S. R. and Hoover, H. H. (1983). Science 221, 951.
9. Friedman, J. M., Fialkow, P. J., Greene, C. L. and Weinberg, M. W. (1982). J. Natl. Cancer Inst. 69, 1289.
10. Baylin, S. B., Gann, D. S. and Hsu, S. H. (1976). Science 193, 321.
11. Baylin, S. B., Hsu, S. H., Gann, D. S., Smallridge, R. C. and Wells, S. A. (1978). Science 199, 429.
12. Wolfe, H. J., Melvin, K. E. W., Cervi-Skinner, G. J., Saadi, A. A. A.. Juliar, J. F., Jackson, C. E. and Tashjian, A. H. (1973). New Eng. J. Med. 289, 437.
13. Cavenee, W. K., Dryja, T. P., Phillips, R. A., Benedict, W. F., Godbaut, R., Gallie, B. L., Murphie, A. L., Strong, L. C. and White, R. L. (1983). Nature 305, 779.
14. Gallie, B. L., Ellsworth, R. M., Abramson, D. H. and Phillips, R. A. (1982). Br. J. Cancer 45, 513.
15. Beutler, E., Collins, Z. and Irwin, L. G. (1967). New Eng. J. Med. 276, 389.
16. McCurdy, P. R. (1968). In Hered. Disorders of Erythrocyte Metab. (Beutler, E., ed.), New York, Grune & Shatton, 121.
17. Fialkow, P. J., Klein, G., Gartler, S. M. and Clifford, P. (1970). Lancet 1, 384.
18. Marczynska, B., Falk, L., Wolfe, L. and Deinhardt, F. (1973). J. Natl. Cancer Inst. 50, 331.
19. Reddy, L. A. and Fialkow, P. J. (1979). J. Exp. Med. 150, 878.
20. Linder, D. and Gartler, S. M. (1966). Proceed. 5th Berkeley Symp. Math. Stat. and Prob. 625.

G-6-PD AS A MARKER FOR TUMORS[1,2]

Ernest Beutler, M.D.

Scripps Clinic & Research Foundation

Department of Basic and Clinical Research

10666 North Torrey Pines Road

La Jolla, California 92037

[1]This is publication #4246 BCR from the Research Institute of Scripps Clinic.

[2]This work is supported in part by Grant #HL 25552 from the National Institutes of Health.

How does a tumor begin? Does it start with one single cell that has gone awry? Does this single cell serve as progenitor of all tumor cells in neoplasm? Or are many cells transformed by a tumorogènic stimulus, be it a chemical or a virus, so that the neoplasm as seen clinically has multiple parentage? These questions are important in our conceptualization of how neoplasms arise, and ultimately in planning strategies for their management. The use of glucose-6-phosphate dehydrogenase (G-6-PD) as a marker for tumors in females has proven to be very useful in helping to answer some of these questions.

The principle upon which this strategy is based is simple and quite straight forward. X-inactivation appears to be fixed in early embryogenesis. After this time only one of the two copies of the G-6-PD gene are expressed by each female cell and by its progeny. If a tumor arises as a single cell, i.e. has a unicellular origin, then only one of the two G-6-PD alleles should be expressed in the tumor. If, on the other hand, more than one cell undergoes malignant transformation the products of both alleles would be found in neoplasms.

This approach to the origin of tumors was first conceptualized by Linder and Gartler (1), who used this strategy as a means of studying the origin of uterine leiomyomas. Investigating black females heterozygous for G-6-PD A and G-6-PD B, they found that uterine myomas, even when very large, manifested only a single G-6-PD phenotype. In contrast, even very small pieces of intervening myometrium could be demonstrated to contain both G-6-PD A and G-6-PD B. This suggested to Linder and Gartler that uterine myomas arose from a single cell.

Subsequently, we (2) extended these investigations to patients with malignant tumors. In a black woman heterozygous for G-6-PD A and B, we found that the primary tumor and metastases from a carcinoma of the colon manifested both the G-6-PD A and G-6-PD B phenotype. In contrast, in a patient with a lymphoma only a single G-6-PD phenotype was found. Over the next two decades since these early studies utilizing G-6-PD as a marker for tumors were published, numerous additional investigations were made (3-11). In this chapter we shall review some of the information which has been gained by the use of this approach, some of the constraints which might limit the usefulness of this technique and some of the alternative approaches which have become available.

Tables I and II summarize many of the reported studies of patients with various types of tumors. A fairly uniform picture emerges: Acquired tumors tend to be unicentric in origin while hereditary tumors, specifically neurofibromata, trichoepithelioma, and hereditary polyposis of the colon tend to be multiple. To a large extent these findings might have been regarded predictable. Chromosomal markers had already made it quite clear that certain forms of leukemia were clonal in origin. Even the fact that erythroid series was involved in the clonal disorder, chronic granulocytic leukemia, had been deduced from chromosomal studies (12). However, the fact that the lymphoid cells were part of the same clone in chronic granulocytic leukemia was surprising (13). Moreover, the involvement of erythroid cells in the etiology of acute granulocytic leukemia is far less obvious; we were able to demonstrate that this was the case by studying the peripheral blood cells of a patient who fortuitiously had one son with a rapidly moving G-6-PD and one son with a slowly moving G-6-PD (5). Since the G-6-PD phenotype of neither this patient nor any other patient with acute leukemia was determined before development of the neoplasm one cannot be certain that only one clone is not expressed even under normal circumstances (14). Indeed, this is not an altogether rare occurrence in normal individuals (15). However, a report of three such cases makes it much more likely that all cell lineages are, indeed, involved in acute leukemia.

TABLE I

Single or multiple cell origin of hematologic disorders determined with glucose-6-phosphate dehydrogenase (G-6-PD) markers.

	Single Cell	Multiple Cells	References
Myeloproliferative disorders			
Chronic myelocytic leukemia	8	0	(3,16,17)
Polycythemia vera	2	0	(18)
Idiopathic myelofibrosis	1	0	(19)
Paroxysmal nocturnal hemoglobinuria	2	0	(20)
Acute myelogenous leukemia	3	0	(5,7,21)
Myelodysplastic syndrome	1	0	(22)
Lymphoproliferative disorders			(22)
Reticulum cell sarcoma	2	0	(3,23)
Burkitt lymphoma	45	1	(3,24)
Non-Hodgkin lymphoma	4	0	(2,3)

Multiple myeloma	2	0	(25,26)
Plasmacytoma	1	0	(23)
Chronic lymphocytic leukemia (B-cells)	2	0	(6)

TABLE II

Single or multiple cell origin of tumors determined with glucose-6-phosphate dehydrogenase (G-6-PD) markers

	Single Cell	Multiple Cells	References
Carcinoma			
Nasopharynx (anaplastic)	7	0	(3,23)
Cervix, pre-invasive	9	0	(27,28)
invasive	8	2	(27,28)
Adrenal cortex	1	0	(3)
Bladder	1	0	(3)
Colon	0	1	(2,29)
Kidney	1	0	(3)
Ovary	3	0	(3)
Palate	4	0	(3,23)
Thyroid	6	0	(3,4,23)
Vulva, Bowen's disease	1	0	(30)
Melanoma	2	0	(3,23)
Neuroblastoma	1	0	(23)
Nephroblastoma	2	0	(3)
Hereditary			
Neurofibroma	0	14	(31)
Trichoepithelioma	0	12	(9)
Viral			
Common wart	6	0	(32)
"Venereal" wart	0	4	(33)
Endocrine			
Solitary thyroid adenoma	22	0	(3,23)
Pheochromocytoma	1	0	(4)
Miscellaneous benign			
Leiomyoma of uterus	184	0	(1,26)
Lipoma	6	0	(3,26)
Salivary gland adenoma	2	0	(3,23)

Ovarian teratoma	39	0	(26,34,35)
Neurofibroma (sporadic)	2	0	(3)

The most serious limitation in the use of G-6-PD phenotypes as markers for the origin of tumors is the fact that the A/B polymorphism is common only among those of African origin. Here some 40% of females may prove to be appropriate for study. Although it also occurs as very high polymorphic frequency, the Mediterranean variant is much less simple to use for this purpose, since it is electrophoretically normal. However, Ferraris, et al (36) have ingeniously used the increased utilization of 2-deoxy-glucose-6-phosphate by this mutant enzyme to distinguish it histochemically from the normal type B enzyme. The recent appreciation of restriction length polymorphisms involving X-linked genes is very promising in this respect. The existence of a retriction polymorphism alone is not enough, however. What is needed, in addition, is a region in which under- or over-methylation clearly distinguishes the active from the inactive X-chromosome. Such a polymorphism has been described in the region of the HGPRT gene (37). The frequency of this polymorphism is such that studies and 27% of females with tumors would be useful. Polymorphisms of this type, detected with G-6-PD probes that are becoming available may extend even further the usefulness of the G-6-PD locus.

REFERENCES

1. Linder D, Gartler SM: Glucose-6-phosphate dehydrogenase mosaicism: Utilization as cell marker in the study of leiomyomas. Science 150:67-69, 1965.

2. Beutler E, Collins Z, Irwin LE: Value of genetic variants of glucose-6-phosphate dehydrogenase in tracing the origin of malignant tumors. N Engl J Med 276:389-391, 1967.

3. Fialkow PJ: Clonal origin of human tumors. Biochem Biophys Acta 458:283-321, 1976.

4. Baylin SB, Ganns DS, Hsu SH: Clonal origin of inherited medullary thyroid carcinoma and pheochromocytoma. Science 193:321-323, 1976.

5. Beutler E, West C, Johnson C: Involvement of the erythroid series in acute myeloid leukemia. Blood 53:1203-1205, 1979.

6. Fialkow PJ, Reddy AL, Najfeld V, Singer J: Chronic lymphocytic leukemia: Clonal origin in a committed β-lymphocyte progenitor. Lancet 2:444-445, 1978.

7. Wiggans RG, Jacobson RJ, Fialkow PJ, Woolley PV, Mac Donald JS, Schein PS: Probable clonal origin of acute myeloblastic leukemia following radiation and chemotherapy of colon cancer. Blood 52:659-663, 1978.

8. Hsu S, Luk GD, Krush AJ, Hamilton SR, Hoover Jr HH: Multiclonal origin of polyps in Gardner Syndrome. Science 221:951-953, 1983.

9. Gartler SM, Ziprkowski L, Krakowski A, Ezra R, Szeinberg A, Adam A: G-6-PD mosaicism as a tracer in the study of hereditary multiple trichoepithelioma. Am J Hum Genet 18:282-287, 1966.

10. Fialkow PJ, Klein G, Gartler SM, Clifford P: Clonal origin for individual Burkitt tumours. Lancet 1:384-386, 1970.

11. Fialkow PJ: The origin and development of human tumors studied with cell markers. N Engl J Med 291:26-35, 1974.

12. Trujillo JM, Ohno S: Chromosomal alteration of erythropoietic cells in chronic myeloid leukemia. Acta Haematol 29:311-316, 1963.

13. Martin PJ, Najfeld V, Hansen JA, Penfold GK, Jacobson RJ, Fialkow PJ: Involvement of the B-lymphoid system in chronic myelogenous leukaemia. Nature 287:49-50, 1980.

14. Beutler E: Clonal remission in acute leukemia. N Engl J Med 311:922-923, 1984.

15. Gandini E, Gartler SM, Angione G, Argiolas N, Dell Acqua G: Developmental implications of multiple tissue studies in glucose-6-phosphate dehydrogenase-deficient heterozygotes. Proc Natl Acad Sci USA 61:945-948, 1968.

16. Fialkow PJ, Gartler SM, Yoshida A: Clonal origin of chronic myelocytic leukemia in man. Proc Natl Acad Sci USA 58:1468-1471, 1967.

17. Barr RD, Fialkow PJ: Clonal origin of chronic myelocytic leukemia. N Engl J Med 289:307-09, 1973.

18. Adamson JW, Fialkow PJ, Murphy S, Prchal JF, Steinmann L: Polycythemia vera: Stem-cell and probable clonal origin of the disease. N Engl J Med 295:913-916, 1977.

19. Jacobson RJ, Fialkow PJ: Idiopathic myelofibrosis: Stem cell abnormality and probable neoplastic origin. Clin Res 24:439A, 1976.

20. Oni SB, Osunkoya BO, Luzzatto L: Paroxysmal nocturnal hemoglobinuria: Evidence for monoclonal origin of abnormal red cells. Blood 36:145-152, 1970.

21. Jacobson RJ, Temple MJ, Singer JW, Raskind W, Powell J, Fialkow PJ: A clonal complete remission in a patient with acute nonlymphocytic leukemia originating in a multipotent stem cell. N Engl J Med 310:1513-1517, 1984.

22. Perona G, Guidi GC, Tummarello D, Mareni C, Battistuzzi G, Luzzato L: A new glucose-6-phosphate dehydrogenase variant (G-6-PD Verona) in a patient with myelodysplastic syndrome. Scand J Haematol 30:407-414, 1983.

23. Fialkow PJ, Martin GM, Klein G, Clifford P, Singh S: Evidence for a clonal origin of head and neck tumors. Int J Cancer 9:133-142, 1972.

24. Fialkow PJ, Klein E, Klein G, Clifford P, Singh S: Immunoglobulin and glucose-6-phosphate dehydrogenase as markers of cellular origin in Burkitt lymphoma. J Exp Med 138:89-102, 1973.

25. McCurdy PR: The genetics of glucose-6-phosphate dehydrogenase deficiency. In: Hereditary disorders of erythrocyte metabolism. Beutler E (eds.): pp. 121-124. Grune & Stratton, New York, NY 1968.

26. Linder D: Gene loss in human teratomas. Proc Natl Acad Sci USA 63:699-704, 1969.

27. Park I-J, Jones HW Jr: Glucose-6-phosphate dehydrogenase and the histogenesis of epidermoid carcinoma of the cervix. Am J Obstet Gynecol 102:106-109, 1968.

28. Smith JW, Townsend DE, Sparkes RS: Genetic variants of glucose-6-phosphate dehydrogenase in the study of carcinoma of the cervix. Cancer 28:529-532, 1971.

29. Burch HB, Bessey OA, Lowry OH: Fluorometric measurements of riboflavin and its natural derivatives in small quantities of blood serum and cells. J Biol Chem 175:457-470, 1948.

30. Smith JW, Townsend DE, Sparkes RS: Glucose-6-phosphate dehydrogenase polymorphism: A valuable tool to study tumor origin. Clin Genet 2:160-162, 1971.

31. Fialkow PJ, Sagebiel RW, Gartler SM, Rimoin DL: Multiple cell origin of hereditary neurofibromas. N Engl J Med 284: 298-300, 1971.

32. Murray RF, Hobbs J, Payne B: Possible clonal origin of common warts (verruca vulgaris). Nature 232:51-52, 1971.

33. Friedman JM, Fialkow PJ: Viral "tumorigenesis" in man: Cell markers in condylomata acuminata. Int J Cancer 17:57-61, 1976.

34. Linder D, Power J: Further evidence for post-meiotic origin of teratomas in the human female. Ann Hum Genet 34:21-30, 1970.

35. Linder D, Kaiser McCaw B, Hecht F: Parthenogenic origin of benign ovarian teratomas. N Engl J Med 292:63-66, 1975.

36. Ferraris AM, Giuntini P, Galiano S, Gaetani GF: 2-Deoxy-glucose-6-phosphate utilization in the study of glucose-6-phosphate dehydrogenase mosaicism. Am J Hum Genet 33:307-313, 1981.

37. Migeon BR, Kaloustian V, Nyhan WL, Young WJ, Childs B: X-linked hypoxanthine-guanine phosphoribosyl transferase deficiency: Heterozygote has two clonal populations. Science 160:425-427, 1968.

GLUCOSE-6-PHOSPHATE DEHYDROGENASE AND MOSAIC ANALYSIS OF HUMAN ATHEROSCLEROTIC LESIONS[1]

Earl P. Benditt

Department of Pathology,
University of Washington,
Seattle, Washington, U.S.A.

I. INTRODUCTION

Recognition of the importance of smooth muscle cell proliferation as a part of the pathologic process involved in atherosclerosis (Moss and Benditt, 1970; Ross and Glomset, 1976) raised questions about the generation of this proliferative component of the atherosclerotic plaque. The work of Linder and Gartler (1965) on the smooth muscle tumors of the uterus suggested a means of examining the question of clonality of the vascular lesions in human beings. The clear cut finding of clonality (Benditt and Benditt, 1973; Benditt, 1976; Benditt and Gown, 1980) in many of the lesions raised a further set of questions relating to development and maintenance of artery walls. I will discuss these and the current state of the analysis of human atherosclerotic lesions.

---

[1]Support for the research reported was provided by National Institutes of Health grants HL-03174 and HL-07312.

Table I
Cell Populations in Arteries of a Newborn Infant

Artery	No. of Samples	% "A" Enzyme
Thoracic aorta	64	68.8 ± 3.6
Abdominal aorta	64	65.8 ± 4.9
Umbilical	40	77.5 ± 3.4

A. The Nature of Cell Populations in Arteries at Birth

In the development of vessels in the embryo, endothelial tubes appear first and then there is recruitment of the smooth muscle coat from the surrounding mesenchymal cell population. This being so, the vessels developing from mesenchyme in different regions of the embryo should show cell population mixtures reflecting those of the local mesenchyme.

In Table I are shown the data obtained from the arteries of a 1-week-old newborn. The population of cells in the thoracic and abdominal segments of the aorta are identical; whereas it is evident that the umbilical artery wall is distinctly different (significant at <0.001), indicating that differences in cell population mixtures do occur in certain tissues during embryonic development.

B. The Character of Cell Populations in Major Arteries of Adults and Changes with Age

We were able to obtain from autopsy specimens multiple samples of non-atherosclerotic thoracic and abdominal aortas in a series of glucose-6-phosphate dehydrogenase (G-6-PD) heterozygotes ranging in age from 1 week to 91 years. The data in Table II show the following: 1) There is no significant difference in

Table II
G-6-PD Mixtures in Aorta Wall: Variation with Age and Site

Age	Thoracic		Abdominal	
Years	Number of Samples	%A Mean ± S.D.	Number of Samples	%A Mean ± S.D.
0.02	64	69 ± 3.6	64	66 ± 4.9
16	284	31 ± 5.7	116	34 ± 10.4
29	25	44 ± 8.2	25	27 ± 8.4
48	25	47 ± 9.3	25	58 ± 11.6
56	74	42 ± 8.0	23	33 ± 15.0
91	--	--	16	38 ± 10.0

population mixtures between thoracic and abdominal aortae in the same person irrespective of age. 2) The abdominal aorta when compared to the thoracic aorta shows a small but distinctly larger variation, as indicated by the standard deviation (mean coefficient of variation ratio A/T = 1.6). 3) In this series no regular increase in the variation with age is evident, except for the fact that the infant vessel population mixture shows less than half the standard deviation of the adult vessel. These data can be interpreted to mean that the artery wall cell population in spite of demands for local repair with age does not show any evidence of either population drift or increased variation which might be expected to occur if there were significant cell selection.

C. Primordial Cell Pool Size for Arteries

Data on a set of individuals drawn from a larger population on the basis of sufficient satisfactory sampling are shown in Table III. The calculations for "primordial" pool size based upon simple binomial statistics are given. The apparent

Table III
Distribution of G-6-PD (A:B) Ratios of Thoracic Aorta and Estimate of Primordial Pool Size

Age	19	25	31	54	62	79	91
"A" Fraction	.67	.36	.52	.27	.65	.42	.38

Mean = 0.43 Variance = 0.0267

$$\text{Primordial Cell Pool} = \frac{0.43\ (1 - 0.43)}{0.0267} = 9$$

95% Confidence Limits 2 - 21

primordial pool size of 9 cells is similar to that found for other tissues by Fialkow (1973).

D. Analysis of Atherosclerotic Plaques

Pearson and co-workers in Baltimore and Lee and Thomas and co-workers in Albany have extended the data on atherosclerotic plaques. The data of Pearson _et al._ (1978a, b) are much like those we originally presented (Benditt and Benditt, 1973). The data of the Albany group likewise reveals that a lesser but significant population of lesions have a clonal character. Lee _et al._ (1981, 1985) and Janakidevi _et al._ (1984) have suggested that a selection process could be at work based on experiments with experimental crosses of two species of hares having G-6-PD enzymes of different mobilities. Thomas _et al._ (1979) and Lee _et al._ (1980) suggested that a similar phenomenon is seen in human lesions based upon an apparent bias in favor of one G-6-PD phenotype in the plaque cell population of individual patients. They presented data on 25 cases; the lesion populations in 20 out of the 25 cases fall within statistical limits of expectation by chi

Table IV
Phenotypes of Normal Vessel Wall and Fibrous Caps of Atherosclerotic Plaques

Ratio A:B	Number of Sites Showing Enzyme Phenotype					
	Normal			Plaque		
	AB	A	B	AB	A	B
1:1	25	0	2	3	4	8
1:3	7	0	0	2	1	6
1:2	12	0	0	1	3	2
2:3	74	0	0	4	13	10
2:3	16	0	0	1	11	0
1:3	56	0	0	0	0	3
1:1	19	0	0	3	5	7
Total	209	0	2	14	37	36

square analyses. In Table IV are shown data on plaque phenotype from a group of our cases with multiple plaques assayed. One out of the seven cases shows an extreme A bias in the single phenotype plaques that is beyond statistical expectation. Thus the large majority of their cases and ours fit within the limits of

Table V
Blood-Derived Cells in Human Atherosclerotic Plaques

	Monocytes/ Macrophages %	T Lymphocytes %	B Lymphocytes %	Total Non-SMC %
Plaque	29 ± 10	11 ± 6	1 ± 1	41 ± 12
Adjacent intima	37 ± 12	8 ± 4	4 ± 3	49 ± 13

Specimens from 16 patients; 1,200 cells counted per specimen. % = Total cell population in carotid endarterectomy specimen. Data modified from L. Jonasson *et al*., J. Clin. Invest. (1985) *76*, 125.

expectation based on the cell population mixtures in the uninvolved vessel. All three groups who have investigated the problem have reported a significant population of atherosclerotic plaques with mixed cell phenotypes. Some of the reasons for this are now emerging as discussed in the next section.

E. Varieties of Cells Populating Human Atherosclerotic Plaques

Atherosclerotic plaques are composed of several different cell types: The relative quantities of smooth muscle and blood-derived cells in carotid artery plaques have been enumerated using immunocytochemical markers (Jonasson et al., 1985; Gordon, personal communication). The data show (Table V) that the macrophages and T and B lymphocytes on the average represent nearly half the cell population. It is also evident that in some instances the blood-derived cell population can be much more than half of the total cell population in the plaque, a factor which could easily lead to "mixed" type plaques. Another feature relates to the distribution of the cells in the plaque. There is a distinct clustering of cells of individual types; there may be quite discrete masses of muscle cells, macrophages or of lymphocytes. The substantial non-smooth muscle cell population and the non-random distribution of different cell populations could well account for at least two of the features of the cell mixtures found by G-6-PD analysis of atherosclerotic plaques: 1) The fact that many plaques appear to be of mixed cell type and 2) the description by Pearson et al. (1978b) of regions of single phenotype and mixed G-6-PD phenotypes in the same plaque.

II. DISCUSSION

The ability to use the mosaic analysis with the X-linked G-6-PD A/B enzyme polymorphism provides a means of asking some cogent questions about development of vessels in human beings. The clonal proliferation found in human atherosclerotic plaques stands out clearly when viewed against the background of the stability of the cell population mixture in the non-atherosclerotic portions of the vessel wall. Recent enumeration of the blood-derived cell populations present in atherosclerotic plaques and the immunohistochemical studies show a strong focal character in the smooth muscle proliferation and a clustering of blood-derived cell populations in plaques. Such data provide a basis for understanding some of the noise in the analytic system and strengthen confidence in the clonal nature of the proliferative response of the smooth muscle cells in human atherosclerotic plaques and pursuit of the implications of this property of the lesions (Benditt and Gown, 1980; Benditt *et al*., 1983; Majesky *et al*., 1985).

ACKNOWLEDGMENTS

The author wishes to acknowledge the dedicated assistance and participation of Dr. Nils Eriksen, Dr. Allen Gown and Ronald Hanson in this work and Virginia Wejak for editorial assistance.

REFERENCES

Benditt, E. P. (1976) Beitr. Pathol. 158, 405.

Benditt, E. P., and Benditt, J. M. (1973) Proc. Natl. Acad. Sci. U.S.A. 70, 1753.

Benditt, E. P., and Gown, A. M. (1980) Int. Rev. Exp. Pathol. 21, 55.

Benditt, E. P., Barrett, T. B., and McDougall, J. K. (1983) Proc. Natl. Acad. Sci. U.S.A. 80, 6386.

Fialkow, P. J. (1973) Ann. Hum. Genet. 37, 39.

Janakidevi, K., Lee, K. T., Kroms, M., Imai, H., and Thomas, W. A. (1984) Exp. Mol. Pathol. 41, 354.

Jonasson, L., Hohn, J., Skalli, O., Gabbiani, G., and Hansson, G. K. (1985) J. Clin. Invest. 76, 125.

Lee, K. T., Janakidevi, K., and Thomas, W. A. (1980) Artery 8, 581.

Lee, K. T., Thomas, W. A., Janakidevi, K., Kroms, M., Reiner, J. M., and Borg, K. Y. (1981) Exp. Mol. Pathol. 34, 191.

Lee, K. T., Janakidevi, K., Kroms, M., Schmee, J., and Thomas, W. A. (1985) Exp. Mol. Pathol. 42, 71.

Linder, D., and Gartler, S. M. (1965) Science 150, 67.

Majesky, M. W., Reidy, M. A., Benditt, E. P., and Juchau, M. R. (1985) Proc. Natl. Acad. Sci. U.S.A. 82, 3450.

Moss, N. S., and Benditt, E. P. (1970) Lab. Invest. 23, 231.

Pearson, T. A., Dillman, J. M., Solez, K., and Heptinstall, R. H. (1978a) Am. J. Pathol. 93, 93.

Pearson, T. A., Dillman, J. M., Solez, K., and Heptinstall, R. H. (1978b) Circ. Res. 43, 10.

Ross, R., and Glomset, J. (1976) N. Engl. J. Med. 295, 369 and 420.

Thomas, W. A., Reiner, J. M., Janakidevi, K., Florentin, R. A., and Lee, K. T. (1979) Exp. Mol. Pathol. 31, 367.

MOLECULAR BIOLOGY OF HUMAN G6PD

STRUCTURE OF HUMAN GLUCOSE-6-PHOSPHATE DEHYDROGENASE[1]

Akira Yoshida and I-Yih Huang[2]

Department of Biochemical Genetics
Beckman Research Institute of the City of Hope
Duarte, California USA

Human G6PD was first purified to homogeneity from red blood cells in 1965 (1). Our knowledge of its molecular size, subunit structure, and primary structure has advanced gradually since then. The partial and tentative amino acid sequence was cited in previous publications (2,3). This paper summarizes the current status of our knowledge about the structure of normal human G6PD.

1) Amino acid sequence: For amino acid sequence determination, about 1.5 g of human G6PD was prepared from about 1400 l of outdated and fresh blood. To be adaptable to the large-scale preparation, the original purification method (1) was modified and simplified. In comparison with the original method, the yield of the enzyme was somewhat reduced, i.e., about 30% versus 50%, after about a 60,000-fold purification by the modified method. Homogeneity of the purified enzyme was checked by polyacrylamide gel electrophoresis in the presence and absence of sodium dodecylsulfate. Since individual blood was not typed in advance, the enzyme preparation could have contained some G6PD variants. Most of the blood came from blood banks in the Boston area, and about

[1]Supported by U.S. Public Health Service Grant HL-29515
[2]Present address: Smith-Kline, Palo Alto, California USA

500 l was supplied by the Amsterdam Central Blood Bank, The Netherlands. Considering the ethnic origin in these areas, more than 90% of the blood donors should be normal B^+ in the Gd locus, and the rest could be variant A^+ and A^-, and fewer could be Mediterranean B^- and other variants. The final enzyme preparation indicated a single enzyme band which matched the normal B^+ enzyme in the starch gel electrophoresis. Thus, the enzyme preparation represents normal human G6PD.

For amino acid sequence determination, the enzyme was dialyzed against water and lyophilized. The enzyme protein was reduced, S-carboxymethylated and digested by trypsin. Fifty-six tryptic peptides, ranging from one to 24 residues, were isolated by chromatography, with Agl-X2, CM-Sephadex, and DEAE-Sephadex; by gel filtration; by two-dimensional paper chromatography and electrophoresis; and by HPLC chromatography. These peptides were sequenced by the modified dansyl-PTH, double-monitored, Edman degradation procedure. The S-carboxymethylated enzyme was cleaved by cyanogen bromide treatment. CNBr peptides thus produced were separated by gel filtration, CM-cellulose chromatography, and HPLC chromatography. Amino acid sequences of isolated CNBr peptides were determined by a combination of automated and manual Edman degradation. Tryptic, chymotryptic, and thermolytic peptides of these CNBr peptides were also characterized. From these analyses, the amino acid sequence of the enzyme was deduced. The amino acid sequence of the enzyme and the summary of the evidence used to establish the structure are shown in Fig. 1A-B.

The sequence of about 150 amino acid residues of the NH_2-terminal region is shown in Fig. 1-A. There are some ambiguities in the structure of CNBr-2, since the alignment of several tryptic peptides is not established. The NH_2-terminal of human G6PD was found to be pyro Glu (4). Therefore, Pro cannot be the NH_2-terminal, and a small peptide with pyro Glu-X---Met probably attached to the terminal Pro of CNBr-1. The NH_2-terminal sequence of G6PD remains to be further clarified. The complete

CNBr-1

Pro-Arg-Ile-Asp-Ala-Asp-Leu-Lys-Leu-Asp-Phe-Lys-Asp-Val-Leu-Leu-Arg-Pro-Lys-Ser-

Ser-Leu-Lys-Ser-Arg-Ala-Glu-Val-Asp-Leu-Glu-Arg-Thr-Phe-Thr-Phe-Arg-Asn-Ser-Lys-

Glu-Thr-Tyr-Ser-Gly-Ile-Pro-Ile-Ile-Val-Ala-Asp-Met

CNBr-2

(Ala-Arg) , (Ser-Arg) , (X-Thr-Ile) (Gly-Ala-Ser-Gly-Asp-Leu-Ala-Lys)

(Leu-Phe-Arg-Asp-Gly-Leu-Leu-Pro-Glu-Asp-Thr-Phe-Ile-Val-Gly-Tyr)

Leu-Thr-Val-Ala-Asp-Ile-Asp-Lys-Gln-Ser-Glu-Pro-Phe-Phe-Lys-Ala-Thr-Pro-Glu-Glu-

Lys-Leu-Lys-Leu-Glu-Asp-Phe-Phe-Ala-Arg-Asn-Ser-Tyr-Val-Ala-Gly-Gln-Tyr-Asp-Ser-

Ala-Ala-Ser-Tyr-Gln-Arg-Leu-Asn-Ser-His-Met- -----------→ to CNBr-3

Figure 1. Summary of Evidence used to establish the amino acid sequence of human G6PD. —T— : tryptic peptides obtained from whole G6PD molecule; —t— : tryptic peptides obtained from individual CNBr peptides; ----c---- : chymotryptic peptides obtained from individual CNBr peptide.

A: Structure of CNBr peptide No. 1, which constitutes the NH_2-terminal region (about 25%) of G6PD. Since overlapping peptides were not obtained, the peptides in parenthesis are not aligned and the peptide chain is not completed.

B: Complete amino acid sequence of 390 amino acid residues which constitute about 75% of G6PD. * indicates the alignments deduced from cDNA sequence, not directly from the protein sequencing.

sequence of 390 amino acid residues of the COOH-terminal region is established (Fig. 1-B). It should be mentioned that the alignments of some peptide fragments (marked * in Fig. 1-B) were deduced from the knowledge of the cDNA sequence, since overlapping peptides were not identified and characterized.

CNBr-2 ←-------- |← CNBr-3
Asn-Ala-Leu-His-Leu-Gly-Ser-Gln-Ala-Asn-Arg-Leu-Phe-Tyr-Leu-

CNBr-3 →| ←CNBr-4
Ala-Leu-Pro-Pro-Thr-Val-Tyr-Glu-Ala-Val-Thr-Lys-Asn-Ile-His-Glu-Ser-Cys-Met-Ser-

Gln-Ile-Gly-Trp-Asn-Arg-Ile-Ile-Val-Glu-Lys-Pro-Phe-Gly-Arg-Asp-Leu-Gln-Ser-Ser-

Asp-Arg-Leu-Ser-Asn-His-Ile-Ser-Ser-Leu-Phe-Arg-Glu-Asp-Gln-Ile-Tyr-Arg-Ile-Asp-

CNBr-4 →|← CNBr-5 →|← CNBr-6
His-Tyr-Leu-Gly-Lys-Glu-Met-Val-Gln-Asn-Leu-Met-Val-Leu-Arg-Phe-Ala-Asn-Arg-Ile-

Phe-Gly-Pro-Ile-Trp-Asn-Arg-Asp-Asn-Ile-Ala-Cys-Val-Ile-Leu-Thr-Phe-Lys-Glu-Pro-

CNBr-6 →|
Phe-Gly-Thr-Glu-Gly-Arg-Gly-Gly-Tyr-Phe-Asp-Glu-Phe-Gly-Ile-Ile-Arg-Asp-Val-Met-

|← CNBr-7 →|← CNBr-8 →|← CNBr-9
Gln-Asn-His-Leu-Leu-Gln-Met-Leu-Cys-Leu-Val-Ala-Met-Gly-Lys-Pro-Ala-Ser-Thr-Asn-

Ser-Asp-Asp-Val-Arg-Asp-Glu-Lys-Val-Lys-Val-Leu-Lys-Cys-Ile-Ser-Glu-Val-Gln-Ala-

Asn-Asn-Val-Val-Leu-Gly-Gln-Tyr-Val-Gly-Asn-Pro-Asp-Gly-Glu-Gly-Glu-Ala-Thr-Lys-

Gly-Tyr-Leu-Asp-Asp-Pro-Thr-Val-Pro-Arg-Gly-Ser-Thr-Thr-Ala-Thr-Phe-Ala-Ala-Val-

Val-Leu-Tyr-Val-Glu-Asn-Glu-Arg-Trp-Asp-Gly-Val-Pro-Phe-Ile-Leu-Arg-Cys-Gly-Lys-

Ala-Leu-Asn-Glu-Arg-Lys-Ala-Glu-Val-Arg-Leu-Gln-Phe-His-Asp-Val-Ala-Gly-Asp-Ile-

Phe-His-Gln-Gln-Cys-Lys-Arg-Asn-Glu-Leu-Val-Ile-Arg-Val-Gln-Pro-Asn-Glu-Ala-Val-

CNBr-9 →| |← CNBr-10 →|← CNBr-11
Tyr-Thr-Lys-Met-Met-Thr-Lys-Lys-Pro-Gly-Met-Phe-Phe-Asn-Pro-Glu-Glu-Ser-Glu-Leu-

Asp-Leu-Thr-Tyr-Gly-Asn-Arg-Tyr-Lys-Asn-Val-Lys-Leu-Pro-Glu-Pro-Tyr-Glu-Arg-Leu-

CNBr-11 →|← CNBr-12
Ile-Leu-Asp-Val-Phe-Cys-Gly-Ser-Gln-Met-His-Pro-Val-Arg-Ser-Asp-Glu-Leu-Arg-Glu-

Ala-Trp-Arg-Ile-Phe-Thr-Pro-Leu-Leu-His-Gln-Ile-Glu-Leu-Glu-Lys-Pro-Lys-Pro-Ile-

CNBr-12 →|← CNBr-13
Pro-Tyr-Ile-Tyr-Gly-Ser-Arg-Gly-Pro-Thr-Glu-Ala-Asp-Glu-Leu-Met-Lys-Arg-Val-Gly-

Phe-Gln-Tyr-Glu-Gly-Thr-Tyr-Lys-Trp-Val-Asn-Pro-His-Lys-Leu-COOH

<u>Figure 1-B.</u>

2) Cysteine residues: The cysteic acid content of the acid hydrolysate of the performic acid-oxidized enzyme is 1.68%; i.e., 16 x 10^{-5} mole of half-cysteine residues per gram, which is compatible to seven cysteine residues per subunit counted from the amino acid sequence (1). The total sulfhydryl content of the enzyme determined by titration with p-chloromercuribenzoate in the presence of 8 M urea was found to be 17 x 10^{-5} mole per gram (5,6). Therefore, the fully active enzyme, prepared in the presence of 1 mM mercaptoethanol, has only cysteine residues and no inter- or intra- S-S bridges in the molecule. In the absence of urea, the -SH content was estimated to be 6.8~7.3 x 10^{-5} mole per gram, indicating that approximately 60% of the sulfhydryl groups of the enzyme are not exposed to react with the SH-reagent in the absence of urea (5,6).

The enzyme was inactivated by p-chloromercuribenzoate (for example, 85% inactivation by 0.5 mM p-chloromercuribenzoate at pH 7.0 and at 25°C for 2 hrs). The activity could be completely restored by dithioerythritol treatment, with or without NADP (6).

When the dialyzed enzyme was exposed to air at 25°C at pH 8.0 for 3 hrs in the absence of reducing reagents, about 70% of sulfhydryl groups were oxidized, but the enzyme remained almost fully active. The result indicates that not all sulfhydryl groups are essential for catalytic activity. When the enzyme was oxidized with 1.5% H_2O_2, a drastic reduction of enzyme activity ocurred, accompanied by the oxidation of sulfhydryl and methionyl groups (5,6).

3) Secondary Structure: The secondary structure of a protein can be predicted from its amino acid sequence. Several empirical methods have been proposed for estimating the probability of the secondary structures of given amino acid sequences. Fig. 2 shows a predicted secondary structure of the COOH-terminal region of human G6PD, which constitutes a functional domain of the enzyme.

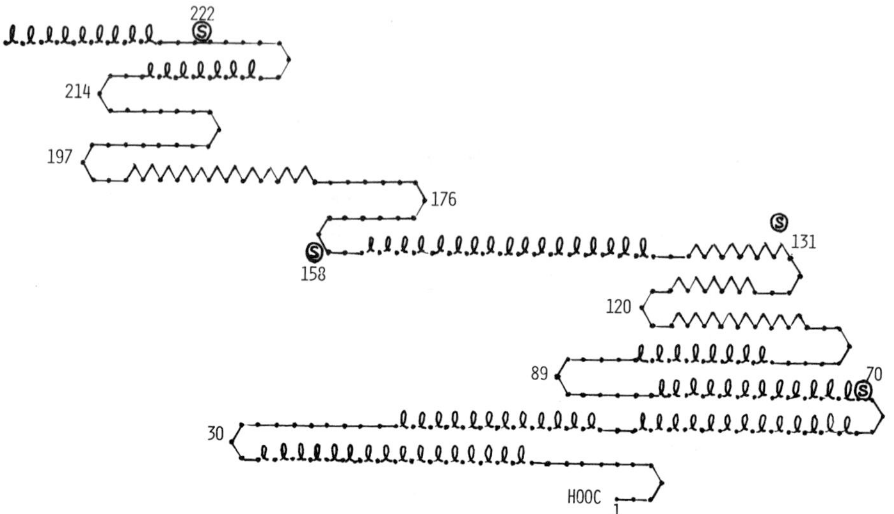

Figure 2. Predicted secondary structure of the COOH-terminal region which is implicated in catalytic activity. ℓ.ℓ.ℓ.ℓ : α-helix; ⋁⋁⋁ : β-sheet; ⋖ : β-turn; •—•—• : other structure. Dots (•) indicate amino acid residues. S : Cysteine residue. (Computer analysis done by Drs. T. Samejima and H. Kaji.)

The computer analysis was carried out by Drs. Samejima and Kaji, Aoyama University, Tokyo, Japan, based on the method of Chou and Fasman (7). It is readily noticeable that parallel β-strands, which commonly exist in nucleotide-binding enzymes (8), are located in the catalytic domain.

4) Oligomeric Structure and Micro Heterogeneity: The human G6PD assumes different oligomeric forms, depending upon protein concentrations, pH, and ionic strength. It assumes predominately tetrameric form at a neutral pH and a high ionic strength (>0.1), and it is predominately dimeric at an alkaline pH (pH 8 to 9) (Table I) (9). At the lower ionic strength of pH 6.0, the enzyme assumes a higher oligomeric form.

Table I

Sedimentation Constant of Human G6PD

	Buffer	Solvent pH	NADP (μM)	G6P (mM)	$S_{20,w}$
A	Acetate	6.0	20	---	9.37
		6.0	---	33	9.67
	Tris-HCl	8.0	25	---	6.26
B	Acetate	6.0	250	2.5	6.85
		6.0	500	5.0	7.12
	Tris-HCl	8.0	500	5.0	6.64
		8.0	250	2.5	6.01
		8.0	500	5.0	6.14

A: The protein concentration used for determination (moving boundary method) was about 0.5 mg/ml.
B: The protein concentrations used for determination ("active enzyme centrifugation" method) ranged from 0.5~5 μg/ml.

In the presence of both NADP and G6P in the solvent (i.e., in the reaction mixture), the sedimentation constant of the enzyme, determined by the "active enzyme centrifugation" method, is about 6.5 S at pH 6.0 and pH 8.0, indicating that the functioning enzyme in red blood cells is predominately dimeric form (8).

Despite the fact that the enzyme consists of a single type subunit, several enzyme bands are observed in thin-layer electrophoresis and isoelectric focusing (10). Usually, a single enzyme band is observed in starch gel electrophoresis. The nature of this micro-heterogeneity is not clear. It is conceivable that binding with NADP, binding with glutathione, internal S-S bridge formation, partial proteolysis, deamination or acetylation of certain amino acid residues, and binding with ampholine could cause the formation of multiple bands. The existence of small molecular weight "G6PD-modifying factors," in leukemic granulocytes was reported by Kahn and co-workers (11). It was suggested that the modifying factors bind to special residues of the enzyme; or they could be non-enzymatic catalysts that cause an alteration of the G6PD molecule itself.

5) NADP Binding Sites: The reports on human G6PD NADP binding sites are controversial. Early reports described that dialyzed normal G6PD contained one molecule of reducible NADP per dimer (1, 12). Based on fluorometric assay of the nucleotides released from the enzyme by proteolysis, De Flora et al concluded that the tetrameric enzyme had four tightly bound NADP, i.e., one NADP per monomer (13). They also reported that, in addition to these four tightly bound NADP, four more NADP could be associated with the tetrameric enzyme. Since their data indicated that the tightly bound NADP was totally reduced to NADPH in the presence of G6P, the bound NADP should be considered as "functional," not merely "structural." We determined NADP and NADPH binding to the enzyme by equilibrium dialysis, using C^{14}-labeled NADP and NADPH. The results indicated that two NADP or two NADPH bind to the tetrameric G6PD, and the dissociation constants for the nucleotides are very low (Table II). It should be pointed out that dissociation constants at 37°C are much greater (three order of magnitude) than those at 4°C. This would imply that a large decrease in entropy (about 30 entropy units) takes place, i.e., the enzyme assumes a more orderly form in the association of the enzyme with the co-enzyme. Using C^{14}-labeled G6P, it was shown that the enzyme-NADPH complex could not associate with G6P; i.e., the binding of the co-enzyme and the substrate to the enzyme should be compulsory order.

The problem of NADP binding to G6PD remains unsettled. The equilibrium dialysis method is straight forward, but it cannot tell us whether or not the enzyme contains additional "structural" NADP. The fluorometric assay, which has some inheritant

Table II

NADP and NADPH Content of G6PD and Dissociation Constants of Enzyme-NADP and Enzyme-NADPH.

Condition	Nucleotide Content (μ mol/g G6PD) NADP	NADPH	Dissociation Constant (M) NADP	NADPH
pH 6.8, 4°C	8.6	8.4	1.5×10^{-9}	3.3×10^{-9}
pH 7.3, 37°C	8.8	8.4	1.7×10^{-7}**	4.9×10^{-7}
pH 8.0, 4°C	9.0 ± 0.3*	8.9	2.0×10^{-9}	1.5×10^{-9}

*Mean values and standard deviation of 6 equilibrium dialysis experiments. NADP concentrations in the buffer ranged from 2.25 to 47.8 μM.

**Mean values of three determinations. Values ranged from 1.4 to 2.1×10^{-7} M.

The nucleotide content of G6PD was determined by equilibrium dialysis. The enzyme (1~2 mg/ml) was incubated with a minute amount of G6P to reduce enzyme-bound NADP to NADPH, and the enzyme was dialyzed against the buffer containing C^{14}-labeled NADP, or C^{14}-labeled NADPH. The amount of enzyme-bound nucleotide was determined from the radioactivities of inside and outside solutions of a dialysis bag.

For determination of dissociation constant, the above dialysis bag was dialyzed against 20 ml of fresh buffer without NADP or NADPH.

methodological problems, cannot tell us whether the added NADP is functional at the site, or it becomes functional after substituting for NADPH at the functional site. X-ray crystallographic analysis may clarify the NADP-binding site of the enzyme.

In conclusion, the subunit of human G6PD consists of about 535 amino acid residues and has a molecular weight of 58000. The nearly complete amino acid sequence of the subunit was determined in the present study by protein sequencing (Fig. 1). It should be mentioned that the tentative sequence data (2,3) included sequence errors and misalignments of peptide fragments. The enzyme prepared from red blood cells in the presence of a reducing reagent, such as mercaptoethanol, does not contain intra- and inter-S-S bridges. The secondary structure of the enzyme can be predicted from the amino acid sequence. However, for further understanding of the functional structure, the determination of the three-dimensional structure is essential. It is hoped that the amino acid sequence data presented in this paper will encourage and stimulate researchers in the field.

ACKNOWLEDGMENTS

We are indebted to Drs. T. Samejima and H. Kaji, Aoyama University, Tokyo, Japan, for computer analysis of the secondary structure of G6PD.

REFERENCES

1. Yoshida, A. (1966). J. Biol. Chem. 241, 4966-4976.
2. Yoshida, A (1971). Acta Biol. Med. Germ. 36, 689-701
3. Huang, I-Y. and Yoshida, A. (1980), cited by Beutler, E. in The Metabolic Basis of Inherited Disease, eds. Stanbury, J.B., Wynegaarden, J.B., and Frederickson, D.S. (McGraw-Hill, N.Y.) p. 1629
4. Yoshida, A. (1972). Anal. Bioch. 49, 320-325
5. Yoshida, A. (1973). Arch. Bioch. Biophys. 159, 82-88
6. Yoshida, A. (1975). In Isozymes, Vol. IV, ed. Markert, C.L. (Academic Press, N.Y.). pp. 853-866.
7. Chou, P.Y., and Fasman, G.D. (1978). Annu. Rev. Biochem. 47, 251-276
8. Rossman, M.G., and Argus, P. (1977). J. Mol. Biol. 109, 99-129.
9. Yoshida, A., and Hoagland, V.D., Jr. (1970). Bioch. Biophys. Res. Commun. 40, 1167-1172.
10. Kahn, A., Vobin, P., Rubinson, H., Cottreau, D., Marie, J., and Dreyfus, J.C. (1977). Bioch. Biophys. Res. Commun. 77, 65-72.
11. Kahn, A., Bovin, P. Rubinson, H., Cottreau, D., Marie, J., and Dreyfus, J.C. (1976). Proc. Natl. Acad. Sci. U.S.A. 73, 77-81.
12. Cohen, P., and Rosemeyer, M.A. (1969). Eur. J. Biochem. 8, 1-7.
13. De Flora, A., Morelli, A., and Giuliano, F. (1974). Bioch. Biophys. Res. Commun. 59, 406-413.

IDENTIFICATION OF A REACTIVE LYSINE RESIDUE IN HUMAN ERYTHROCYTE GLUCOSE-6-PHOSPHATE DEHYDROGENASE*

Laura Camardella

Carla Caruso

Bruno Rutigliano

Mario Romano

Guido di Prisco

Fiorella Descalzi

International Institute of Genetics and Biophysica, CNR, Naples, Italy

I. INTRODUCTION

Detailed studies on the human G6PD structure have led to the elucidation of the subunit structure (1) and to an almost complete knowledge of the amino acid sequence (2).

Supported by the Progetto Finalizzato Ingegneria Genetica e Basi Molecolari delle Malattie Ereditarie of C.N.R.

Abbreviations: G6PD, Glucose-6-Phosphate dehydrogenase; G6P, Glucose-6-phosphate; PLP, Pyridoxal-5'-phosphate.

Surprisingly, little work has so far been done on the identification of essential amino acids and the structure of the active sites.

Studies on the effect of pH and ionic strength on the kinetic parameters reported for G6PD from Saccaromyces carlsbergensis (3) and Leuconostoc mesenteroides (4) have suggested the involvement of a lysine residue in the catalytic activity confirmed by modification studies in the G6PD from L. mesenteroides (5). In this case a reactive lysine residue was labeled with PLP and an eight-residue sequence including the labeled lysine was established (6).

The existence of an essential lysine residue identified by inactivation by acetyl-salicylic acid and its surrounding amino acid sequence has recently been reported also for G6PD from S. cerevisiae (7).

We have shown also that human G6PD is inactivated by PLP (8). This inhibition is reversible upon dilution, but becomes irreversible after treatment with $NaBH_4$. This means that a mechanism of Schiff-base formation between PLP and a lysine residue is at the basis of this inactivation, as already shown for other enzymes. The G6P substrate fully protects the enzyme against this inhibition suggesting that the reaction site with PLP may be close to the G6P binding site. In this paper we report the characterization of a peptide containing the PLP-binding lysine.

II. RESULTS

A. Enzyme purification. The enzyme used was prepared according to the method of De Flora (9) with slight modifications (10). It consisted of one step of affinity chromatography on 2',5'-ADP-Sepharose followed by gel filtration on Sephadex G100. Figure 1 shows the elution pattern of the G100 column; the electrophoresis pattern on SDS gel is shown at the bottom: the first lane represents the pool loaded onto the column, the second lane represents pool I from the G100 column containing homogeneous G6PD; the third, fourth and fifth lanes represent pools II, III and IV containing the lower molecular weight contaminating bands. The enzyme prepared by this method appeared to be homogeneous and suitable for structural studies. The apparent molecular weight was 58,000; the total amino acid composition did not show appreciable differences compared with that reported by Yoshida (11).

B. PLP inactivation. The enzyme was incubated in the presence of different PLP concentrations and the time course of the activity was followed. After an initial drastic loss the activity reached a plateau corresponding to 60-80% of inactivation. The binding of PLP to the enzyme was then stabilized by the addition of $NaBH_4$ and the extent of bound PLP

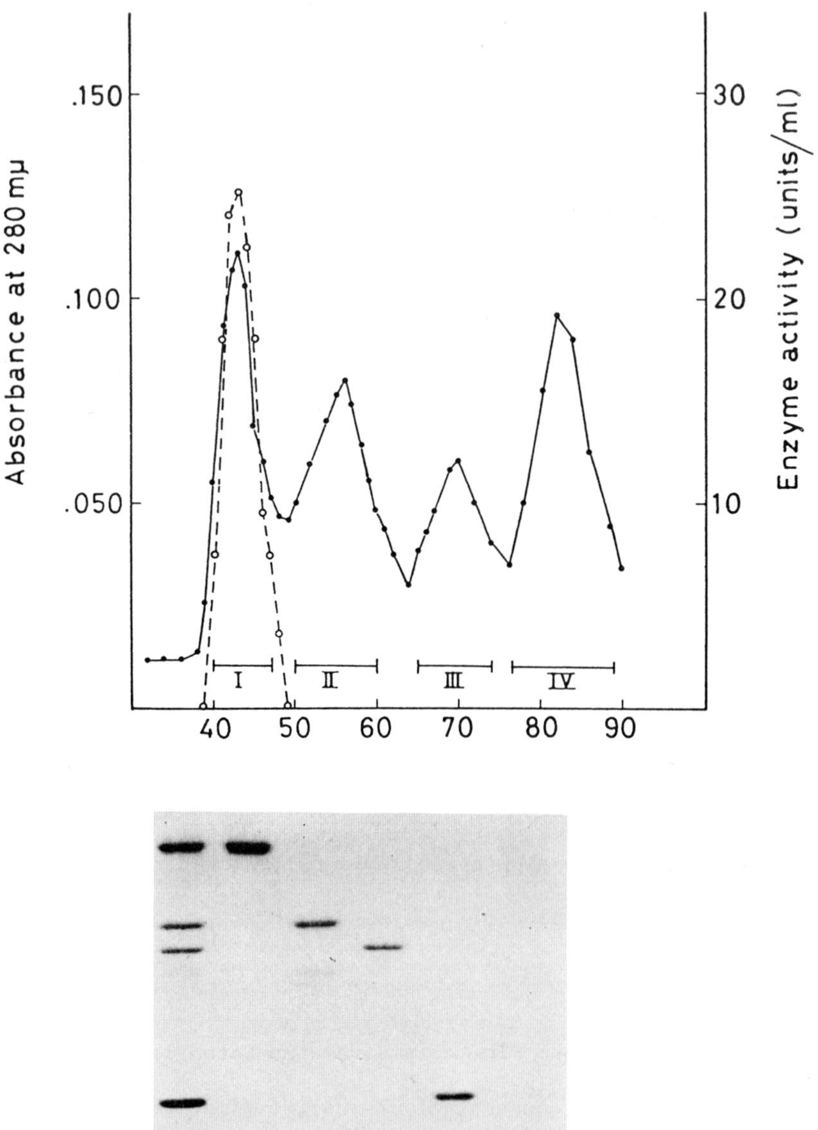

Fig. 1: Upper section: Chromatography on Sephadex G100 (1x100 cm) in 50 mM phosphate buffer pH 7.5, containing 25 mM NaCl, 1 mM EDTA, 0.2% β-mercaptoethanol and 10 μM NADP of the concentrated pool from the 2',5'-ADP-Sepharose column. Lower section: 10% polyacrylamide gel electrophoresis in SDS. First lane: sample before G100 column; second, third, fourth and fifth lanes correspond to pools I, II, III and IV from the G100 column, respectively.

was calculated from the extinction at 325 nm. When the residual activity was 30-35% of the initial one, only one mole of PLP was incorporated per mole of enzyme subunit. By using tritiated $NaBH_4$ we were able to label only one PLP-binding lysine.

C. Peptide analysis. The PLP labeled enzyme was subjected to tryptic digestion and the resulting peptides were chromatographed on a Sephadex G50 column. Almost all the labeled material was eluted in one peak corresponding to peptides of about twenty residues in size which were further purified by reverse-phase high-performance liquid chromatography. A single peak was found to contain the radioactivity associated with a pure peptide having the amino acid composition shown in Table 1. It was only possible to determine the amino acid sequence of the first five residues, after which the Edman degradation was blocked. Sub-digestion of the peptide with CNBr produced three peptides, revealing the presence of two methionine residues. The peptide derived from the amino-terminal end contained the labeled lysine. The overall situation is outlined in Table 2 where the first line shows the amino acid composition and the partial sequence of the isolated peptide. The amino acids in brackets have not been sequenced but are present in the amino acid composition. The second line shows a fragment deduced from the nucleotide sequence (M.G. Persico et al., this volume). A total correspondence is observed. The third line shows the amino acid sequence of the analogous peptide from S. cerevisiae G6PD (7) where only two differences, Met-->Leu and Gln-->Lys, are observed.

Table 1. Aminoacid composition of the tryptic peptide containing the PLP-labeled lysine residue.

Aminoacid	mol/mol of peptide[a]	
Asp	2	(2)
Glu	2	(2)
Gly	1.3	(1)
Val	1.5	(2)
Met	0.8	(2)[b]
Ile	0.9	(1)
Leu	2.6	(3)
Tyr	0.9	(1)
His	0.7	(1)
Lys (Pxy)[c]	+	(1)
Arg	0.6	(1)

[a]The numbers in parentheses are the nearest integer values.

[b]Subdigestion of the peptide with cyanogen bromide revealed the presence of two methionine residues.

[c]Pyridoxyllysine.

Table 2. Data on the aminoacid sequence of the labeled tryptic peptide from human G6PD compared with the aminoacid sequence derived from nucleotide sequence and with the corresponding peptide from yeast G6PD

Ile/Leu - Asp - His - X - Ile/Leu -(Gly, Lys*, Glx, Met)-(Val, Glx, Asx, Leu, Met)-Val - Leu - Arg[a]
Ile - Asp - His - Tyr - Leu - Gly - Lys* - Glu - Met - Val - Gln - Asn - Leu - Met -Val - Leu - Arg[b]
Ile - Asp - His - Tyr - Leu - Gly - Lys* - Glu - Leu - Val - Lys[c]

[a]Human G6PD: peptide containing the reactive lysine (ε-PLP)

[b]Human G6PD: corresponding fragments deduced from nucleotide sequence

[c]Yeast G6PD: peptide containing the reactive lysine (ε-acetyl)

*Labeled lysine

III. DISCUSSION

The existence in human G6PD of an essential lysine which can be labeled by reaction with PLP has been demonstrated and the amino acid structure surrounding this residue has been characterized, allowing the positioning of the reactive residue in the total amino acid sequence of the protein.

A high homology is observed with the analogous amino acid sequence of G6PD from S. verevisiae, indicating that the enzymes have related structure and that the reactive lysine occupies a similar position in both enzymes. In contrast, a low homology is observed with the sequence containing the essential lysine of G6PD from L. mesenteroides. This may indicate that the peculiarity of the L. mesenteroides enzyme, capable of using both NAD^+ or $NADP^+$ as coenzymes according to a complex regulation mechanism, is associated with differences at the primary structure level.

It has been suggested for G6PD from L. mesenteroides that the lysine labeled by PLP is directly involved in the binding of G6P. A structural analogy exists between PLP and aldohexose-6-phosphates. Therefore the lysine residue involved in the Schiffbase formation with the aldehyde group of PLP might be the same one that normally binds to the oxygen ring of the α-D-glucopyranose-6-phosphate (5,12).

Some authors (13) have compared the amino acid sequences around the PLP binding lysine in several enzymes. They believe that this lysine residue is not essential for catalytic activity but that the binding of PLP to this residue makes a phosphate (anion) binding site inaccessible to the natural anionic substrate.

The involvement of the labeled lysine in the human G6P binding site is not clear. This study establishes the functional importance of the segment containing the labeled lysine present without substantial differences in the G6PD from organisms very distant on the evolutionary scale such as yeast and man.

The specific role of this residue in the enzyme mechanism requires further clarification.

REFERENCES

1. Bonsignore, A. and De Flora, A. (1972) in Current Topics in Cellular Regulation (Horecker, B.L. and Stadtman, E.R. eds.) vol. 6, p. 21-62, Academic Press, New York.

2. Beutler, E. (1983) in The Metabolic Basis of Inherited Diseases (Stanbury, J.B., Wyngaarden, J.B., Fredrickson, D.S., Goldstein, J.L. and Brown M.S. eds) 5th ed. p. 1629-1653, McGraw-Hill, New York.

3. Kuby, S.A. and Roy, R.N. (1976) Biochemistry 15, 1975-1987.

4. Olive,C., Geroch,M.E. and Levy, H.R. (1971) J. Biol. Chem. 246, 2047-2057.

5. Milhausen, M. and Levy, H.R. (1975) Eur. J. Biochem. 50, 453-461.

6. Haghighi, B., Flynn, T.G. and Levy, H.R. (1982) Biochemistry 21, 6415-6420.

7. Jeffery, J., Hobbs, L. and Jornvall,H. (1985) Biochemistry 24, 666-671.

8. Camardella, L., Romano, M., di Prisco, G. and Descalzi-Cancedda, F. (1981) Biochem. Biophys. Res. Commun. 103, 1384-1389.

9. De Flora, A., Morelli, A., Benatti, U. and Giuliano, F. (1975) Arch. Biochem. Biophys. 169, 362-363.

10. Descalzi-Cancedda, F., Caruso, C., Romano, M., di Prisco, G. and Camardella, L. (1984) Biochem. Biophys. Res. Commun. 118, 332-338.

11. Yoshida, A. (1966) J. Biol. Chem. 241, 4966-4976.

12. Levy, H.R. (1979) in Advances in Enzymology (Meister, A. ed.) vol. 48, p. 97-192, John Wiley and Sons, Inc. New York.

13., Minchiotti, L., Ronchi, S. and Rippa, M. (1981) Biochim. Biophys. Acta 657, 232-242.

KINETICS AND MOLECULAR ABNORMALITIES OF HUMAN G6PD VARIANTS[1]

Akira Yoshida

Department of Biochemical Genetics
Beckman Research Institute of the City of Hope
Duarte, California USA

INTRODUCTION

At the present time, about 280 human G6PD variants, which are distinguishable from each other by the standard WHO characterization method (1), have been reported. In addition, about 50 variants, characterized by non-standard methods, were also reported. The G6PD variants reported up to 1982 are listed in previous publications (2,3). The original standard characterization methods include: 1) red cell enzyme activity; 2) starch gel electrophoresis; 3) Km for G6P; 4) relative rate of utilization of 2-deoxy G6P; and 5) thermal stability. In addition, determination of pH optima, the Km for NADP, the heat of activation, the response to inhibitors, and the migration on chromatographic media were also included in characterizing G6PD variants. Since so many variants have already been reported, it becomes increasingly difficult to differentiate a newly found variant from those variants previously reported. Ultimately, as established in the case of hemoglobin variants, each variant can be identified by its specific amino acid substitution at the protein level. Each variant can also be identified by determining its nucleotide base substitution at the

[1]Supported by U.S. Public Health Service Grant HL-29515

gene level. Because of its extremely low concentration, it is technically difficult to identify the amino acid substitutions of G6PD variants at the protein level. Utilizing G6PD cDNA to detect base substitutions in DNA samples would be a more practical approach for identifying the molecular abnormality of G6PD variants. This paper reviews the nature of some G6PD abnormalities.

1. Nature of the G6PD Deficiency: "Red cell G6PD deficiency" could be due to: a) rapid inactivation of variant enzymes during red cell maturation and/or aging; b) lower catalytic activity of variant enzymes due to molecular defect; c) suppressed biosynthesis of variant enzymes; and d) a combination of these causes.

The degree of G6PD deficiency is more severe in red blood cells than in leukocytes and other nucleated tissues, and in older red cells than younger red cells of variant subjects. The very rapid inactivation of some G6PD variants during the early stage of red cell maturation was observed by De Flora. (See Chapter "G6PD Stability in Normal and Variant Red Blood Cells.")

Specific enzyme activity can be directly estimated by measuring the enzyme activity of fully purified enzyme preparations, assuming that no inactivation occurred during the purification process. Alternatively, it can be indirectly estimated from the ratio between enzyme activity and immunologically estimated G6PD protein. Immuno-precipitation, rocket immuno-electrophoresis, and immuno-titration were used for the quantification of G6PD protein, using rabbit antibody against purified normal human G6PD.

Presumption for the indirect immunological quantification, i.e., immunological equivalence of variant protein, was not held in certain G6PD variants. The problems and uncertainty involved in the indirect assay were previously discussed (4). Nevertheless, the immunological method would provide a rough estimation of the specific activity of several G6PD variants. The relative specific activities of G6PD variants are listed in Table I.

TABLE I

Relative Specific Activity of G6PD Variants

Variant	Red Cell Activity (% of normal)	Specific Activity (% of normal)		References
A+	80~100	~100	(100)	11
A-	10~20	~100	(80)	16
Mediterranean	<5	30~50		17
Ube	33-45	77~91		4
Union	<3	~10		4
Manchester	25-30	55	(45)	4
Toronto	10~20	60	(40)	18
West Bengal	10	~100		19
Seattle	13	80		19
Worcester-like	10	85		19
Mali	0~10	72		19
Fort de France	18	72		19
Ankara	5	84		19
Matam	~0	37		19
Hektoen	400	~100	(100)	6

Relative specific activities (values not in parentheses) are estimated by the indirect immunological method.

Values in parentheses are estimated by the direct method (unpublished observation).

These results indicate that the red cell enzyme activity is not proportional to the specific enzyme activity (Table I). Several variants with severe red cell enzyme deficiency had normal or near-normal specific activity. It was recently reported that G6PD Nukus, Tashkent, and El Fayoum showed severe red cell enzyme deficiency and normal G6PD protein levels in the red cells, i.e., very low specific activities (5); however, the quantification of the variant G6PD protein by immuno-titration which they report is questionable.

The specific activities of several variant enzymes were directly determined. Specific activities of A^+, a common Negro variant with normal or near-normal red cell enzyme activity, and Hektoen, a rare variant with several-fold increased red cell enzyme activity, are found to be similar to normal. The specific activity of A^-, a common Negro variant with 10~15% normal red cell enzyme activity, is about 80% of normal. Toronto and Manchester variants have less than 50% of normal specific activity (Table I).

The rate of biosynthesis of normal and variant G6PD has not been well studied. The rate of increase in G6PD activity in fibroblast cultures of normal (B^+) and Hektoen variant, a variant associated with several-fold increased red cell enzyme activity (6), was previously compared. The rate of increase in G6PD activity (relative to reference enzymes such as lactate dehydrogenase and 6-phosphogluconate dehydrogenase) was several times higher in the Gd Hektoen fibroblast than in the control cells. Since the catalytic activity of G6PD Hektoen is similar to normal, the finding implies that either the rate of synthesis is increased or the rate of degradation is decreased in Gd Hektoen cells.

It was reported that the half-life of G6PD is about two days in a normal fibroblast and 60 days in normal red cells (7). The same authors observed that variants with decreased half-life in the red cells had normal half-life in their fibroblasts. Since the level of G6PD (and other enzymes) is determined by the rate of synthesis and the rate of degradation, the causes of enzyme deficiency in various G6PD variants would be diverse.

Thus far, there is only one true null G6PD variant. Three male siblings of a Canadian family showed G6PD activity in their red cells and leukocytes that was too low to be detected (less than 0.1% of normal levels); and immunologically cross-reacting material was also absent (less than 0.2% of normal) (8). Other so-called "null" G6PD variants exhibited residual activity in nucleated tissues, and their kinetic characteristics were measurable; therefore, these were not truly null for G6PD. The null-G6PD-variant subjects suffered from chronic non-spherocytic hemolytic anemia and chronic granulomatous disease. The cause of this unique variant remains to be further examined.

Since G6PD cDNA probes and genomic DNA probes are now available, we shall be able to gather more information regarding the biosynthetic regulation of normal and variant G6PD.

2. Kinetic Abnormalities: In reviewing the reported kinetic properties of G6PD variants, it is noticeable that some variants exhibit unusual substrate specificity. Normal human G6PD has strict substrate specificity. It oxidizes G6P, but not Gal6P or 2-deoxy G6PD; and uses NADP, but not NAD, as a co-enzyme. Several G6PD variants have more relaxed substrate specificity. These oxidize Gal6P and 2-deoxy G6P, as well as G6P; and use NAD, as well as NADP, as a co-enzyme (Table II).

TABLE II

Substrate Specificity of Variant G6PD

Variant	d-deoxy G6P	Gal6P	deamino NADP	NAD
B^{+} (normal)	<4	<4	55-60	<1
Bat-Yam	40-45	---	---	---
Benevento	245	---	---	---
Hualien	72.2	---	---	---
Lifta	60	---	---	---
Manus	<4	---	300-400	60-130
Markham	180	66	300	130
Mediterranean	25	20	350	---
Mexico	26-33	---	130-160	---
Orchomenos	105	---	350	---
Union	180	80	370	<3
Worcester	<4	---	21	---
B^{+}, oxidized	68	---	222	3.6

Relative rate of utilization of substrate analogues in comparison to the primary substrates (i.e., G6P and NADP).

Deamino NADP is a very useful analogue for distinguishing many variants, since the rate of utilization ranges from <50% to 400% of NADP in G6PD variants. Another NADP analogue, 3-acetylpyridine adenine dinucleotide phosphate, was found to be less discriminatory.

At the present time, the structural abnormality of any G6PD variant with unusual substrate specificity has not been identified, and the molecular region that governs enzyme specificity is not clear. The mildly oxidized normal B^+ enzyme exhibited altered substrate specificity in accompanying the oxidation of cysteine and methionine residues (9). The modified G6PD oxidized deoxy G6P, and exhibited a high utilization of deamino NADP and a lower catalytic activity (Table II).

Some variants have a several-fold lower affinity to NADP and/or G6P than do the normal-type enzyme (Table III). Some variants are very sensitive to inhibition by NADPH (i.e., low K_I for NADPH). Since the red cell concentrations of G6P and, particularly, NADP are lower than the K_m values, and since the concentration of NADPH is much higher than that of NADP, variants with high K_m for NADP and/or G6P and low K_I for NADPH are expected to exhibit disproportionately lower physiological activity than their "red cell G6PD" activities, which are determined in the presence of high concentrations of NADP and G6P, and in the absence of NADPH. The degree of red cell enzyme deficiency and hemolytic severity are not always parallel. For example, Gd Manchester, Alhambra, and Tripler, associated with relatively mild (10~30% of normal) red cell enzyme deficiency, exhibit chronic hemolytic symptoms; whereas Gd Union, Markham, and Mediterranean, associated with severe red cell enzyme deficiency, do not cause a chronic hemolytic problem. The former group of variants with low K_I is very sensitive to NADPH inhibition, while the latter group with high K_I is tolerant of the inhibition. Part of NADP and NADPH is bound to catalase and other proteins in red blood cells (see Chapter "Regulation of Glucose-6-Phosphate Dehydrogenase

Activity in Normal and Variant Red Blood Cells"), and it is difficult to accurately estimate physiological G6PD activity in variant red cells at the present time. Nevertheless, it is likely that one of the major factors distinguishing hemolytic manifestation is the sensitivity of G6PD variants to NADPH inhibition.

TABLE III

Kinetic Properties of Hemolytic and Non-Hemolytic Variants

Variant	Activity (% of normal)	KM (μM) G6P	KM (μM) NADP	Inhibition by NADPH*	Heat Stability
Variants not associated with chromic hemolytic anemia:					
B^+ (normal)	100	50-70	2.9-4.4	1.38	Normal
A^+	80	Normal	Normal	1.2	Normal
A^-	8-20	Normal	Normal	0.62	Normal
Union	<3	8-12	3.6-5.2	0.22	Low
Markham	1.5-10	4.4-6.3	---	0.375	Low
Mediterranean	<5	19-26	1.2-1.6	0.4	Low
Variants associated with chronic hemolytic anemia:					
Bat-Yam	~0	27	---	---	Very Low
Ramat-Gan	~0	35	---	---	Very Low
Worcester	~0	11.6	61	---	Very Low
Oklahoma	4-10	127-200	20	---	Low
Ashdod	10	100	---	---	Low
Freiburg	10-20	87-118	4	---	---
Albuquerque	~1	115	11	---	Very Low
Milwaukee	0.5	224	---	---	---
Clichy	2	178	---	---	---
Strasbourg	6	96	13	---	Low
Torrance	2.4	48-60	2.4	3.1	Very Low
Manchester	25-30	64	6	23	Low
Alhambra	9-20	55	2.6	5.9	Low
Tripler	35	30	---	3.1	Very Low

*Ratio of K_m and K_I at pH 7.3. Variants with higher values are more sensitive to NADPH inhibition.

3) Structural Abnormalities: All reported variants, except for the "null" variant mentioned above, are considered to be structural variants. On the analogy of other mutant proteins, most G6PD variants are presumably associated with single amino acid substitutions. Some variants, such as A+, A-, and Mediterranean, are prevalent in certain populations; and variants with double amino acid substitutions could exist, as in the cases of hemoglobin and α_1-antitrypsin variants.

Thus far, amino acid substitution was identified only in G6PD A+; i.e., one of the Asn residues in the NH_2-terminal tryptic peptide of CNBr-3 of the normal enzyme was substituted by Asp in the A+ enzyme (see Chapter "Structure of Human Glucose-6-Phosphate Dehydrogenase"). From the fingerprint patterns of their tryptic and chymotryptic digests, single substitutions were suggested in Mediterranean, Toronto, Manchester, and Hektoen variants (unpublished observations).

Modiano et al claimed that, based on kinetic and chromatographic criteria, the common B+ and common A+ could be subdivided into "B_1" and "B_2" types, and into "A_1" and "A_2" types (10). Since the methodology used to distinguish these "subtypes" is not totally reliable, comparison of peptide maps between "B_1" and "B_2," and between "A_1" and "A_2," or direct genomic analysis, is required to establish their claim. As previously demonstrated, a single peptide substitution exists in peptide maps between the "pooled" B+ and "pooled" A+ G6PD preparations (11). This fact favors the concept of the existence of one, and not two, types of A+.

The usual human G6PD has a tetrameric form at a neutral pH and a dimeric form at an alkaline pH (>pH 8.0) in a relatively high concentrated solution (~1mg/ml). In highly diluted solutions, such as in a starch gel for electrophoresis, the enzyme resumes dimeric form, even at a neutral pH, as indicated by the formation of one hybrid enzyme, in addition to the two original bands, after the hybridization of normal and variant human G6PDs

(12). By contrast, several variants, such as Capetown (13), Yamaguchi (14), and Manchester (15), have tetrameric forms at the neutral pH, even in a very dilute solution, as demonstrated by the extremely slow electrophoretic mobility (less than 50% normal mobility) at pH 7.0 and by the estimation of the molecular size by gel filtration (unpublished observation).

CONCLUSION

When compared with hemoglobin variants, our knowledge of G6PD variants is still scanty and superficial. Since the quantity of G6PD, particularly in deficient variants, is extremely low (less than 0.02% in red cell proteins), it is difficult to obtain sufficient amounts of material for examination at the molecular level. Lack of knowledge about the structure of normal G6PD has also been a major obstacle. The study of the primary structure of the normal G6PD, and the cloning of cDNA and genomic DNA for G6PD, reported in this meeting, would facilitate further study of G6PD variants at the protein and gene levels. Determination of the three-dimensional structure of G6PD is essential for clarifying the structure-function relationships of G6PD variants.

REFERENCES

1. Standardization of procedures for the study of glucose-6-phosphate dehydrogenase. Wld. Hlth. Org. Techn. Rep., Ser. No. 366, 1967
2. Yoshida, A., and Beutler, E. (1983). Ann. Hum. Genet. 47, 25-38
3. McKusick, V.A. (1983). "Mendelian Inheritance in Man," 6th Ed. (Johns Hopkins Univ. Press, Baltimore)
4. Nakashima, K., and Yoshida, A. (1977). J4.. Lab. Clin. Med. 89, 446-454
5. Yermakov, N., Tokarev, Ju., Chernjak, N., Schönian, Ct., Grieger, M., Guckler, A., Jacobasch, Ct., Mahmudova, M., and Bahramov, S. (1984). Acta Biol. Med. Germ. 40, 563-570
6. Dern, R.J., McCurdy, P.R., and Yoshida, A. (1969). J. Lab. Clin. Med. 73, 283-290
7. Luzzatto, L., and Battistuzzi, G. (1984). Advan. Hum. Genet. 14, 217-329

8. Gray, G.R., Stamatoyannopoulos, G., Naiman, S.C., Kliman, M.R., Klebanoff, S.J., Astin, T., Yoshida, A., and Robinson, G.C.F. (1973). Lancet, Sept. 8, 530-534
9. Yoshida, A. (1973). Arch. Biochem. Biophys. 159, 82-88
10. Modiano, G., Battistuzzi, G., Esan, G.J.F., Testa, U., and Luzzatto, L. (1979). Proc. Natl. Acad. Sci. USA 76, 852-856
11. Yoshida, A. (1967). Proc. Natl. Acad. Sci. USA 57, 835-840
12. Yoshida, A., Steinman, L., and Harbert, P. (1967). Nature 216, 275-276
13. Botha, M.C., Dern, R.J., Mitchell, M., West, C., and Beutler, E. (1969). Am. J. Hum. Genet. 21, 547-551
14. Miwa, S., Fujii, H., Nakatsuji, T., Ishida, Y., Oda, E., Kaneto, A., Motokawa, M., Ariga, Y., Fukuchi, S., Sasai, S., Hiraoka, K., Kashii, H., Kodama, T., and Miwa, Y. (1978). Am. J. Hum. Genet. 5, 131-138
15. Milner, G., Delamore, I.W., and Yoshida, A. (1974). Blood 43, 271-276

ANALYSIS OF THE PRIMARY STRUCTURE OF HUMAN G6PD DEDUCED FROM THE cDNA SEQUENCE

M.G. Persico[o]*, G. Viglietto*, G. Martini*, R. Dono*, M. D'Urso*, D. Toniolo*, T. Vulliamy+ and L. Luzzatto+

*International Institute of Genetics and Biophysics, CNR, Naples, Italy, and +Department of Haematology, Royal Postgraduate Medical School, London, UK

The first information on primary structure of human G6PD was published nearly 20 years ago by A. Yoshida, who obtained tryptic digest fingerprints of both B and A variants (Yoshida, 1967). In the meantime physico-chemical and enzymatic studies provided ample evidence that most and possibly all other common and rare variants of G6PD, numbering now over 300 (E.Beutler, personal communication), were structurally different from G6PD B.

However, complete elucidation of the primary structure proved extremely difficult, and therefore amino acid replacement in variant types could neither be identified, nor less located. Partly for this purpose, and also in order to investigate the regulation of G6PD synthesis in normal and mutant cells (see D'Urso et al., 1983), we embarked in 1980 on cloning the human gene.

The first *Gd*-specific clone was obtained from a cDNA library prepared by us in pBR322, from human diploid fibroblasts, after about 50-fold enrichment of G6PD-specific mRNA. The criterion for identifying the clones was positive selection of G6PD mRNA translatable in a cell-free system (Persico et al., 1981). One of the clones, pGD6405, yielded signals proportional to the number of X-chromosomes in Southern blots of human genomic DNA, and it was subsequently mapped to Xq26-Xqter by using somatic cell hybrids (Martin-De Leon, 1985). The plasmid pGD6405 was used to isolate genomic clones from phage libraries (Toniolo et al., 1984a), and a unique fragment subcloned from one of the phages was mapped to Xq28 by *in situ* hybridization (Szabo et al., 1984). All these data were consistent with pGD6405 and its cognate genomic clones belonging to the *Gd* gene. Sequencing of pGD6405 revealed features of a 3'-untranslated region of mRNA (Toniolo et al., 1984b). However, sequencing of about 5 kb of cDNA and genomic clones upstream of pGD6405 failed to reveal an open reading frame compatible with a G6PD-coding region.

An almost complete amino acid sequence of the protein became available to us when it was published in 1983 (Beutler, 1983). This made it possible to construct oligonucleotide probes, which were used to screen another G6PD-enriched cDNA library, prepared this time from HeLa cell RNA. This approach led eventually to the elucidation of the mRNA sequence and of its relationship to the genomic _Gd_ gene.

The G6PD mRNA

Overlapping cDNA clones were isolated not only from our original HeLa cell library, but also from other cDNA libraries from SV 40 transformed fibroblasts, from placenta and from a teratocarcinoma cell line. The six longest inserts were sequenced, and each segment was sequenced at least twice in order to minimize errors and to avoid possible cloning artefacts. The sequence of an mRNA molecule of about 2500 nucleotides was determined (Fig. 1). This exhibited two interesting features. First, we have 254 nucleotides in the 5'-untranslated region: this is longer than in most other known eukaryotic mRNA species, and it is likely to be still an understimate, since we have not yet identified the transcription initiation site. Second, the sequence is GC-rich: the GC fraction is 59.5% in the coding region (slightly higher than the 53.4% which is the average value of a sample of human proteins compiled by Lathe, 1985); it

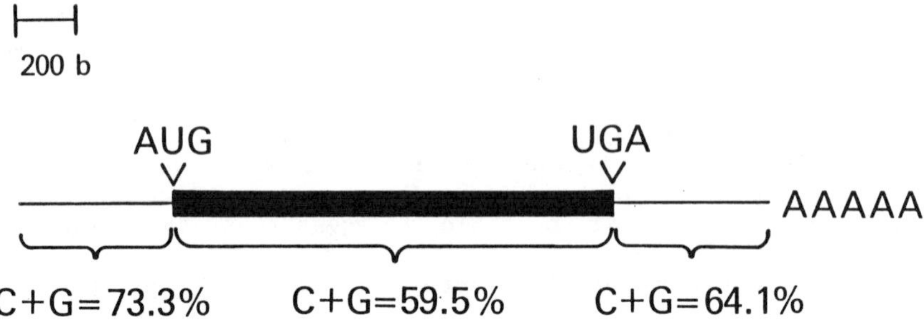

FIG. 1 Diagram of G6PD/mRNA
The nucleotides sequence of the cDNA clones was determined using both chemical (Maxam and Gilbert, 1980) and enzymatic (Sanger et al., 1977) sequencing methods

is 64.1% in the 3'-untranslated region, and as high as 73.3% in the 5'-untranslated region.

<u>The coding sequence</u>

Within the mRNA molecule, an open reading frame of 1437 nucleotides (479 codons) was identified (Fig. 2). The primary structure deduced from the cDNA sequence yields a molecular weight of 55,176. This is intermediate between previous estimates of about 52 and about 59 kd (Shreve and Levy, 1977) The amino acid composition (table I) is in reasonable agreement with two previously published sets of data (De Flora et al., 1977; Yoshida & Huang, quoted by Beutler, 1978). The number of acidic amino acids is 62, and the number of basic amino acids is 61. The number of hydrophobic amino acids is 168 (35%), and a hydrophobicity profile is shown in Fig.3.

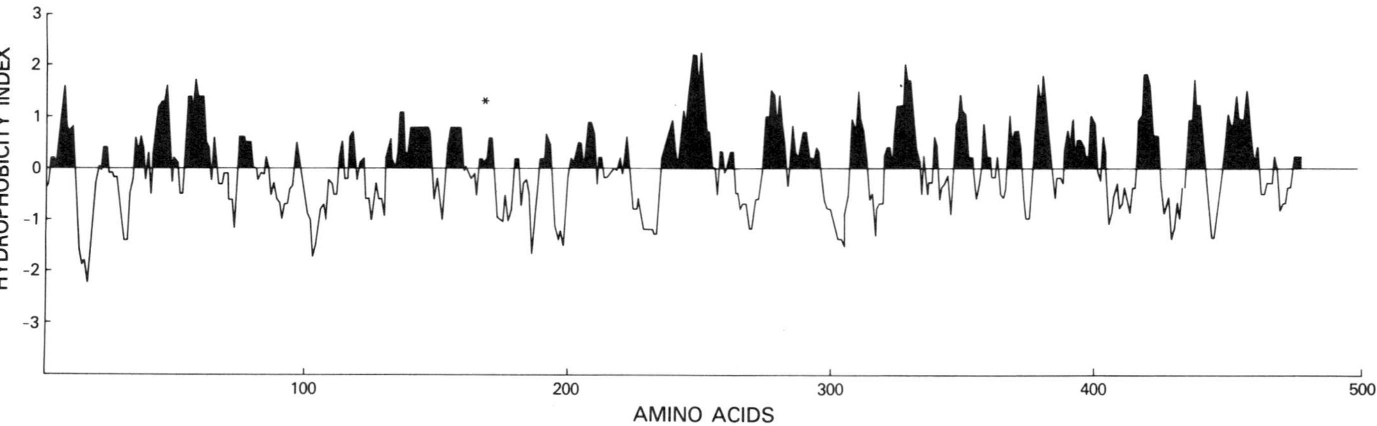

FIG. 2 Hydropathy Plot of the Amino Acid Sequence of human G6PD using the Algorithm and hydropathy values of Hopp and Woods (1974)

Table I. AMINO ACID COMPOSITION OF HUMAN G6PD

	Number of residues previously reported from protein analysis		Deduced from CDNA sequence	
Aminoacid	a	b	n	%
Ala	29	37	29	6.1
Arg	29	30	32	6.7
Asn	48'	47'	25	5.2
Asp			27	5.6
Cys	7	9	7	1.5
Gln	55"	52"	19	4.0
Glu			35	7.3
Gly	32	37	31	6.5
His	11	14	11	2.3
Ile	22	26	26	5.4
Leu	39	43	43	9.0
Lys	26	31	29	6.1
Met	9	12	13	2.7
Phe	22	23	25	5.2
Pro	21	28	24	5.0
Ser	23	29	22	4.6
Thr	20	26	20	4.2
Trp	7	7	7	1.5
Tyr	17	21	21	4.4
Val	29	30	33	6.9

(a) is from De Flora et al.(1977); (b) is from Yoshida & Huang (Beutler, 1978). ': Asn + Asp; "Gln + Glu.

The last four peptides of our sequence are in complete agreement with those of the protein sequence previously published (Beutler, 1983). The C-terminal peptide is also identical to that previously identified in red cell G6PD (De Scalzi-Cancedda et al. 1984), and this finding does not support post-synthetic cleavage of C-terminal amino acids (Kahn et al., 1977). Because the methionine corresponding to the initiation codon is often cleaved after translation, one can never be certain of the N-terminus deduced from a cDNA sequence. However, a peptide identical to our N-terminal peptide, including the N-terminal methionine, was part of the published protein sequence (Beutler, 1983), although it was placed in position 14. We believe therefore that it is the very first peptide and that methionine is the N-terminal amino acid, the amino group of which has been reported to be blocked to Edman degradation (Yoshida, 1972; Kahn et al., 1976). Numerous other discrepancies between this and the previous sequence are apparent (Fig. 3). Because of the different methodologies used in DNA and protein sequencing, we consider the DNA-based sequence more likely to be correct.

Special features of G6PD-sequence

In analyzing the amino acid sequence, we have observed two internal repeats: a pentapeptide Ala-Asn-Arg-$^{\text{Leu}}_{\text{Ile}}$-Phe (residues 98-102 and 181-185), and a decapeptide Leu-Arg-$^{\text{Ile}}_{\text{Glu}}$-Ala-$^{\text{Asn}}_{\text{Trp}}$-Arg-Ile-Phe-$^{\text{Gly}}_{\text{Thr}}$-Phe (residues 178-187 and 422-431) (fig.4). These repeats have 70 and 80% homology at the amino acid level and 50

```
M G A S G D L A K K K I Y P T I W W L F R D G L L P E N T F I V G Y A R S R L T V A D I R K Q S E P   50
      14
F F K A T P E E K L K L E D F F A R N S Y V A G Q Y D D A A S Y Q R L N S H M N A L H L G S Q A N R  100
          12  *           11                  7                           1
L F Y L A L P P T V Y E A V T K N I H E S C M S Q I G W N R I I V E K P F G R D L Q S S D R L S N H  150
              2                                               3       4           5
I S S L F R E D Q I Y R I D H Y L G K E M V Q N L M V L R F A N R I F G P I W N R D N I A C V I L T  200
   8          10        9                                  36
F K E P F G T E G R G G Y F D E F G I I R D V M Q N H L L Q M L C L V A M E K P A S T N S D D V R D  250
        24                21
E K V K V L K C I S E V Q A N N V V L G Q Y V G N P D G E G E A T K G Y L D D P T V P R G S T T A T  300
29       17                                    20              *          18
F A A V V L Y V E N E R W D G V P F I L R C G K A L N E R K A E V R L Q F H D V A G D I F H Q Q C K  350
                                                              16            * *  22
R N E L V I R V Q P N E A V Y T K M M T K K P G M F F N P E E S E L D L T Y G N R Y K N V K L P D A  400
                        28      31                      32                *         42        41
Y E R L I L D V F C G S Q M H F V R S D E L R E A W R I F T P L L H Q I E L E K P K P I P Y I Y G S  450
                                          38    39              *40 *     *                    43
R G P T E A D E L M K R V G F Q Y E G T Y K W V N P H K L                                            479
          44                  45      *     46
```

FIG. 3 Amino Acid Sequences of Human G6PD

We have numbered the 46 tryptic peptides of the partial A.A. sequence published (Yoshida, 1983) and compared them with the A.A. sequence deduced from the nucleotide sequence. The asterisks denote the differences between individual peptides.
The drawing of this figure has been made using the computer program developed by Shapiro and Senapathy (1986)

```
ATGGGTGCATCGGGTGACCTGGCCAAGAAGAAGATCTACCCCACCATCTGGTGGCTGTTCCGGGATGGCCTTCTGCCCGAAAACACCTTCATCGTGGGCT   100
 M  G  A  S  G  D  L  A  K  K  K  I  Y  P  T  I  W  W  L  F  R  D  G  L  L  P  E  N  T  F  I  V  G
ATGCCCGTTCCCGCCTCACAGTGGCTGACATCCGCAAACAGAGTGAGCCCTTCTTCAAGGCCACCCCAGAGGAGAAGCTCAAGCTGGAGGACTTCTTTGC   200
Y  A  R  S  R  L  T  V  A  D  I  R  K  Q  S  E  P  F  F  K  A  T  P  E  E  K  L  K  L  E  D  F  F  A
CCGCAACTCCTATGTGGCTGGCCAGTACGATGATGCAGCCTCCTACCAGCGCCTCAACAGCCACATGAATGCCCTCCACCTGGGGTCACAGGCCAACCGC   300
  R  N  S  Y  V  A  G  Q  Y  D  D  A  A  S  Y  Q  R  L  N  S  H  M  N  A  L  H  L  G  S  Q  A  N  R
CTCTTCTACCTGGCCTTGCCCCCGACCGTCTACGAGGCCGTCACCAAGAACATTCACGAGTCCTGCATGAGCCAGATAGGCTGGAACCGCATCATCGTGG   400
 L  F  Y  L  A  L  P  P  T  V  Y  E  A  V  T  K  N  I  H  E  S  C  M  S  Q  I  G  W  N  R  I  I  V
AGAAGCCCTTCGGGAGGGACCTGCAGAGCTCTGACCGGCTGTCCAACCACATCTCCTCCCTGTTCCGTGAGGACCAGATCTACCGCATCGACCACTACCT   500
E  K  P  F  G  R  D  L  Q  S  S  D  R  L  S  N  H  I  S  S  L  F  R  E  D  Q  I  Y  R  I  D  H  Y  L
GGGCAAGGAGATGGTGCAGAACCTCATGGTGCTGAGATTTGCCAACAGGATCTTCGGCCCCATCTGGAACCGGGACAACATCGCCTGCGTTATCCTCACC   600
  G  K  E  M  V  Q  N  L  M  V  L  R  F  A  N  R  I  F  G  P  I  W  N  R  D  N  I  A  C  V  I  L  T
TTCAAGGAGCCCTTTGGCACTGAGGGTCGCGGGGGCTATTTCGATGAATTTGGGATCATCCGGGACGTGATGCAGAACCACCTACTGCAGATGCTGTGTC   700
 F  K  E  P  F  G  T  E  G  R  G  G  Y  F  D  E  F  G  I  I  R  D  V  M  Q  N  H  L  L  Q  M  L  C
TGGTGGCCATGGAGAAGCCCGCCTCCACCAACTCAGATGACGTCCGTGATGAGAAGGTCAAGGTGTTGAAATGCATCTCAGAGGTGCAGGCCAACAATGT   800
L  V  A  M  E  K  P  A  S  T  N  S  D  D  V  R  D  E  K  V  K  V  L  K  C  I  S  E  V  Q  A  N  N  V
GGTCCTGGGCCAGTACGTGGGGAACCCCGATGGAGAGGGCGAGGCCACCAAAGGGTACCTGGACGACCCCACGGTGCCCCGCGGGTCCACCACCGCCACT   900
  V  L  G  Q  Y  V  G  N  P  D  G  E  G  E  A  T  K  G  Y  L  D  D  P  T  V  P  R  G  S  T  T  A  T
TTTGCAGCCGTCGTCCTCTATGTGGAGAATGAGAGGTGGGATGGGGTGCCCTTCATCCTGCGCTGCGGCAAGGCCCTGAACGAGCGCAAGGCCGAGGTGA  1000
 F  A  A  V  V  L  Y  V  E  N  E  R  W  D  G  V  P  F  I  L  R  C  G  K  A  L  N  E  R  K  A  E  V
GGCTGCAGTTCCATGATGTGGCCGGCGACATCTTCCACCAGCAGTGCAAGCGCAACGAGCTGGTGATCCGCGTGCAGCCCAACGAGGCCGTGTACACCAA  1100
R  L  Q  F  H  D  V  A  G  D  I  F  H  Q  Q  C  K  R  N  E  L  V  I  R  V  Q  P  N  E  A  V  Y  T  K
GATGATGACCAAGAAGCCGGGCATGTTCTTCAACCCCGAGGAGTCGGAGCTGGACCTGACCTACGGCAACAGATACAAGAACGTGAAGCTCCCTGACGCC  1200
  M  M  T  K  K  P  G  M  F  F  N  P  E  E  S  E  L  D  L  T  Y  G  N  R  Y  K  N  V  K  L  P  D  A
TACGAGCGCCTCATCCTGGACGTCTTCTGCGGGAGCCAGATGCACTTCGTGCGCAGCGACGAGCTCCGTGAGGCTTGGCGTATTTTCACCCCACTGCTGC  1300
 Y  E  R  L  I  L  D  V  F  C  G  S  Q  M  H  F  V  R  S  D  E  L  R  E  A  W  R  I  F  T  P  L  L
ACCAGATTGAGCTGGAGAAGCCCAAGCCCATCCCCTATATTTATGGCAGCCGAGGCCCCACGGAGGCAGACGAGCTGATGAAGAGAGTGGGTTTCCAGTA  1400
H  Q  I  E  L  E  K  P  K  P  I  P  Y  I  Y  G  S  R  G  P  T  E  A  D  E  L  M  K  R  V  G  F  Q  Y
TGAGGGCACCTACAAGTGGGTGAACCCCCACAAGCTC                                                                 1437
  E  G  T  Y  K  W  V  N  P  H  K  L
```

Fig. 4 Nucleotide and amino acid sequence of human G6PD

Nucleotide residues are numbered in the 5' to 3' direction beginning with the first residues at the ATG triplet encoding the initiation methionine. The number of nucleotide residue at the right end of each line is given. The deduced amino acid sequence in one letter notation is shown below the nucleotide sequence.

The peptides homologous in human and yeast G6PD (see table) is underlined (continuous line). The dashed and wavy lines indicate the aminoacid repeats discussed in the text.

and 80% homology at the nucleotide level respectively. Their significance is not clear yet.

Codon usage was analyzed by employing the microgenie programme (Beckman). We have observed considerable deviations (Table II) from the average reported by Lathe (1985), based on a sample of human proteins totalling 4285 amino acid residues. Although in that compilation of data a preference for codons ending in C and G is apparent, this bias is much more extreme in the case of G6PD. For instance, CAC is the His codon in 90% of cases (60% in the other proteins); GAG is the condon for Glu in 94% of cases (60% in other proteins); TTA for Leu and GTA for Val are not used at all in G6PD.

Sequence homologies

The deduced amino acid composition is remarkably similar to that previously reported for G6PD from Saccharmomyces cerevisiae (Jeffery et al., 1985) suggesting considerable sequence conservation throughout evolution of eukaryotes. This is made evident from inspection of a yeast peptide which differs in only two residues from the homologous human peptide (Table III). This peptide is of special interest because it contains a lysine residue essential in the yeast enzyme for catalytic activity, which may be near the glucose-6-phosphate binding site (Jeffery et al., 1985), and which has been found previously in the human enzyme to be reactive with pyridoxal phosphate (Camardella et al., 1981). A search of the Dayhoff data base has been carried

Table II

a	b	c	d	e	a	b	c	d	e	a	b	c	d	e	a	b	c	d	e
TTT	Phe	5	35	17	TCT	Ser	1	17	4	TAT	Tyr	7	47	33	TGT	Cys	1	30	14
TTC	Phe	20	65	83	TCC	Ser	9	26	40	TAC	Tyr	14	53	77	TGG	Cys	6	70	85
TTA	Leu	0	5	0	TCA	Ser	3	11	13	TAA	End	0	0	0	TGA	End	1	100	100
TTG	Leu	2	9	4	TCG	Ser	2	7	9	TAG	End	0	0	0	TGG	Trp	7	100	100
CTT	Leu	1	11	2	CCT	Pro	1	24	4	CAT	His	1	42	9	CGT	Arg	5	9	18
CTC	Leu	12	22	28	CCC	Pro	19	41	79	CAC	His	10	58	90	CGC	Arg	15	19	55
CTA	Leu	1	7	2	CCA	Pro	2	24	8	CAA	Gln	0	26	0	CGA	Arg	1	10	3
CTG	Leu	27	46	63	CCG	Pro	2	11	8	CAG	Gln	19	74	100	CGG	Arg	4	15	14
ATT	Ile	4	23	15	ACT	Thr	2	20	10	AAT	Asn	3	34	12	AGT	Ser	1	11	4
ATC	Ile	21	64	80	ACC	Thr	15	47	75	AAC	Asn	22	66	88	AGC	Ser	6	29	27
ATA	Ile	1	13	3	ACA	Thr	1	21	5	AAA	Lys	3	45	10	AGA	Arg	3	24	11
ATG	Met	13	100	100	ACG	Thr	2	12	10	AAG	Lys	26	55	89	AGG	Arg	4	23	14
GTT	Val	1	13	3	GCT	Ala	3	31	10	GAT	Asp	9	38	33	GGT	Gly	4	15	13
GTC	Val	8	27	24	GCC	Ala	22	44	75	GAC	Asp	18	62	77	GGC	Gly	17	44	54
GTA	Val	0	9	0	GCA	Ala	4	17	13	GAA	Glu	2	40	5	GGA	Gly	1	17	3
GTG	Val	24	50	72	GCG	Ala	0	12	0	GAG	Glu	33	60	94	GGG	Gly	9	24	29

a, codon; b, amino acid; c, number of times codon occurs in human G6PD;

d, per cent codon utilization in a sample of human proteins (Lathe, 1985);

e, per cent codon utilization in human G6PD.

Table III

YEAST	Iso -	Asp -	His -	Tyr -	Leu -	Gly -	Lys -	Glu -	Leu -	Val -	Lys
HUMAN	Iso -	Asp -	His -	Tyr -	Leu -	Gly -	Lys -	Glu -	Met -	Val -	Gln

out. Homologies have been detected with several other dehydrogenases. However, the matches observed never comprise more than 4 or 5 consecutive amino acids, and therefore these are at the moment difficult to interpret.

Conclusion

Gd is of special interest in human genetics for its high rate of genetic heterogeneity and for its clinical implications. G6PD deficiency is a well-characterized example of balanced polymorphism which has evolved by virtue of the selective advantage it confers to female heterozygous mosaic with respect to *P.falciparum* malaria. The sequence provided here for normal G6PD will serve as a basis to solve the entire structure of the gene, and for identifying structural changes in G6PD variants. This in turn will help to understand which are the critical sites for enzyme activity, and why different mutations entail different clinical expression. Since human cDNA probes of the coding sequence cross-hybridize with a variety of mammalian and non-mammalian DNA (data now shown) it should be also possible soon to investigate fully the evolution of this gene and how its expression is controlled in parasites such as *P.falciparum*.

REFERENCES

Beutler, E. (1983). In "The Metabolic Basis of Inherited Disease", 3rd ed. (J.B. Stanbury, J.B. Wyngaarden, D.S. Fredrickson, J.L. Goldstein, M.S. Brown, Eds). McGraw-Hill, New York.

Camardella, L., Romano, M., di Prisco, G., Descalzi-Cancedda, F. (1981). Biochem. Biophys. Res. Commun. 103, 1384

De Flora, A., Morelli, A., Frascio, M., Corte, G., Curti, B., Galliano, M. Gozzer, C., Minchiotti, L., Mareni, C., Gaetani, G.F. (1977). Biochim. Biophys. Acta, 500, 109.

Descalzi-Cancedda, F., Caruso, C., Romano, M., di Prisco, G., Camardella, L. (1984). Biochem. Biophys. Res. Commun., 118, 332.

D'Urso, M., Mareni, C., Toniolo, D., Piscopo, M., Schlessinger,D. Luzzatto, L. (1983). Somatic Cell Genetics, 9, 429.

Jeffery, J., Hobbs, L., Jornvall, H. (1985). Biochemistry, 24, 666

Hopp and Woods (1974) PNAS, 78, 3824

Kahn, A. et al. (1976). Biochim. Biophys. Acta, 445, 537.

Kahn, A. et al. (1977). Biochem. Biophys. Res. Commun., 77, 65

Lathe, R. (1985). J. Mol. Biol., 183, 1.

Luzzatto, L., O'Brien, S., Usanga, E., Wanachiwanawin, W., this Symposium.

Martin-De Leon, P.A., et al. (1985). Cytogenet. Cell. Genet., 39, 87.

Persico, M.G., Toniolo, D., Nobile, C., D'Urso, M., Luzzatto, L. (1981). Nature, 294, 778

Shapiro, M., Senapathy, B. (1986). Nucleic Acids Res., (in press).

Shreve, D.S., Levy, H.R. (1977). Biochim. Biophys. Res. Commun., 78, 1369

Szabo, P., Purrello, M., Rocchi, M., Archidiacono, N., Alhadeff, B., Filippi, G., Toniolo, D., Martini, G., Luzzatto, L., Siniscalco, M. (1984). Proc. Natl. Acad. Sci. USA, 81, 7855

Toniolo, D., D'Urso, M., Martini, G., Persico, M., Tufano, V., Battistuzzi, G., Luzzatto, L. (1984a). EMBO Journal, 3, 1987

Toniolo, D., Persico, M.G., Battistuzzi, G., Luzzatto, L. (1984b). Mol. Biol. Med., 2, 89

Yoshida, A. (1967). Biochem. Genet., 1, 81

Yoshida, A. (1972). Anal. Biochem., 49, 320

Yoshida, A., Huang, L.Y. (1978). Quoted by Beutler, E. In "the Metabolic Basis of Inherited Disease." 4th ed. (Stanbury, J.B., Wyngaarden, J.B., Fredrickson, D.S., eds.), p. 1430, McGraw-Hill, New York.

MOLECULAR CLONING OF cDNA FOR G6PD[1]

Akira Yoshida and Takenori Takizawa

Department of Biochemical Genetics
Beckman Research Institute of the City of Hope
Duarte, California USA

cDNA for a protein can be cloned by screening a cDNA library constructed in a plasmid vector with synthetic oligonucleotide probes which match the protein. Alternatively, a cDNA library constructed in phage λgt 11 can be screened with a specific antibody and/or with synthetic oligonucleotide probes. We attempted to clone G6PD cDNA by following these two approaches.

A human liver cDNA library constructed in plasmid pUC13 was screened by the colony hybridization method, using mixed synthetic oligonucleotides (ranging from 17- to 20-mers), which corresponded to the amino acid sequence of G6PD, as probes. Several positive clones were obtained, but all of them proved to be false-positive by partial nucleotide sequencing. Subsequently, two λgt11 libraries, i.e., an adult human liver cDNA library constructed by Dr. S. Woo (Howard Hughes Medical Institute, Houston, TX), and a human hepatoma Li-7 cDNA library constructed by J.R. DeWet (Univ. of Calif., San Diego, CA) were screened. We found that the hepatoma Li-7 cell line expressed G6PD far more strongly than did usual human livers; i.e., 1.1 mg and 0.037 mg of G6PD per g of soluble protein in these two tissues, respectively. The

[1]Supported by U.S. Public Health Service Grant HL-29515

use of the hepatoma Li-7 library would be advantageous for cloning G6PD cDNA.

The two λgt11 libraries were screened with a specific antibody against human G6PD and/or with synthetic 17- to 20-mer nucleotide probes. Many hybridization-positive clones were obtained, and some of them were hybridized with both the antibody and the synthetic nucleotide. However, all of them were found to be false-positive by further characterization of their nucleotide sequences.

In order to avoid the problem of non-specific cross-hybridization, a 41-mer synthetic nucleotide, with an unique sequence matching the amino acid sequence of human G6PD (position 99 to 112 from the COOH-terminal), was used as a hybridization probe (Fig. 1).

Amino Acid:(N)-Met-Met-Thr-Lys-Lys-Pro-GLy-Met-Phe-Phe-Asn-Pro-Glu-Glu-(C)

mRNA: 5' AUG·AUG·ACC·AAG·AAG·CCC·GGC·AUG·UUC·UUC·AAC·CCC·GAG·GA 3'

cDNA: 3' TAC·TAC·TGG·TTC·TTC·GGG·CCG·TAC·AAG·AAC·TTG·GGG·CTC·CT 5'
(41 mer)

Figure 1. Synthetic oligonucleotide probe. The probe consisted of 41 nucleotide residues corresponding to the amino acid sequence of human G6PD.

Approximately $5x10^5$ recombinant phage plaques from the two libraries were screened under stringent conditions (at 60°C). Twelve positive clones were obtained from the hepatoma Li-7 cDNA library and four positive clones were obtained from the liver cDNA library. Examples of positive and negative plaques are shown in Fig. 2. At the present time, three clones, λ G6PD-19, λ G6PD-25, and λ G6PD-26 have been characterized.

Restriction endo-nuclease maps of these clones are shown in Fig. 3. Clone λ G6PD-19 contained an insertion of 2.0 kbp. The

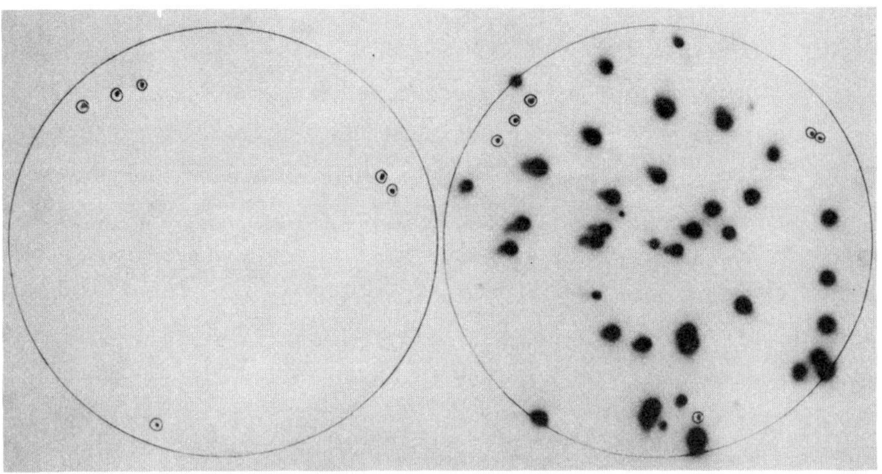

Figure 2. Autoradiogram of phage plaques hybridized with the nucleotide probe. Left: negative hybridization; Right: positive hybridization.

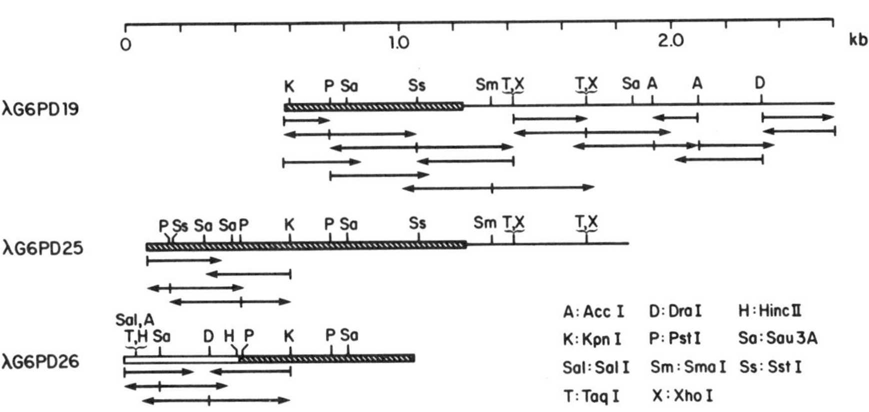

Figure 3. Restriction maps of three positive clones and sequence determination strategy. Horizontal arrows indicate the direction and extent of sequencing. [hatched bar]: coding region compatible with G6PD; [open bar]: region incompatible with G6PD.

insert encoded 204 amino acid residues whose sequences were completely compatible with the COOH-terminal part of the enzyme (see Chapter "Structure of Human Glucose-6-Phosphate Dehydrogenase"), and had a 3'-non-coding region of 1.36 kbp (Fig. 4). The non-coding region does not contain a poly (A) sequence or the AATAAA polyadenylation signal sequence. Another clone, λ G6PD-25, contained an insertion of 1.8 kbp, and encoded 362 amino acid residues of the COOH-terminal part of G6PD and a 3'-non-coding region of 0.7 kbp (Fig. 4). Clone λ G6PD-26, containing an insertion of 1.0 kbp, was found to be a defective cDNA clone. The insertion of λ G6PD-26 contained a coding region encoding about 200 amino acid residues of G6PD. However, the upstream (5'-side) of the insert contained about 400 nucleotide sequences which were not compatible to G6PD. Several in-phase termination codons exist in this region. Clone λ G6PD-26 might have been produced artificially during the construction of the cDNA library.

Cloning of cDNA for human G6PD was previously reported (1). However, the cDNA reported appears to lack the coding sequence (2,3), and its authenticity was not verified. The nucleotide sequence of the present clones is completely compatible with the amino acid sequence of human G6PD; thus, the authenticity of the cDNA clones was established.

Utilizing G6PD cDNA (insertion of clone λG6PD-19) as a probe, human G6PD loci were examined. Dot-blot hybridization of DNA samples of flow-sorted human chromosomes indicated that chromosome 17, as well as the X-chromosome, contained DNA sequences compatible to the G6PD probe (done by Dr. Y.W. Kan, Howard Hughes Medical Institute, San Francisco, CA). An existence of the G6PD-like locus on chromosome 17 was further confirmed by the Southern hybridization analysis of DNA from the mouse-human hybrid cell line containing human chromosome 17 only. It is known that in addition to the X-linked G6PD, an autosomal isozyme, commonly named glucose dehydrogenase or hexose-6-phosphate dehydrogenase, exists in microsomes of human and other mammals (see Chapter

```
                                                         EcoRI
                                                       GAATTCCC·ATT·CAC·GAG
                                                                Ile His Glu

TCC·TGC·ATG·AGC·CAG·ATA·GGC·TGG·AAC·CGC·ATC·ATC·GTG·GAG·AAG·CCC·TTC·GGG·AGG·GAC
Ser Cys Met Ser Gln Ile Gly Trp Asn Arg Ile Ile Val Glu Lys Pro Phe Gly Arg Asp

CTG·CAG·AGC·TCT·GAC·CGG·CTG·TCC·AAC·CAC·ATC·TCC·TCC·CTG·TTC·CGT·GAG·GAC·CAG·ATC
Leu Gln Ser Ser Asp Arg Leu Ser Asn His Ile Ser Ser Leu Phe Arg Glu Asp Gln Ile

TAC·CGC·ATC·GAC·CAC·TAC·CTG·GGC·AAG·GAG·ATG·GTG·CAG·AAC·CTC·ATG·GTG·CTG·AGA·TTT
Tyr Arg Ile Asp His Tyr Leu Gly Lys Glu Met Val Gln Asn Leu Met Val Leu Arg Phe

GCC·AAC·AGG·ATC·TTC·GGC·CCC·ATC·TGG·AAC·CGG·GAC·AAC·ATC·GCC·TGC·GTT·ATC·CTC·ACC
Ala Asn Arg Ile Phe Gly Pro Ile Trp Asn Arg Asp Asn Ile Ala Cys Val Ile Leu Thr

TTC·AAG·GAG·CCC·TTT·GGC·ACT·GAG·GGT·CGC·GGG·GGC·TAT·TTC·GAT·GAA·TTT·GGG·ATC·ATC
Phe Lys Glu Pro Phe Gly Thr Glu Gly Arg Gly Gly Tyr Phe Asp Glu Phe Gly Ile Ile

CGG·GAC·GTG·ATG·CAG·ACC·CAC·CTA·CTG·CAG·ATG·CTG·TGT·CTG·GTG·GCC·ATG·GAG·AAG·CCC
Arg Asp Val Met Gln Asn His Leu Leu Gln Met Leu Cys Leu Val Ala Met Glu Lys Pro

GCC·TCC·ACC·AAC·TCA·GAT·GAC·GTC·CGT·GAT·GAG·AAG·GTC·AAG·GTG·TTG·AAA·TGC·ATC·TCA
Ala Ser Thr Asn Ser Asp Asp Val Arg Asp Glu Lys Val Lys Val Leu Lys Cys Ile Ser

GAG·GTG·CAG·GCC·AAC·AAT·GTG·GTC·CTG·GGC·CAG·TAC·GTG·GGG·AAC·CCC·GAT·GGA·GAG·GGC
Glu Val Gln Ala Asn Asn Val Val Leu Gly Gln Tyr Val Gly Asn Pro Asp Gly Glu Gly

GAG·GGC·ACC·AAA·GGG·TAC·CTG·GAC·GAC·CCC·ACG·GTG·CCC·CGC·GGG·TCC·ACC·ACC·GCC·ACT
Glu Ala Thr Lys Gly Tyr Leu Asp Asp Pro Thr Val Pro Arg Gly Ser Thr Thr Ala Thr

TTT·GCA·GCC·GTC·GTC·CTC·TAT·GTG·GAG·AAT·GAG·AGG·TGG·GAT·GGG·GTG·CCC·TTC·ATC·CTG
Phe Ala Ala Val Val Leu Tyr Val Glu Asn Glu Arg Trp Asp Gly Val Pro Phe Ile Leu

CGC·TGC·GGC·AAG·GCC·CTG·AAC·GAG·CGC·AAG·GCC·GAG·GTG·AGG·CTG·CAG·TTC·CAT·GAT·GTG
Arg Cys Gly Lys Ala Leu Asn Glu Arg Lys Ala Glu Val Arg Leu Gln Phe His Asp Val

GCC·GGC·GAC·ATC·TTC·CAC·CAG·CAG·TGC·AAG·CGC·AAC·GAG·CTG·GTG·ATC·CGC·GTG·CAG·CCC
Ala Gly Asp Ile Phe His Gln Gln Cys Lys Arg Asn Glu Leu Val Ile Arg Val Gln Pro

AAC·GAG·GCC·GTG·TAC·ACC·AAG·ATG·ATG·ACC·AAG·AAG·CGC·GGC·ATG·TTC·TTC·AAC·CCC·GAG
Asn Glu Ala Val Tyr Thr Lys Met Met Thr Lys Lys Pro Gly Met Phe Phe Asn Pro Glu

GAG·TCG·GAG·CTG·GAC·CTG·ACC·TAC·GCC·AAC·AGA·TAC·AAG·AAC·GTG·AAG·CTC·CCT·GAG·CCC
Glu Ser Glu Leu Asp Leu Thr Tyr Gly Asn Arg Tyr Lys Asn Val Lys Leu Pro Glu Pro

TAC·GAG·CGC·CTC·ATC·CTG·GAC·GTC·TTC.TGC·GGG·AGC·CAG·ATG·CAC·TTC·GTG·CGC·AGC·GAC
Tyr Glu Arg Leu Ile Leu Asp Val Phe Cys Gly Ser Gln Met His Phe Val Arg Ser Asp

GAG·CTC·CGT·GAG·GCC·TGG·CGT·ATT·TTC·ACC·CCA·CTG·CTG·CAC·CAG·ATT·GAG·CTG·GAG·AAG
Glu Leu Arg Glu Ala Trp Arg Ile Phe Thr Pro Leu Leu His Gln Ile Glu Leu Glu Lys

CCC·AAG·CCC·ATC·CCC·TAT·ATT·TAT·GGC·AGC·CGA·GGC·CCC·ACG·GAG·GCA·GAC·GAG·CTG·ATG
Pro Lys Pro Ile Pro Tyr Ile Tyr Gly Ser Arg Gly Pro Thr Glu Ala Asp Glu Leu Met

AAG·AGA·GTG·GGT·TTC·GAG·TTT·GAG·GGC·ACC·TAC·AAG·TGG·GTG·AAC·CCC·CAC·AAG·CTC·TGA
Lys Arg Val Gly Phe Gln Tyr Glu Gly Thr Tyr Lys Trp Val Asn Pro His Lys Leu ***
```

Figure 4-A.

TGA GCC CTG GGC ACC CAC
***

CTC CAC CCC CGC CAC GGC CAC CCT CCT TCC CGC CGC CGA CCC GAG TCG GGA GGA CTC CGG

GAC CAT TGA CCT CAG CTG CAC ATT CCT GGC CCC GGG CTC TGG CAC CCT GGC CGC CCC TCG

CTG CTG CTA CTA CCC GAG CCC AGC TAC ATT CCT CAG CTG CCA AGC ACT CGA GAC CAT CCT

GGC CCC TCC AGA CCC TGC CTG AGC CCA GGA GCT GAG TCA CCT CCT CCA CTC ACT CCA GCC

CAA CAG AAG GAA GGA GGA GGG CGC CCA TTC GTC TGT CCC AGA GCT TAT TGG CCA CTG GGT

CTC ACT CCT GAG TGG GGC CAG GGT GGG AGG GAG GGA CAA GGG GGA GGA AAG GGG CGA GCA

CCC ACG TGA GAG AAT CTG CTG TGC CTT GCC CGC CAG CCT CAG TGC CAC TTG ACA TTC CTT

GTC ACC AGC AAC ATC TCG AGC CCC CTG GAT GTC CCC TGT CCC ACC AAC TCT GCA CTC CAT

GGC CAC CCC GTG CCA CCC GTA GGC AGC CTC TCT GCT ATA AGA AAA GCA GAC GCA GCA GCT

GGG ACC CCT CCC AAC CTC AAT GCC CTG TAA GGA TTT TGC CCA CTG GGC AAC TCA GAA ATA

CTT CGA TCT CCC AAG ATA TAA GAG GCA GCA GCA AAC GTG CCT ATT GAC GTC TGT TTC ATA

GTT ACC ACT TAC GCG AGT AGA CAG AAC TCG GCT TTT CAG AAA ATA GGT GTC AAG TCC ACT

TTA TAA GAA CCT TTT TTT CTA AAA TAA GAT AAA AGG TGG CTT TGC ATT TTC TGA TTA AAC

GAC TGT GTC TTT GTC ACC TCT GCT TAA CTT TAG GAG TAT CCA TTC CTG TGA TTG TAG ACT

TTT GTT GAT ATT CTT CCT GGA AGA ATA TCA TTC TTT TCT TGA AGG GTT GGT TTA CTA GAA

TAT TCA AAA TCA ATC ATG AAG GCA GTT ACT ATT TTG AGT CTA AAG GTT TTC TAA AAA TTA

ACC TCA CAT CCC TTC TGT TAG GGT CTT TCA GAA TAT CTT TTA TAA ACA GAA GCA TTT GAA

GTC ATT GCT TTT GCT ACA TGA TTT GTG TGT GTG AAG GAC ATA CCA CGT TTA AAT CAT TAA

TTG AAA AAC ATC ATA TAA GCC CCA ACT TTG TTT GGA GGA AGA GAC GGA GGT TGA GGT TTT

TCC TTC TGT ATA AGC ACC TAC TGA CAA AAT GTA GAG GCC ATT CAA CCG TCA AAC ACC ATT

TGG TTA TAT CGC AGA GGA GAC GGA TGT GTA AAT TAC TGC ATT GCT TTT TTT TTC AGT TTG

TAT AAC CTC TAA TCT CCG TTT GCA TGA TAC GCT TTG TTA GAA ACA TTA ATT GTA GTT TGG

AAG CAA GTG TGT ATG GGAATTC
EcoR I

Figure 4. Nucleotide sequence of G6PD cDNA. Solid underlining indicates the region corresponding to the synthetic nucleotide probe. Termination codon is marked ***.
A: Coding region which encodes 362 amino acid residues.
B: Non-coding 3'-region of 1.36 kbp.

"Hexose-6-Phosphate Dehydrogenase"). The locus for this autosomal isozyme was assigned to chromosome 1 (4). Therefore, the G6PD-like locus on chromosome 17 cannot be for hexose-6-phosphate dehydrogenase. The G6PD-like locus could be related to a putative pseudogene for G6PD. Alternatively, it could be for the G6PD isozyme which was reported to exist in fetal human brain (5). This isozyme has not yet been well characterized and it would be interesting to further characterize the fetal-brain-specific G6PD isozyme and the G6PD-like locus on chromosome 17.

About 50 X-chromosomes were haplotyped using the G6PD cDNA as a probe. Several restriction-site-length polymorphisms were found in Caucasians and Orientals. Details of genomic studies of the G6PD locus described above will be reported elsewhere.

Recent technological developments greatly facilitate the elucidation of specific mutations of variant G6PD at the DNA level by using the G6PD cDNA probe. Structural abnormalities of the variant G6PD at the protein level can be readily deduced from their DNA abnormalities. Our knowledge about the structure of variant G6PD at the nucleotide and protein levels is expected to increase rapidly in the near future. Examining restriction-site-length polymorphism in various populations with common G6PD variants, such as Mediterranean, A+, A-, Mahidol, Markham, and Union, would contribute to revealing the root of these variants.

ACKNOWLEDGMENTS

We are indebted to Drs. S. Woo and J.R. DeWet for providing the cDNA libraries, and to Dr. T. Mohandas for providing the DNA of human-rodent hybrid cell lines. We are also grateful to Dr. Y.W. Kan for carrying out dot-blot hybridization of the flow-sorted chromosome DNA.

REFERENCES

1. Persico, M.G., Toniolo, D., Nobile, C., D'Urso, M., and Luzzatto, L. (1981). Nature 241, 4966-6976.
2. Luzzatto, L., and Battistuzzi, G. (1984). Advan. Hum. Genet. 14, 217-329.
3. Toniolo, D., Persico, M.G., Battisuzzi, G., and Luzzatto, L. (1984). Mol. Biol. Med. 2, 89-103.
4. Hameister, H., Rogers, H.-H., Grzeschik, K.-H. (1978). Cytogenet. Cell Genet. 22, 200-202.
5. Toncheva, D., Evres, T., Tzoneva, M. (1982). Hum. Hered. 32, 193-196.

They came to America:
Some from the East,
The others from West.
It was like a feast,
They wanted the best.

We heard ancient tales
Of dangerous beans
We heard about messages
And cloning of genes:
Whether in *Drosophila*
Or *Leuconostoc*
G6PD
Must be studied *ad hoc.*

You have seen the slides
Coming in tides;
Enzyme kinetics
And human genetics,
Geographical maps
Then filling the gaps:
Of course you might say
—It is haematology;
Or perhaps, nay,
—It is enzymology:
It was like a flood
Dear students of blood.
Who cares about labels,
Who cares about disciplines
As long as we discipline
The science that must be?
In every session
There was one obsession
It's G6PD.

But where can it be?
One minute just wait—
Yes, now I can see
On Xq28.

Some variants deficient
—Not all in one class—
Some are spectacular,
Some serious—halas—
Is our locus really a hot spot?
Is it mutation
Hypermethylation
It is GC, the clusters, the lot!

Specific
Terrific
And yet it's housekeeping:
The fruits we are now reaping
Of work that was done
For G6PD is second to none.

Starving, refeeding
Transcription increases;
Selection, inbreeding
The mutants releases;
Molecular studies
—Listen, my buddies—
The scientists' selection
Has brought to perfection:
But when evolution
Has mounted the podium
The only explanation
Must be the *Plasmodium.*
Whatever thou thinkest,
Whatever it may be,
I tell you, the greatest
Is G6PD.

It's clear, Pasadena
Was worthy an arena;
With such a crowd,
You can be proud:
You only can cope
From the City of Hope.
Akira, we thank you,
You brought us together
With insight and zest
Your meeting for ever
Was ever at its best.
You've earned it
Dear Ernie
You've done it again
You spared us pain.
We thank Dian
And Vibha, you too:
One X may be inactive
But women are active!
This was a wonderful jubilee
For all of us and for G6PD.

ANONYMOUS

INDEX*

*Entries followed by "f" denote figures; entries followed by "t" denote tables.

F

I

J